AF780274

Wildlife Conservation: Managing Resources for a Sustainable World

Wildlife Conservation: Managing Resources for a Sustainable World

Editor

Vasilios Liordos

MDPI

Basel • Beijing • Wuhan • Barcelona • Belgrade • Novi Sad • Cluj • Manchester

Editor
Vasilios Liordos
Forest and Natural
Environment Sciences
International Hellenic
University
Drama
Greece

Editorial Office
MDPI
St. Alban-Anlage 66
4052 Basel, Switzerland

This is a reprint of articles from the Special Issue published online in the open access journal *Sustainability* (ISSN 2071-1050) (available at: https://www.mdpi.com/journal/sustainability/special_issues/Wildlife_Conservation_Managing_Resources).

For citation purposes, cite each article independently as indicated on the article page online and as indicated below:

Lastname, A.A.; Lastname, B.B. Article Title. *Journal Name* **Year**, *Volume Number*, Page Range.

ISBN 978-3-7258-0974-5 (Hbk)
ISBN 978-3-7258-0973-8 (PDF)
doi.org/10.3390/books978-3-7258-0973-8

Cover image courtesy of Panos Vafeidis
Eurasian blue tit *Cyanistes caeruleus*

© 2024 by the authors. Articles in this book are Open Access and distributed under the Creative Commons Attribution (CC BY) license. The book as a whole is distributed by MDPI under the terms and conditions of the Creative Commons Attribution-NonCommercial-NoDerivs (CC BY-NC-ND) license.

Contents

About the Editor

Vasilios Liordos

Vasilios Liordos is a Professor of Wildlife Management and the Director of the Wildlife Ecology and Management Laboratory at the Department of Forest and Natural Environment Sciences of the International Hellenic University at Drama, Greece. He has been active in wildlife ecology and management fields, both in Greece and abroad. He has published over 50 scientific articles in many international journals. His main interests include wildlife conservation, wildlife management, human dimensions of wildlife, avian ecology, community ecology, spatial ecology, movement ecology, animal behavior, and ecotoxicology.

Preface

The unprecedented rate of vertebrate species loss over the last century, up to 100 times higher than the pre-human background rate, indicates that a sixth mass extinction is already underway. Human activities, particularly habitat loss, overexploitation for economic gain, and climate change, are considered responsible for the exceptionally rapid loss of animal species.

Sustainability is achieved when the goals of its three basic components, society, environment, and economy, are satisfied. The loss and decline of wildlife species and populations destabilize trophic webs, thus compromising ecosystem function and health. Such consequences also negatively affect society and the economy. Wildlife conservation and management research therefore focuses on efforts to avert a true sixth mass extinction, maintain healthy wildlife populations, and thus contribute toward a sustainable world for future generations.

In this Special Issue, we present a collection of state-of-the-art studies on the conservation and management of wildlife species under the framework of sustainability. In this volume, wildlife conservationists and managers will find a holistic view of the current problems and proposed solutions that could be useful as a guide for successfully designing and implementing conservation and management plans.

Vasilios Liordos
Editor

sustainability

Editorial

Conservation, Sustainability, Conflict and Coexistence: Key Themes in Wildlife Management

Vasilios Liordos

Department of Forest and Natural Environment Sciences, International Hellenic University, P.O. Box 172, 66100 Drama, Greece; liordos@for.ihu.gr

1. Introduction

Human activities have shaped the Earth's biotic and abiotic elements for millennia [1]. These activities have become so significant, both in their extent and intensity, that negative anthropogenic impacts have affected even the most remote places where human presence is scarce or even non-existent [2,3]. This profound impact of humans on Earth has led scientists to propose the dawn of a new geological epoch, aptly named the Anthropocene [2,3]. These impacts, the most important among which are habitat destruction and degradation, overexploitation, and climate change, have caused an unprecedented rate of vertebrate species loss over the last century, up to 100 times higher than the pre-human background rate [4–6]. These rates of extinction indicate that a sixth phase of mass extinction is already in progress [4].

The world's population exceeded 8 billion people in 2022, and the latest projections by the United Nations suggested that it could grow to around 8.5 billion in 2030, 9.7 billion in 2050, 10.4 billion in the 2080s and remain stable until 2100 [7]. This increase in the global population size is accompanied by an increase in urbanization: the global urban population was 50% for the first time in 2007, and 57.5% of people currently live in cities, which is a percentage that is projected to reach 68.4% by 2050 [8]. The large human population and the ever-increasing rates of urbanization are detrimental to biodiversity because the need for natural resources also proportionately increases and these cities now occupy former natural ecosystems [9,10]. This situation has called for our better use of resources and led to the formulation of the concept of sustainability. In 1987, the United Nations Brundtland Commission defined sustainable development as "development that meets the needs of the present without compromising the ability of future generations to meet their own needs" [11]. Sustainability is achieved when the goals of its three basic components—society, environment, and economy—are satisfied (Figure 1). The integration of all environmental, social, and economic interests into all aspects of decision making in a bearable, equitable, and viable way is key to achieving sustainability [12].

The ever-increasing rates of global population size and urbanization increase the pressures on wildlife. These pressures often result in human–wildlife conflicts (HWC) that occur "when the needs and behavior of wildlife impact negatively on the goals of humans or when the goals of humans negatively impact the needs of wildlife" [13]. These conflicts are better perceived as conservation conflicts, "situations that occur when two or more parties with strongly held opinions clash over conservation objectives and when one party is perceived to assert its interests at the expense of another" [14,15]. Conservation conflicts can be separated into two components: (a) human–wildlife impacts, i.e., the direct interactions between humans and wildlife [16], and (b) the inherent human–human conflicts between those interested in wildlife conservation and those with other priorities [17].

Citation: Liordos, V. Conservation, Sustainability, Conflict and Coexistence: Key Themes in Wildlife Management. *Sustainability* **2024**, *16*, 3271. https://doi.org/10.3390/su16083271

Received: 3 March 2024
Accepted: 9 April 2024
Published: 14 April 2024

Copyright: © 2024 by the author. Licensee MDPI, Basel, Switzerland. This article is an open access article distributed under the terms and conditions of the Creative Commons Attribution (CC BY) license (https://creativecommons.org/licenses/by/4.0/).

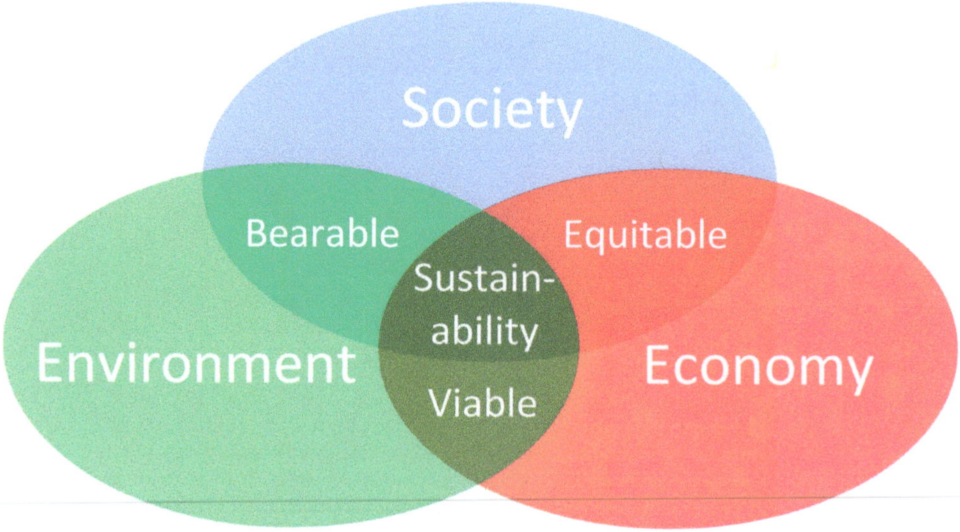

Figure 1. The sustainable development concept.

Wildlife management seeks to satisfy three basic societal desires: (a) species conservation, (b) sustainable use, and (c) addressing the negative impacts of certain behaviors. Wildlife management is the science of applying methods and practices to regulate the interactions between society, wildlife, and their habitats to satisfy society's wishes [18,19]. Wildlife managers need to apply suitable management strategies to improve habitats, address conflicts, and promote coexistence, following the principles of sustainability, to achieve wildlife conservation. Only in this way will it be possible to simultaneously satisfy the needs of wildlife species and humans. This volume includes relevant research comprising the Special Issue "Wildlife Conservation: Managing Resources for a Sustainable World". Overall, 11 articles are presented, showcasing research regarding the conservation management of wildlife species across the globe under the prism of sustainability.

2. Geography, Taxa, Management Themes

Contributions to our Special Issue span the globe: three came from Africa (Shiweda and colleagues, Zyambo and colleagues, Lines and colleagues), three from Asia (Sun and colleagues, Yin and colleagues, Abd Rahman and Matthew), four from Europe (Katselis and colleagues, Raftogianni and colleagues, Sultana and colleagues, Liordos and colleagues), and one covers global wildlife (Prokop and colleagues) (Figure 2). Furthermore, they involved species from many taxa: six articles involved mammals (Shiweda and colleagues, Zyambo and colleagues, Lines and colleagues, Sun and colleagues, Raftogianni and colleagues, Prokop and colleagues), six articles involved birds (Yin and colleagues, Katselis and colleagues, Raftogianni and colleagues, Sultana and colleagues, Liordos and colleagues, Prokop and colleagues), two articles involved reptiles (Raftogianni and colleagues, Prokop and colleagues), two articles involved fish (Abd Rahman and Matthew Prokop and colleagues), one article involved amphibians (Prokop and colleagues), and one article involved invertebrates (Prokop and colleagues). A variety of wildlife management themes, often more than one per study, were identified from the contributed research, highlighting similarities and differences in hot contemporary issues across continents: eight articles tackled conservation issues (Shiweda and colleagues, Lines and colleagues, Yin and colleagues, Abd Rahman and Matthew, Sultana and colleagues, Liordos and colleagues, Prokop and colleagues), five articles tackled sustainability issues (Zyambo and colleagues, Lines and colleagues, Abd Rahman and Matthew, Katselis and colleagues, Raftogianni and colleagues), four articles addressed human–wildlife coexistence (Shiweda and colleagues, Yin and colleagues, Sultana and colleagues, Liordos and colleagues), two articles tackled human–wildlife conflict issues (Shiweda and colleagues, Katselis and colleagues), and two articles examined wildlife use issues (Zyambo and colleagues, Raftogianni and colleagues).

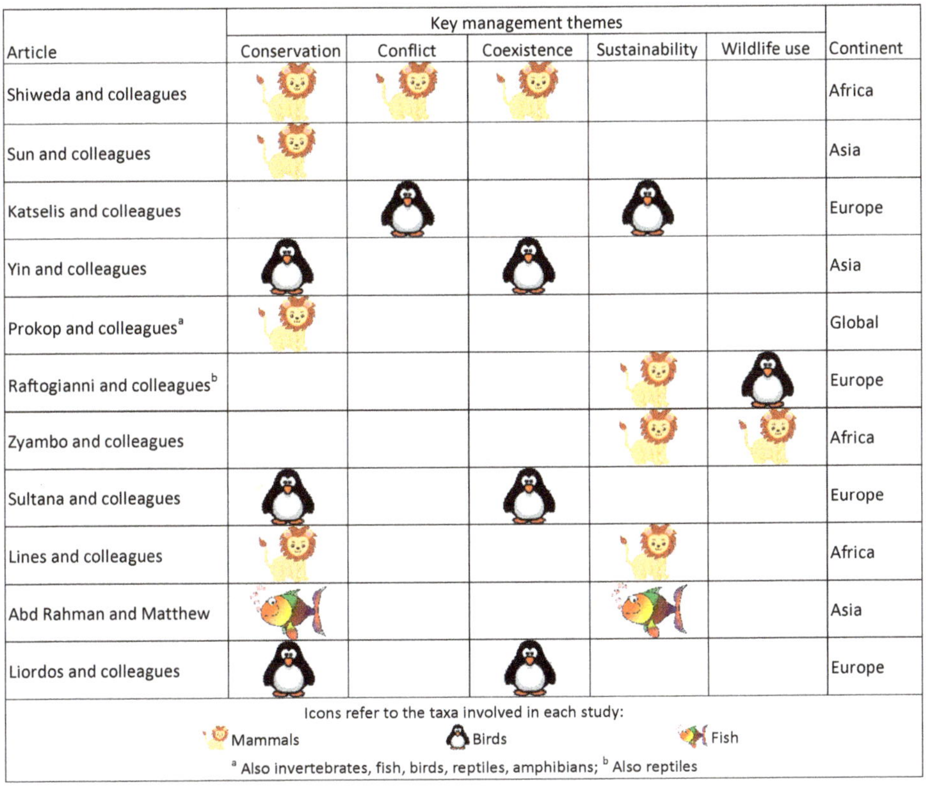

Figure 2. Management themes, taxa, and continents involved in each article of the Special Issue.

3. Research from Africa

In their article "Climate change and anthropogenic factors are influencing the loss of habitats and emerging human–elephant conflict in the Namib desert", Shiweda and colleagues studied how climate change and other human pressures affected the habitat of African elephants (*Loxodonta africana*) and induced human–elephant conflicts in the Ugab River basin in Namibia. They used geographical information systems to analyze data on land cover, African elephant movement, rainfall, and temperature. They found that farming activities, poor rainfall, and frequent droughts were responsible for the loss of around 73.0% of habitat in the lower river basin consisting of ephemeral streams. They also determined human–elephant conflict hotspots. Furthermore, they proposed a reduction in livestock numbers to favor vegetation growth and water conservation and the financial support of local organizations to provide local farmers with the skills to coexist with African elephants. They concluded that habitat conservation and farmer education would help protect African elephants and reduce conflicts.

In their article "Conceptualising drivers of illegal hunting by local hunters living in or adjacent to African protected areas: A scoping review", Zyambo and colleagues reviewed the drivers of illegal hunting among hunters living in protected areas in Africa. Illegal hunting activities endanger wildlife populations and are highly unsustainable. The need to generate income and a lack of an income source, the need or preference for bushmeat consumption, cultural needs and rights, preventative or retaliatory killing, poverty, and weak or inadequate law enforcement were the main drivers of illegal hunting among African societies. They concluded that wildlife conservation and sustainability would not be achieved without the alleviation of deep-rooted social problems, public participation, and proper law enforcement.

Lines and colleagues integrated modeling and field observations of human pressure to generate a Human Footprint Pressure map at the Kafue–Zambezi interface in their article "Utility of human footprint pressure mapping for large carnivore conservation: The Kafue-Zambezi interface". Then, they used this model along with occurrence data for lions (*Panthera leo*), leopards (*Panthera pardus*), and spotted hyenas (*Crocuta crocuta*) to generate threshold ranges at which the concerned species can persist in their habitats. Model performance suggested great potential for such predictions in the study area, but it can also have implications for local and region-wide conservation planning. Mapping of human disturbances and the estimation of carnivore species' tolerance thresholds would help managers select and apply suitable strategies for their conservation.

4. Research from Asia

In their article "Habitat selection: Autumn and winter behavioral preferences of water deer (*Hydropotes inermis*) in Northeast China", Sun and colleagues investigated the habitat selection of the endangered water deer in the Baishan Musk Deer National Nature Reserve in Northeast China. They used 11 vegetation and matrix characteristics and found that the water deer preferred grasslands at sunny and middle slopes in both autumn and winter. The height of dominant herbage, hiding cover, distance from water, and distance to human settlements predicted habitat selection in winter. They concluded that this new information would be valuable for the restoration and conservation of the endangered water deer.

In their article "Birds' flight initiation distance in residential areas of Beijing are lower than in pristine environments: Implications for the conservation of urban bird diversity", Yin and colleagues determined the flight initiation distance of birds in both urban and natural areas in Beijing, China. They found that urban adapters, including ground foragers, insectivores, and omnivores, displayed shorter flight initiation distances in urban rather than natural areas. Additionally, tree canopy cover positively affected flight initiation distance and floor area ratio negatively affected it. They concluded that their findings would help urban managers promote the coexistence of urban birds and Beijing residents for both parties' benefit.

In their article "Fish hobbyists' willingness to donate for wild fighting fish (*Betta livida*) conservation in Klang valley", Abd Rahman and Matthew presented an econometric model to infer the fish hobbyists' willingness to donate for the conservation of the endangered endemic fighting fish in Klang Valley, Malaysia. They used single- and double-bounded contingent valuation methods that revealed a mean annual willingness to donate of MYR 9.04 among the fish hobbyists. Older fish hobbyists and those who owned fighting fish were more willing to donate for their conservation. These findings implied the preferences of fish hobbyists and provided insights into the potential for funding fighting fish conservation.

5. Research from Europe

In their article "Estimation of fishery losses from great cormorants during the wintering period in Greek lagoons (Ionian Sea, W. Greece)", Katselis and colleagues estimated fish losses from great cormorants (*Phalacrocorax carbo sinensis*) in lagoons of the Amvrakikos Gulf in Greece during the winter. They incorporated several parameters into their modeling approach, including fish population growth, age and size of fish prey, great cormorant numbers, and fishing strategies. Mugilids dominated both the lagoon fisheries and the great cormorant's diet, leading to the conclusion that there is a high level of conflict between great cormorants and fisheries in the lagoons of the Amvrakikos Gulf.

In their article "Wildlife knowledge and attitudes toward hunting: A comparative hunter–non-hunter analysis", Raftogianni and colleagues compared the attitudes toward hunting and the knowledge about wildlife species between hunters and non-hunters in Greece. The hunters stated significant motivations for hunting, especially as a recreational social activity, and considered hunting a valuable management tool. Non-hunters believed that hunting is a source of pride for hunters. They were critical or even negative toward hunting, mostly as an activity but also as a management tool. Interestingly, the hunters had greater knowledge about the biology and ecology of wildlife species, both game and non-game, than the non-hunters. These findings revealed a conflict between hunters and non-hunters and suggested that engaging in outdoor activities would increase public awareness about wildlife. Policies that could reduce this conflict include educating both hunters and the public about good hunting practices and promoting outdoor recreational activities.

Sultana and colleagues examined the factors that affected urban bird diversity in small green spaces in Freiburg and Regensburg, Germany, in their article "Neighboring green network and landscape metrics explain biodiversity within small urban green areas—A case study on birds". They found that variations in species richness and composition were explained by green networks in Freiburg and by green networks and landscape metrics in Regensburg. They concluded that bird communities in small urban green spaces could benefit from the spatial configuration that includes water bodies and other green areas in their vicinity. These findings could be used by urban managers to improve the habitat quality for urban birds and thus promote biodiversity, human well-being, and, ultimately, sustainability.

Liordos and colleagues investigated the ecological requirements of birds in small green spaces in the urban core of Kavala, Greece, and Rovaniemi, Finland, in their article "Niche analysis and conservation of bird species using urban core areas". They determined species niches using the outlying mean index. Niche characteristics differed between cities, while bird species occupied different niches, showing significant niche specialization. In both cities, bird species could be grouped into urban adapters (mostly found in larger green spaces, with high vegetation cover and far from the city center) and urban exploiters (mostly found in green spaces close to the densely built and noisy city center). The conservation priority of urban bird species was also determined based on niche specialization and conservation status. These findings allowed suggestions of measures for enhancing urban bird diversity. The creation and maintenance of large green spaces, the increase in tree and shrub cover, and the retention of mature trees and dense shrubberies would benefit adapter species. Urban exploiters would benefit from the retention of balconies, holes in roofs, and artificial nest boxes. The enhancement of bird diversity in urban areas would promote both the conservation of biodiversity and human well-being.

6. Global Wildlife Conservation

In their article "Prioritisation of charismatic animals in major conservation journals measured by the altmetric attention score", Prokop and colleagues investigated the representation of animals and plants in the covers and articles of three major conservation journals: *Conservation Biology*, *Journal of Applied Ecology*, and *Conservation Letters*. They found that the covers more often depicted mammals, reptiles, and amphibians, and depicted fish less often. Most published articles also concerned mammals, while articles about mammals, invertebrates, and amphibians received the most citations. These results suggested a taxonomic bias in scientific research, with scientists and the public preferring large mammals over other species. Policies are needed to promote support for taxa neglected by all parts of society to promote conservation and achieve sustainability.

7. Conclusions

This Special Issue included research concerning sustainable wildlife management in both natural and urban areas. Research was carried out in many countries around the globe, thus providing an overall snapshot of hot contemporary issues. Key management themes were identified, such as wildlife conservation, human–wildlife conflict and coexistence, and sustainable resource management and use. The research presented would also appeal and be useful to scientists studying a variety of taxa, including mammals, birds, amphibians, reptiles, fish, and invertebrates. We hope that the research presented in this Special Issue will help wildlife managers design successful management plans and provide insights for promoting much-needed relevant research in the field.

Funding: This research received no external funding.

Data Availability Statement: No new data were created or analyzed in this study. Data sharing is not applicable to this article.

Conflicts of Interest: The author declares no conflicts of interest.

References

1. Brooks, T.M.; Mittermeier, R.A.; da Fonseca, G.A.B.; Gerlach, J.; Hoffmann, M.; Lamoreux, J.F.; Mittermeier, C.G.; Pilgrim, J.D.; Rodrigues, A.S.L. Global biodiversity conservation priorities. *Science* **2006**, *313*, 58–61. [CrossRef] [PubMed]
2. Steffen, W.; Grinevald, J.; Crutzen, P.; McNeill, J. The Anthropocene: Conceptual and historical perspectives. *Philos. Trans. R. Soc. Lond. Ser. A* **2011**, *369*, 842–867. [CrossRef] [PubMed]
3. Waters, C.N.; Zalasiewicz, J.; Summerhayes, C.; Barnosky, A.D.; Poirier, C.; Galuszka, A.; Cearreta, A.; Edgeworth, M.; Ellis, E.C.; Ellis, M.; et al. The Anthropocene is functionally and stratigraphically distinct from the Holocene. *Science* **2016**, *351*, aad2622. [CrossRef] [PubMed]
4. Ceballos, G.; Ehrlich, P.R.; Barnosky, A.D.; García, A.; Pringle, R.M.; Palmer, T.M. Accelerated modern human–induced species losses: Entering the sixth mass extinction. *Sci. Adv.* **2015**, *5*, e1400253. [CrossRef] [PubMed]
5. Dirzo, R.; Isaac, N.J.B.; Young, H.S.; Collen, B. Defaunation in the Anthropocene. *Science* **2014**, *345*, 401–406. [CrossRef] [PubMed]
6. Young, H.S.; McCauley, D.J.; Galetti, M.; Dirzo, R. Patterns, causes, and consequences of anthropocene defaunation. *Annu. Rev. Ecol. Evol. Syst.* **2016**, *47*, 333–358. [CrossRef]
7. United Nations Department of Economic and Social Affairs, Population Division. World Population Prospects 2022: Summary of Results. United Nations. 2022. Available online: https://www.un.org/development/desa/pd/sites/www.un.org.development.desa.pd/files/wpp2022_summary_of_results.pdf (accessed on 1 March 2024).
8. United Nations, Department of Economic and Social Affairs, Population Division. World Urbanization Prospects: The 2018 Revision (ST/ESA/SER.A/420). United Nations. 2018. Available online: https://population.un.org/wup/Publications/Files/WUP2018-Report.pdf (accessed on 1 March 2024).
9. Liordos, V.; Jokimäki, J.; Kaisanlahti-Jokimäki, M.-L.; Valsamidis, E.; Kontsiotis, V.J. Patch, matrix and disturbance variables negatively influence bird community structure in small-sized managed green spaces located in urban core areas. *Sci. Total Environ.* **2021**, *801*, 149617. [CrossRef] [PubMed]
10. Grimm, N.B.; Faeth, S.H.; Golubiewski, N.E.; Redman, C.L.; Wu, J.G.; Bai, X.M.; Briggs, J.M. Global change and the ecology of cities. *Science* **2008**, *319*, 756–760. [CrossRef] [PubMed]
11. World Commission on Environment and Development. Our Common Future. United Nations. 1987. Available online: https://digitallibrary.un.org/record/139811 (accessed on 1 March 2024).
12. Dernbach, J.C. Achieving sustainable development: The centrality and multiple facets of integrated decision making. *Indiana J. Glob. Leg. Stud.* **2003**, *10*, 247–285. [CrossRef]
13. Madden, F. Creating coexistence between humans and wildlife: Global perspectives on local efforts to address human–wildlife conflict. *Hum. Dimens. Wildl.* **2004**, *9*, 247–257. [CrossRef]
14. Redpath, S.M.; Young, J.; Evely, A.; Adams, W.M.; Sutherland, W.J.; Whitehouse, A.; Amar, A.; Lambert, R.A.; Linnell, J.D.C.; Watt, A.; et al. Understanding and managing conservation conflicts. *Trends Ecol. Evol.* **2013**, *28*, 100–109. [CrossRef] [PubMed]
15. Peterson, M.N.; Birckhead, J.L.; Leong, K.; Peterson, M.J.; Peterson, T.R. Rearticulating the myth of human-wildlife conflict. *Conserv. Lett.* **2010**, *3*, 74–82. [CrossRef]
16. Woodroffe, R.; Thirgood, S.; Rabinowitz, A. (Eds.) People and Wildlife: Conflict or Coexistence? Cambridge University Press: Cambridge, UK, 2005.
17. Young, J.C.; Marzano, M.; White, R.M.; McCracken, D.I.; Redpath, S.M.; Carss, D.N.; Quine, C.P.; Watt, A.D. The emergence of biodiversity conflicts from biodiversity impacts: Characteristics and management strategies. *Biodivers. Conserv.* **2010**, *19*, 3973–3990. [CrossRef]
18. Caughley, G.; Sinclair, A.R.E. *Wildlife Ecology and Management*; Blackwell Scientific Publications: Oxford, UK, 1994.
19. Bookhout, T.A. (Ed.) *Research and Management Techniques for Wildlife and Habitats*, 5th ed.; Wildlife Society: Bethesda, MD, USA, 1994.

Disclaimer/Publisher's Note: The statements, opinions and data contained in all publications are solely those of the individual author(s) and contributor(s) and not of MDPI and/or the editor(s). MDPI and/or the editor(s) disclaim responsibility for any injury to people or property resulting from any ideas, methods, instructions or products referred to in the content.

sustainability MDPI

Article

Climate Change and Anthropogenic Factors Are Influencing the Loss of Habitats and Emerging Human–Elephant Conflict in the Namib Desert

Markus Shiweda [1,*], Fillipus Shivute [2], Ana Raquel Sales [1] and Mário J. Pereira [3]

1 Department of Biology, University of Aveiro, 3810-193 Aveiro, Portugal
2 Apex Geospatial Intelligence, Khomasdal, Windhoek P.O. Box 10614, Namibia
3 Department of Biology and CESAM, University of Aveiro, 3810-193 Aveiro, Portugal
* Correspondence: shiweda@ua.pt

Abstract: Climate change and anthropogenic factors' impact on habitat loss is a growing problem that is influencing unsustainable wildlife local-population home range shifts and triggering an increase in human–wildlife conflict (HWC). Yet, keystone species involved in HWC such as elephants play a vital role in nature-based ecosystem services and have important economic and cultural value to the people that are living with them. To understand how climate change and anthropogenic factors affect habitat loss and elephants' home range shift, the movement of Namib desert-dwelling elephants was monitored and observed in the Ugab River basin between February 2018 and November 2020 at fortnight intervals. There are 87 elephants in the Ugab River basin that are distributed into two subpopulations: desert-dwelling elephants (N = 28) and semi-desert-dwelling elephants (N = 59). To achieve the objective of the study, land cover change, elephant movement, rainfall, and temperature data were analysed using ArcGIS spatial and statistical tools, such as image analysis, optimised hot spot analysis (OHSA), and cost distance analysis, to distinguish habitat vegetation changes and home range shifts and how these link to emerging human–elephant conflict (HEC) hot spots. Human farming activities, poor rainfall, and frequent droughts are responsible for the loss of habitat of around 73.0% in the lower catchment of the ephemeral river streams; therefore, the urgency of conserving and sustaining these habitats and desert-dwelling elephants is discussed here.

Keywords: desert-dwelling elephants; home range shift; habitat loss; Namib Desert; ephemeral rivers; vegetation cover; NDVI; human–elephant conflict

Citation: Shiweda, M.; Shivute, F.; Sales, A.R.; Pereira, M.J. Climate Change and Anthropogenic Factors Are Influencing the Loss of Habitats and Emerging Human–Elephant Conflict in the Namib Desert. *Sustainability* **2023**, *15*, 12400. https://doi.org/10.3390/su151612400

Academic Editor: Vasilios Liordos

Received: 28 June 2023
Revised: 24 July 2023
Accepted: 27 July 2023
Published: 15 August 2023

Copyright: © 2023 by the authors. Licensee MDPI, Basel, Switzerland. This article is an open access article distributed under the terms and conditions of the Creative Commons Attribution (CC BY) license (https://creativecommons.org/licenses/by/4.0/).

1. Introduction

The overall African elephant population continues to decrease, which might leave one of the world's most charismatic species in jeopardy of extinction [1]. The literature indicated that habitat loss and fragmentation, poaching for ivory, and human–elephant conflict (HEC) are among the most researched factors contributing to the elephant population's decline [2–4]. It is further established that the impacts of human activities [5,6] and climate change undoubtedly lead to habitat and biodiversity loss [7–9]. Targets to limit extinction are not easily achievable [10], affecting species and habitats at local and regional scales [11–13], such as desert-dwelling elephants and their habitats in the Namib Desert. Recently, these worrying observations have led to the reclassification of the African savanna elephant (*Loxodonta africana*) and the forest elephant (*L. cyclotis*) from vulnerable to endangered and critically endangered, respectively, by the IUCN [14–16]. Yet, this keystone species plays a vital role in nature-based ecosystem services and has important economic and cultural value to human populations [17].

The world population of elephants was estimated to be over 1.3 million in 1979 [18], but it has drastically decreased over the last four decades, as some local and regional populations have declined significantly. By 2007, the elephant population in Africa was estimated at 470,000 to 690,000 [19]. A further decline of 144,000 elephants was recorded in 2014, and the population has continued to decline at an unprecedented rate every year [4]. Until 2021, there were approximately 415,400 elephants (*L. africana* and *L. cyclotis*) in Africa [20,21], but population estimates can be questionable, and the number of elephants roaming the African continent may be smaller than that presented in many studies. Data from the Great Elephant Census (GEC) indicate that only 16.0% of the surveyed African savanna elephants roam out of protected areas. However, these areas represent 80.0% of the elephant distribution range in Africa [22]. Thus, most of the elephant's distribution area does not have formal protection, and they are more exposed to the possibility of losing major habitats.

Namibia is one of the countries with a high population of free-roaming elephants outside of national parks. The Ministry of Environment and Tourism (MEFT) reports that 31.0% of 23,736 elephants in Namibia roam outside protected areas [23]. In the 1900s, these elephants had a known home range—an area over which an animal (or group of animals) regularly lives and traverses in its normal activities—of 22.1% (calculated from the report of De Villiers (1975), as cited in MEFT (2021) [23], and they may have had a distribution area—a geographical area where a species can be found—greater than 56.0% of the country's total area of 823,000 km^2 (Supplementary Figure S1). However, this area declined to 13.7% by the 1990s before increasing again to 21.3% in 2020. Nevertheless, not all local populations are growing despite the elephant's national population growth, especially in the northwest part of the Namib Desert, where fragmented local populations within ephemeral river basins, including the Ugab River basin, are reported to be declining [24,25].

1.1. The Desert Elephants

Namib Desert elephants were once close to extinction, and their population has been classified as local and restored since the early 1980s [23]. A genetic distinctiveness between these elephants and other populations of *L. africana* has not yet been established, and studies indicate that what enabled them to survive in the extremely arid environment of the desert was their high learning capacity and adaptive behaviour [26,27]. Thus, they are an ecotype instead of a different species.

This ecotype is divided into two subpopulations: desert-dwelling elephants and semi-desert-dwelling elephants; both are widely distributed throughout communal land and community conservancy. Desert-dwelling elephants roam freely within the ephemeral river basins of the Erongo and Kunene regions (the latter is in the northwest of the country and partially hosts our study area) and can be found below the 200 mm isohyet of each basin [28–31]. More specifically, the herds of this population include less than 250 elephants and are found at the lower catchments of major basins of the Ugab, Huab, Hoanib, Hoarusib, and Uniab Rivers. The adjacent populations of over 200 elephants found further inland in the semi-arid savanna grasslands at the upper catchments of the Ugab and Huab River Basins are classified as transitional or semi-desert elephant populations [23,32]. Namib Desert is characterised by frequent severe droughts, especially in the last 10 years [33], and an increase in the average maximum temperature recorded during the hot months from October to January [34,35], potentially making the Namib Desert one of the harsh environments home to the largest land mammals that may become inhabitable in the near future.

1.2. Namibia as One of the Most Vulnerable Countries to Climate Change and HWC

Namibia is classified as being among the most vulnerable countries relative to climate change, and it is characterised by reduced rainfall; a rapid increase in the number of consecutive dry days, flash floods (Table S1), and frequent droughts; and high-temperature increases [13,36–39], a situation predicted to be getting worse for tropical regions in a recent IPCC Assessment Report (AR6) [40]. The country has an average annual rainfall of 340 mm, with these values ranging from 0 mm (for example, in the western desert) to over than 600 mm (as in the northeast savanna woodland ecosystem) [41].

Namibia is projected to experience a radical temperature increase of 1.0 °C to 3.5 °C in the summer and 1.0 °C to 4.0 °C in the winter by 2050–2065 [34,38,39], further contributing to evaporation that is already exceeding annual rainfall [34]. Moreover, only 1.0% of the annual rainfall infiltrates to recharge the underground aquifers [38], but the communal farms present in the regions of ephemeral rivers (as with the Ugab River communal farms) depend mostly on borehole water throughout the year [23]. Farmers with limited income rely more on few natural springs for human and livestock water consumption [42], in addition to these springs being the main water source for wildlife. Furthermore, the temperature increase is projected to negatively affect ecosystems and local populations of endemic species, possibly leading to their extinction [13].

The observed trends from the World Bank Group country's climate portal [43] show an increase in the annual maximum mean temperature (AMaMT) (Figure 1A), a moderate rise in the annual mean temperature (AMT) (Figure 1B), and a rapid increase in the observed number of consecutive dry days (CDD) (Table S2) over the past 30 years [44]. The 95-confidence interval equation line referring to the average of the national annual precipitation (NAP) (Figure 1C) and the Kunene region annual precipitation (KAP) (Figure 1D), over this time interval, tends towards values indicating a slight increase in annual mean rainfall. However, a loess line with a 95-confidence interval displays an average decrease in NAP since 2008 and in KAP since 2006 (Figure 1E,F).

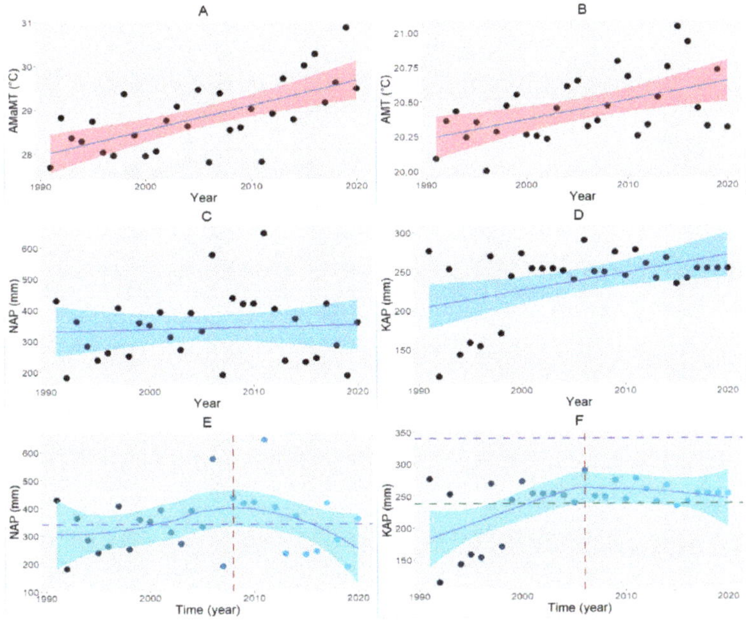

Figure 1. Climate data observation in Namibia over 30 years. Figure shows linear model trend lines for the annual maximum mean temperature (AMaMT) in °C (**A**), the annual mean temperature (AMT) (**B**), the average of the national annual precipitation (NAP) (**C**), and the average of the Kunene region annual precipitation (KAP) (**D**). The average observed precipitation between 1991 and 2020 regarding to the NAP (**E**) and KAP (**F**) data is indicated by the blue and green dashed abline, respectively, whereas the red dashed abline indicates the highest average observation point. Figure is constructed with the climate data in Table S2, adapted with permission from the World Bank Group Climate Change Knowledge Portal (CCKP) (2022, the World Bank Group) [43], using a an Open Access ggplot2 V3.4.2 R package by Wickham (2016) (Accessed from: https://ggplot2.tidyverse.org, accessed on 20 April 2023) [45] in R 4.3.0 Open Access software (accessed from: https://www.r-project.org/, accessed on 20 April 2023) [46].

As the arid transitional areas are more vulnerable to climate change [40], and Namibia's human population is widely scattered in these rural areas—for example, in 2011, more than half of the national population was dispersed in these areas, and almost 75.0% of the inhabitants of the Kunene region are within rural areas—the people and wildlife of Namibia may be facing an unprecedented challenge with advancing climate change. As competition for resources intensify, the interaction between people and wildlife across the landscape is believed to increase HWC [47]. Uncontrolled livestock herds grazing on unfenced land led to an overlap of livestock grazing land and wildlife home ranges [48], but the problem does not end there. Human settlement growth and increasing farmland are leading to deforestation and desertification [49–51], threatening vulnerable ecosystems and biodiversity [13,38,39].

The government encourages communal farmers to register unoccupied land as community conservation units to care for wildlife [52]. In addition, the government and non-governmental organisations (NGOs) develop strategies to empower the residents on how to coexist with wildlife [48,52–54]. Still, the efforts seem not to be enough, with HWC increasing [55] and expanding into new areas [23].

1.3. Namib Desert Ephemeral Rivers and Aims of the Study

The ephemeral rivers of the Namib Desert cut through igneous rocks [56] and may flow for a few days during the summer across this desert [57]. They are mainly dominated by *Faidherbia albida*, *Vachellia erioloba*, and *V. karroo* trees (previously belonging to the genus *Acacia*) and *Salvadora persica* bushes [58–62]. A high abundance of *Phragmites australis* and *Tamarix usneoides* were also observed at natural springs within the riverbed. The importance of these habitats should not be underestimated, as a few perennial springs in ephemeral desert river catchments are the primary source of water for wildlife and riparian vegetation, and both domestic and wildlife animals depend on this vegetation. For example, riparian forests on the riverbanks are essential for elephants, attracting them during dry seasons for food and shade (shelter) [30,63].

Findings outlined by previous studies conducted on dry lands and ephemeral rivers of the Namib Desert on drivers of underground water decline [64] and observations of large trees die-off [61,62,65] supported the development of the objectives of this study. This work established that climate change indicators (such as temperature rise, frequent droughts, and decreasing rainfall [40,66]) and anthropogenic factors (such as overgrazing and over-abstraction of groundwater [57]) negatively affect desert habitats and the availability of vegetation, and such changes are leading to altered elephant behaviour and risks to their ability to survive in this region.

The general aims of this study were to analyse the potential threats to desert-dwelling elephants' habitats, especially climate change and anthropogenic factors, and to understand how these threats influence elephant home range shifts. Spatiotemporal analysis of elephant movement was used to track home range shift patterns, helping to understand what is causing this, where elephants are going, and how such events might impact human populations and, consequently, the elephants themselves.

2. Methods

2.1. Study Area

The study area is in the northwest part of the country, where the Ugab River basin (Figure 2A) is located, and where the maximum temperature exceeds 40 °C in the summer [29]. The Ugab River is one of the ephemeral rivers present in the interior of Namibia that characterises the country's aridness [67]; it has a catchment area of 28,000 km^2, a length of over 540.0 km, and it flows east-to-west into the Atlantic Ocean. There are no perennial rivers in this region [67], and the situation of Namibia's arid landscapes, which already lack surface water during the dry season [57], is worsening with advancing climate change. Consequently, pressures on these ecosystems increase, as do concerns about the future of wildlife and human livelihoods [44,68]. The Ugab River desert-dwelling elephants historically roamed within the lower catchment, between 20.6°–21.1° S and 13.9°–15.2° E. Therefore, the previously known home range for the population was demarcated based on descriptive and observational studies [27–31,63,69] so that it could later be compared with the home range observed in this study.

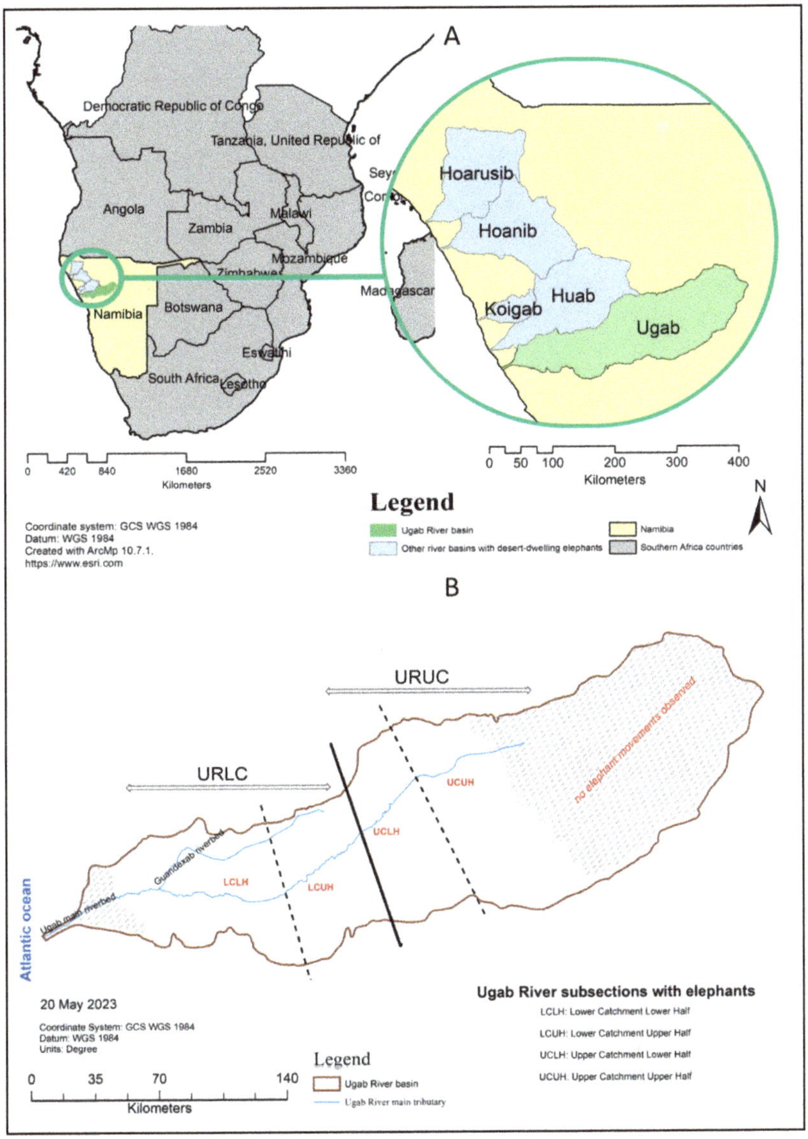

Figure 2. (**A**) Location of the Namib Desert major ephemeral rivers inhabited by elephants in northwest Namibia. (**B**) Division of Ugab River basin into the Ugab River Lower Catchment (URLC)—subdivided into Lower Catchment Lower Half (LCLH) and Lower Catchment Upper Half (LCUH)—and the Ugab River Upper Catchment (URUC)—subdivided into Upper Catchment Lower Half (UCLH) and Upper Catchment Upper Half (UCUH). The border and watershed shapefiles used to construct the figure were obtained from the Ministry of Agriculture, Water and Land Reform, Luther Street, Windhoek, Namibia (2022).

Given the isolation of the two populations in the past, the Ugab River basin was divided into two sections: the Ugab River Lower Catchment (URLC) and the Ugab River Upper Catchment (URUC) (Figure 2B). The URLC was further subdivided into two sections, the Lower Catchment Lower Half (LCLH) and the Lower Catchment Upper Half (LCUH). The two sections of the URLC are the historic home range for the desert-dwelling elephants, with the LCLH representing the area mainly used in the winter and the LCUH in the summer [28,58,60,70]. The URUC was split into the Upper Catchment Lower Half (UCLH) and the Upper Catchment Upper Half (UCUH). The URUC represents the historic home range of the semi-desert elephant population. The two populations' home ranges have not been overlapping in the past based on information obtained through ad hoc interviews on historical observations made by farmers, researchers, and local conservation institutions.

2.2. *Data Collection*

2.2.1. Elephant Observation Movements and Population Structure

The data used in this study were collected in the field and from open-source databases and previous studies. Data from other studies are presented as part of analytical methods and include vegetation raster, rainfall, temperature, historical movement, and home range data. Regarding the latter, it was considered from the information present in Brown et al. (2020), Enzerink and Liefferink (2017), Garstang et al. (2014), Ishida et al. (2016), MEFT (2007, 2020, 2021), Viljoen (1987, 1989a, 1989b), and Viljoen and Bothma (1990) [23,25,27–29,31,47,63,69,71,72]. Additional data from other sources were used: Landsat satellite images (United States Geological Survey) [73], precipitation (World Bank Group) [43], records of drought events cited in Hitila (2019), the Office of Prime Minister's Directorate of Disaster Risk Management (2021), and the Cooperation in International Waters in Africa (CIWA) (2021) that make use of the Emergency Events Database (EM-DAT), Thomson (2021) [33,68,74,75].

Between February 2018 and September 2020, 28 elephants in the URLC (3 herds and 3 breeding males) and 59 in the URUC (3 herds and 7 breeding males) were observed at fortnight intervals. By the end of data collection, 5 and 7 of these individuals, respectively, had died. Daily tracking was conducted from 08h00 to 18h00 by foot and car. The elephants were not collared during the project, requiring careful tracking along small trails between farms (usually 3.0 to 5.0 km apart). Other field observations (such as assessing boreholes conditions and talking to farmers) were carried out until November 2020.

The guidelines of Princeton University [76] on identifying and ageing animals' markings by animal tracking were adopted and modified. An elephant tracking chart was developed covering topics such as the ageing of elephants' footprints and dung, the softness of leaves fed and dropped on the ground, and the classification of individual imprints. Data collectors aged elephant dung as: recent, if it was less than 3 h old (occurred during the day) or 8 h old (from the previous night); not so recent, if it was between 1 to 2 days; and old, if it was between 3 to 5 days. These data were important to ascertain what was the tracking time required to find the herd or individual males.

Even though savanna elephant dung has been fairly studied by many researchers, such as Nchanji and Plumptre (2001) and Barnes et al. (1997) [77,78], the studies did not focus on dung ageing for elephants tracking purposes. However, Masunga et al. (2006) [79] presented an important perspective, mentioning how dung ageing monitoring gaps vary widely and discussing the role of moisture, shade from trees, and the season in the dung decay time-frame. The study presented here also used dung colour to age elephant dung. The researchers relied on their own tracking knowledge of elephants and other species supplemented with information from the literature to generate their own elephant dung ageing chart. A dung was identified as recent if it was wet on the surface and varied in colour from light to dark green (depending on the colour of the plant the elephant ate, which varies with the seasons)—information supported by 40 years of experience tracking elephants and other mammals in the Namib Desert of Mattias Kangumbe (27 February 2018, personal communication).

Leaves that fell on the ground while the elephants were feeding were used to age elephant tracks. For this, it was important to consider the influence of weather conditions on the ageing rate. Leaves get drier as they age but can remain fresh for up to 24 h if it rains, is cloudy or humid, or when it is cold in the winter. In addition, ageing also depends on the plant species, with the leaves of succulent plants staying wet longer than plants with thinner, softer leaves.

To avoid disturbance by invading vehicles, and considering that herds and individual males have shown different levels of tolerance to human presence depending on distance, non-invasive methods and distance restrictions were set up based on animals' behaviours and reactions. Desert-dwelling elephants were not shy as semi-desert-dwelling elephants since data collectors could approach the animals closer without elephants showing signs of being disturbed. Therefore, we maintained a minimum 50 m distance if desert-dwelling elephants were relaxed and a 100 m distance if they were not. When approaching the elephants at the URUC, researchers maintained a 100 m minimum distance when the animals were relaxed (except when feeding in bushy areas) and 150 m if not settled.

When an individual male or a herd was found, the location coordinates were recorded. Individual elephants were counted in every herd spotted, and the elephants' ages were estimated following existing methods [80,81]. The physical characteristics of individuals (males or members of a herd) were identified at every new sighting location, as described in Viljoen (1989) [28], adopted from Douglas–Hamilton (1972) [82]. The data recorded for each individual were sex, age, and behaviour, and, in the case of herds, the number of adults and the total number of elephants that constituted them were also recorded [71,83]. Identification of matriarchs and relationships between calves, juveniles, and adult cows was carried out carefully, following Elephant Voices guidelines [83]. Each individual and each herd were assigned a unique code (e.g., ULH1 for Ugab River Lower Catchment Herd 1 and UUH1 for Ugab River Upper Catchment Herd 1) for future reference and identification.

The previous photographs taken of the elephants, in parallel with the information presented in the guidelines [83] and in an unpublished report from 2019 [84], to verify if a certain individual (or herd) was known or not. For cases where it was the first time that the elephant was sighted, a new individual identification code was created. Specific details such as cuts in ears and tusk shapes helped to identify individuals, which made it possible to know which elephants were found dead, broke facilities at a farm, or gave birth. Elephant sightings were recorded more than once a day if the same individual or herd changed the position for a distance longer than three km from the previous sighting site. This record made it possible to know the direction in which the individual/herd was going, also facilitating its observation and tracking in the next day.

2.2.2. Human–Elephant Conflict Events

Human–elephant conflict (HEC) events, such as damage to water infrastructure, were recorded through ad hoc interviews with farmers and community leaders. In order to guarantee the veracity of the information provided, the project data collector always observed the damage to the infrastructure accompanied by the farmers. Elephant sightings and conflict events were also reported via phone calls. However, mobile cell phone network limitations, together with COVID-19 restrictions implemented at the time of the collection of this data, affected visits to farms and communication with the people involved, and may have led to limitations in the collected data.

The coordinates of the locations of the events (also referred here as HEC hot spots) were recorded on-site and through ArcGIS 10.7.1 software. Records included the types of infrastructure damaged and the frequency of elephant visits to the farm or water point (monthly, weekly, or seasonally). For reasons of animal safety and the ongoing fieldwork, the coordinates and physical description of the sites will not be provided in the attached supplementary materials (Table S3).

2.3. *Data Analysis*

2.3.1. Spatiotemporal Analysis of Elephant Movement Data to Trace Home Range Shift Patterns

The weighted optimised hot spot analysis (OHSA) tool of ArcMap 10.7.1 was used to map the desert-dwelling elephant distribution shift and HEC hot spots based on the number of sightings observed within specific areas. Hot spot analysis is widely used in many research areas to assist in identifying areas of interest (e.g., fire management areas [85]), but it can also be applied in the study of spatial sciences and ecology [86]. The tool identifies and cluster statistically significant high values (hot spots) and low values (cold spots) [87], using an input feature class that contains data points at a 95.0% confidence level ($p = 0.05$) at Gi_Bin cluster levels of +3 and −3 (ESRI https://pro.arcgis.com/en/pro-app/latest/tool-reference/spatial-statistics/optimized-hot-spot-analysis.htm, accessed on 1 February 2021). Input data must meet OHSA requirements, i.e., data have to be spatially autocorrelated (and not randomly occurring), and consideration must be given to an appropriate projection and screening for outliers (an observation that lies an abnormal distance from other values) and any missing values. The spatial autocorrelation of the data was measured using Moran's Index, and it obtained a p-value < 0.01, indicating that the values in the dataset tended to cluster.

The OHSA tool filtered the data and excluded outliers from the analysis. Thus, this tool was initially used to identify the core distribution areas of the populations, and then the kernel density tool could be used to map and visualize the seasonal distribution density of the elephant population and HEC hot spots.

Finally, 30.0 km buffer of all sightings of males and herds from each catchment section were created to calculate the population's home range. The use of buffers to study animal movement and distribution has been used before to answer questions of interest about species [88], and the radius size of a buffer can vary between study areas, depending on climate and types of vegetation and habitat [89]. The 30.0 km buffers were used to estimate a potential distance to waterpoints, and the choice of this value was made considering the midpoint of the values presented in Viljoen (1989) [28], which indicated that elephants in the Namib Desert could travel far from waterpoints in the dry season in a range of 20.0–40.0 km/day. For data analysis, it was important to consider that the daily distance covered by elephants was greater in the desert than in wetter areas [90], even though the main area used (core area) in this arid habitat was usually smaller. Random visits to farms and settlements within the 30.0 km buffer also allowed us to gather information about the presence of elephants in villages and water points. Subsequently, to identify newly established home range and habitats, the data collected in this study were compared with data presented by Viljoen (1989) [28], referring to elephants in the distribution range below the 200 mm isohyet. Random visits to farms and settlements within the 30.0 km buffer also allowed us to gather information about the presence of elephants in villages and water points.

2.3.2. Identification of Historic Vegetation Cover Change and Habitat Modification

Vegetation health was analysed within the historic home range of desert-dwelling elephants at LCLH. The area included natural springs, *F. albida* trees, and several bush species. The vegetation at this site was mainly found in and near the riverbed, and there were patches of dead trees within the riverine, covering areas larger than 500 × 500 m. Therefore, it was intended to determine whether tree mortality was being influenced by the severe drought events and reduced rainfall in the region (Table S1) [33–35].

The use of multispectral satellite imagery bands from Landsat 7 and 8, accessed through the United States Geological Survey Earth Explorer (USGS https://earthexplorer.usgs.gov/, accessed on 1 February 2021), made it possible to examine vegetation cover and its health. The analysed vegetation images came from a location whose coordinates were 20.9°–20.9° S and 14.4°–14.7° E and had a length of 38.5 km, 1.0 km at the widest point, and an area of 164.5 km^2. The best images (those with less than 5.0% cloud cover) were downloaded separately for the wet seasons of 2000, 2006, 2010, 2015, and 2020. Afterwards, these images were inspected again (to look in more detail at cloud cover and dust), projected to the WGS_1984_UTM_Zone_33N, and merged using the mosaic raster function [91]. To render these raster datasets together, the bands 1–7 for Landsat 7 Enhanced Thematic Mapper (ETM+) and 1–8 for the Landsat 8 Operational Land Imager (OLI) were composted to form a single raster image for each year [91,92], using the ArcMap 10.7.1's Image Analysis Raster Composite tool. Bands combinations of 4, 3, and 2 and 5, 4, and 3 were selected for Landsat 7 ETM and 8 OLI surface reflectance data, respectively [93–95]. However, the OLI reflectance is greater than the ETM reflectance band, thus there could be some differences for the near-infrared (NIR) and visible light (VIS) produced [96]. The selected bands combination was suitable for classifying and quantifying vegetation and for driving the Normalised Difference Vegetation Index (NDVI), using the formula [97,98]:

$$NDVI = (NIR - VIS)/(NIR + VIS)$$

The land surface cover types were reclassified [91] to differentiate between seasonal vegetation (SVeg) cover, perennial unhealthy vegetation (PUVeg) cover, and perennial health vegetation (PHVeg) cover. An NDVI index score of zero indicated no vegetation, and the closer this value was to 1 the greater the density of green leaves [93]. Thus, vegetation with a score greater than 0.6 was considered healthy, and the condition of vegetation was verified in the field and using the application Google Earth Pro (for example, through the observation of cells coming from fully green or trees in bad condition). Other classes, such as bare sandy areas, rocky plains, mountainous areas, and human infrastructure (mainly small, corrugated iron houses), were difficult to differentiate, especially due to the homogeneity of the colours of the rocky areas. Therefore, these classes were reclassified into two categories: rocky and bare plains and mountainous areas, which were not reported in this study but contributed to the area calculation. Furthermore, the raster layers of classified attributes, which included vegetation classes, were converted into vector layers to calculate the total area. The percentage of vegetation cover (Table S5) was calculated from these data, using ArcGIS 10.7.1 raster geometry and calculator function. Finally, a linear estimation of the missing values of the years that were not calculated was carried out using a combination of the "mutate" and "na.approx" functions of the Dplyr (https://www.rdocumentation.org/packages/dplyr/versions/1.0.10/topics/mutate) and Zoo (https://www.rdocumentation.org/packages/zoo/versions/1.8-12/topics/na.approx) packages, respectively (packages accessed on 10 February 2021).

2.3.3. Identification and Mapping of Migration Corridors

To map the commuting and migratory corridors that connected elephant habitats, a least cost method was chosen, among several modelling approaches, including factorial least-cost paths analysis, circuit theory, and the resistant kernel described in Rudnick et al. (2012) [99]. A catchment polygon was generated from an Ugab River watershed and was used as extraction extent for the images from the Landsat 8 OLI satellite. These images were taken in April 2020 and were processed for the corridor analysis. A method similar to the one declared in 2.3.2. and used in other studies [97,98] was further used to generate the vegetation index. Viljoen (1989, 1990) [28,29] pointed out that suitable vegetation and riverbeds scan attract elephants, serving as a migration corridor or as a foraging area. This information contributed to the decision of the parameters used for corridor mapping.

A signature was created to reclassify and extract vegetation with an NDVI score value above 0.6 using a table query selection in ArcMap. Considering that the high vegetation density followed river streams [29], the slope and vegetation layers (as described in Evans et al. (2020) and Hazen et al. (2021) [88,89]) and the habitat patches (generated from vegetation cover) were used to model the suitable corridors. The cells with the highest vegetation index were assigned a low value, meaning to be least cost, and the same was repeated for the elevation raster file, with the lower elevation in river streams being preferred over mountainous areas [30]. The two classified raster files were converted into an accumulative cost distance raster file (determined based on the raster cell values) using the Cost Distance tool (ESRI https://pro.arcgis.com/en/pro-app/latest/tool-reference/spatial-statistics/optimized-hot-spot-analysis.htm, accessed on 30 August 2022). Another copy of a raster extraction was converted into a vector file that later served as feature polygons defining the habitat regions. This one and the cost distance file were used, as documented in the ArcGIS Pro 3.0 spatial cost connectivity tool, to generate the least cost paths. A 500 m buffer was created around the paths and then resampled with the original high-value vegetation index within the demarcated corridor buffer. The locations for the corridors were verified in the field for ground truthing purpose. Subsequently, the layers of "Habitat gain", based on observed movements, "Unoccupied Historic Habitat", and "Occupied Historic Habitat" were overlaid to create the final map.

3. Results

3.1. Elephant Movement, Home Range Shift, and Potential Emerging Human–Elephant Conflict Areas

Figure 3 shows the distribution of sightings of elephant movements from February 2018 to September 2020, made within communal conservancies (blue area) and the non-conservation regions. The average annual rainfall is indicated in mm within each home range section. This figure indicates the high prevalence of desert-dwelling elephants (in brown) in the URLC and semi-desert elephants (in blue) in the URUC. Still, the occurrence of semi-desert elephants in the lower catchment is visible, especially in the LCUH, and the presence of desert-dwelling elephants in the upper catchment, with a greater preference for the UCUH.

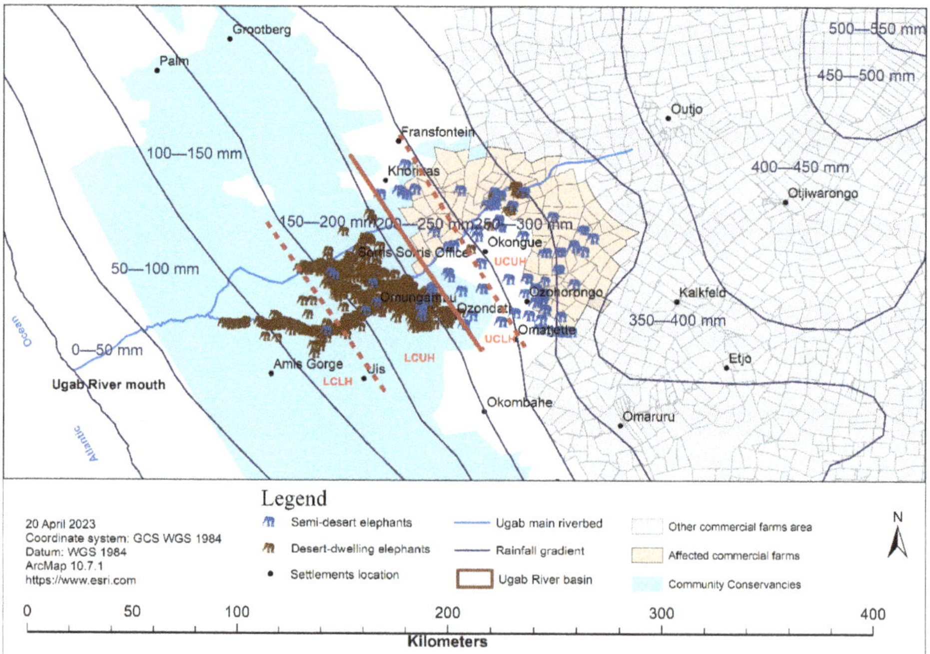

Figure 3. Distribution of elephant movement sightings from February 2018 to September 2020. A single elephant symbol refers to either an individual male, a group of males, a family group, or males and families groups. The brown symbols represent the desert-dwelling elephants, whereas the blue symbols represent the semi-desert (transitional) elephants. The land use shapefiles used to construct the figure were obtained from the Ministry of Agriculture, Water and Land Reform, Luther Street, Windhoek, Namibia (2022).

Overall, half of the sightings of elephants residing in the upper catchment home range moved within 35.0 km and 50.0 km north and south of the river, and the other half occurred within 15.0 km from the main URUC riverbed. However, sightings away from the main riverbed mostly occurred along the tributaries of the Ugab River, which runs through commercial farms in mountainous areas and rocky plains. The main range for the desert-dwelling elephants falls within community conservation areas (the light blue areas) that are managed by the residents as community conservation areas.

The Moran's I p-value < 0.01 obtained indicated that the data regarding elephants' movement were autocorrelated, and this was also confirmed by OHSA, which identified cold spots mainly in the URLC and hot spots emerging at the URUC, especially in the UCUH.

3.2. *Habitats, Home Range Shift, and Potential Emerging Human–Elephant Conflict Areas*

The desert-dwelling elephants have been shifting their distribution range eastward between 2018 and 2020. The overall yearly mean centre shifted by 3.8 km, between 2018 and 2019, and 60.4 km, between 2018 and 2020. Figure 4 shows that the sightings area occurred mainly along the riverbed (especially in the dry season since they move through bush and grass plains in the wet season) and that the distribution area is spreading wide-out towards the upper catchment as more major tributaries are formed. The figure demonstrates the density of desert-dwelling elephants mainly being distributed within the LCUH and LCLH in 2018 and 2019, with the latter section being more relevant during the dry season. In those years, in the wet season, elephants approached the UCLH. However, it was in 2020 that they spent a lot of time at the URUC, where they were not seen before, mostly within the UCUH from April to August. During that time, they were mainly at the commercial farms and used the UCLH section for migration between the URLC and UCUH. Most farmers located at the URUC complained that they were seeing more elephants than before, which was due to the combination of the local semi-desert elephant population and the desert-dwelling elephant population that moved into this area. As a result, damage caused by elephants at the farms increased.

The seasonal distribution of the semi-desert-dwelling elephant population (Figure 5) was mapped with the combination of sightings made in 2018, 2019, and 2020. This figure does not show data by season and year because fewer observations were recorded for this population and due to the fact that there were no major changes in annual home range. Map A indicates that elephants used the UCLH and UCUH sections during the dry season in a similar way, but, when at the UCLH, they agglomerate in the main riverbed. In the wet season, both sections of the URUC were used, but there was a tendency for elephants to concentrate in the UCUH, moving away from the main riverbed and overlapping commercial farms.

Conflict events in the lower catchment section were more frequent in areas closer to the main riverbed, and the animals seemed to spread out from it towards commercial farms in the upper catchment (Figure 6). Patterns of movement and conflict relate to the conditions of the waterpoints, i.e., the constant presence/absence of water and the accessibility to the waterpoint (both for adults and juveniles) influences the frequency it is visited (Table S3 and Supplementary Figure S2). Supplementary Figure S2 indicates the location of water points (drilled boreholes for groundwater abstraction) and the frequency with which elephants visit them (Table S3). Half of the visits recorded in the URUC occurred 15.0 km from the main riverbed, the other half 50.0 km from it, and the vast majority took place below the main riverbed (to the south). The URUC area has more accessible waterpoints, and they are being visited more frequently than those in the URLC (Supplementary Figure S2). These upper catchment waterpoints appear to be attracting more elephants, leading them to establish new habitats upstream.

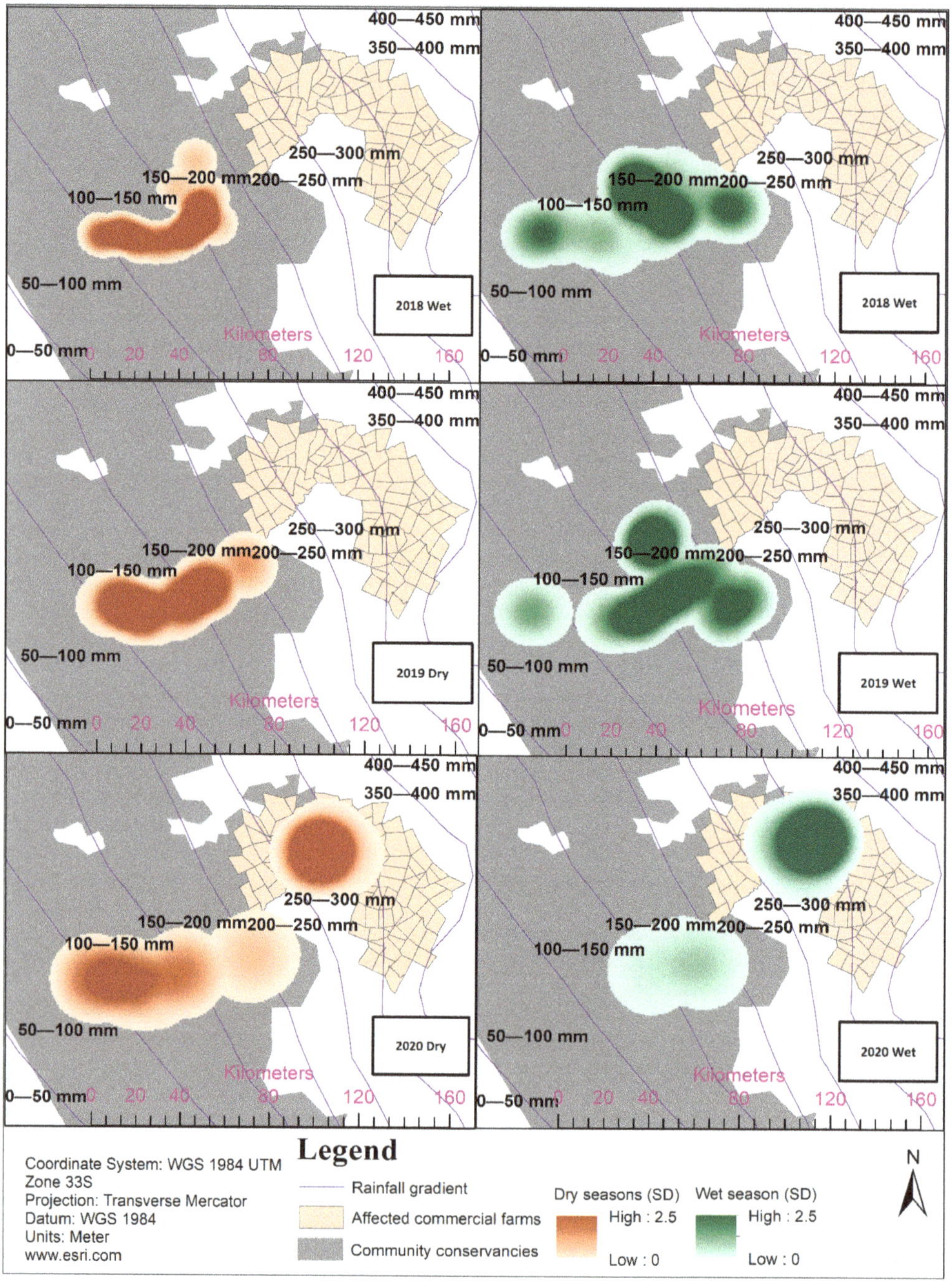

Figure 4. The seasonal distribution of the desert-dwelling elephant population during the dry and wet season in the years 2018, 2019, and 2020.

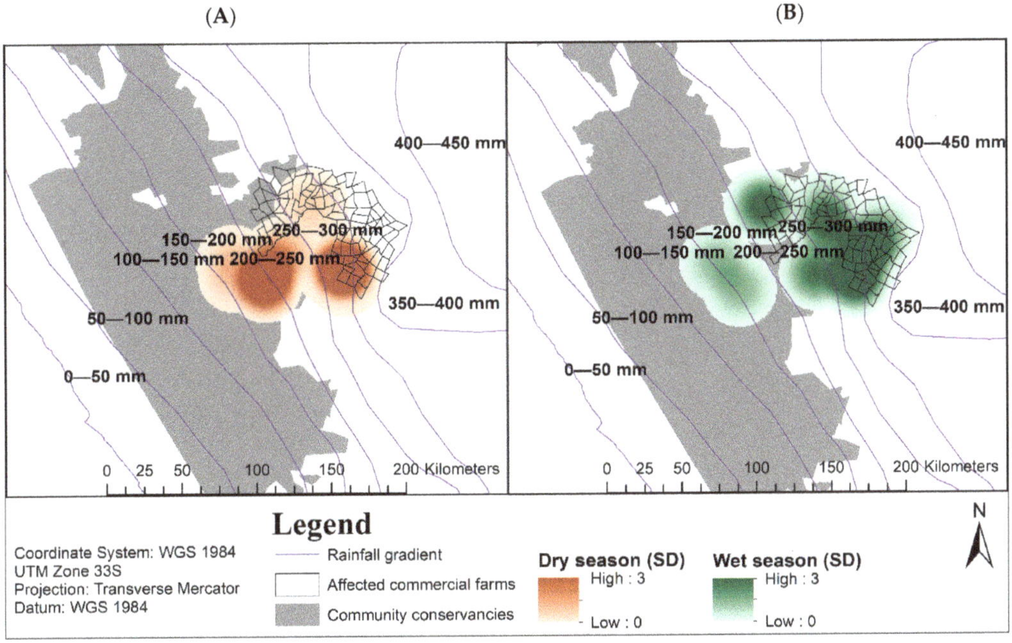

Figure 5. The seasonal distribution of the semi-desert-dwelling elephant population over the wet season (January to June) (**A**) and dry season (July to December) (**B**), combined for the years 2018–2020.

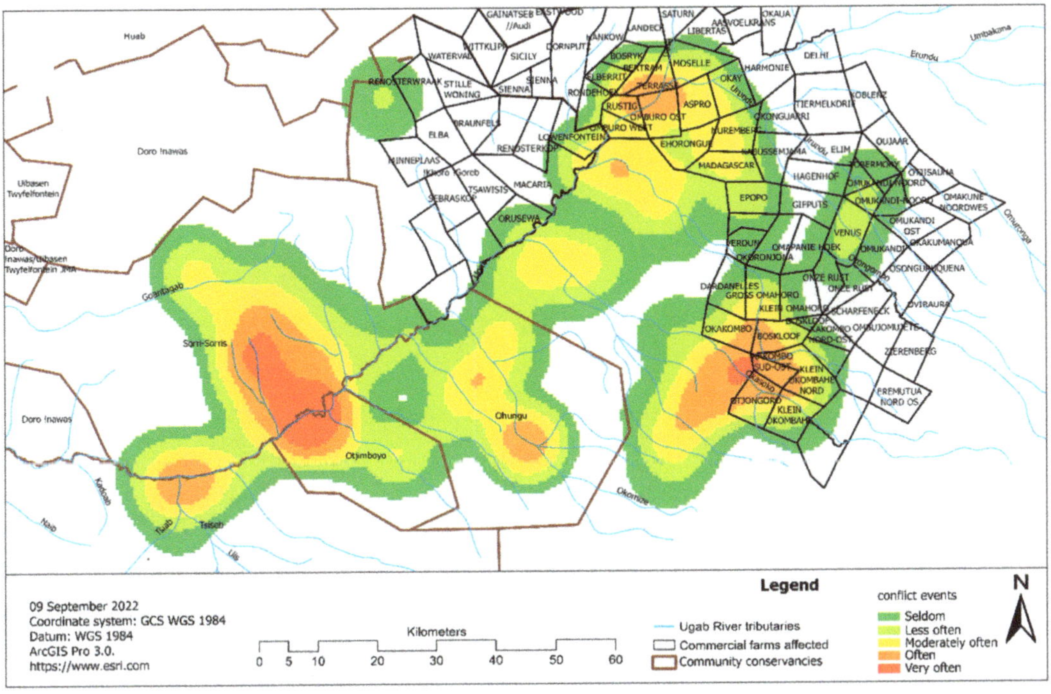

Figure 6. Distribution density of the human–elephant conflict events.

Only four waterpoints were often visited in the URLC, as some of them were damaged by elephants or stopped working and were not repaired. Five of the twelve boreholes within a 5.0 km of the main riverbed of this section, as well as two others further away, ran out of water due to the depletion of the groundwater table, eventually drying up.

3.3. Habitat Loss, Gain, Connectivity, and Migration Corridors

A "suitable historic habitat" was considered as a historic habitat that is still occupied by elephants and a "suitable new habitat" as areas that have been recently occupied by the desert-dwelling elephants (Figure 7) but that were previously occupied by the transitional herds. These areas have a high density of vegetation and suitable vegetation cover for elephants (mostly *F. albida*, *V. erioloba*, and *Colophospermum mopane*).

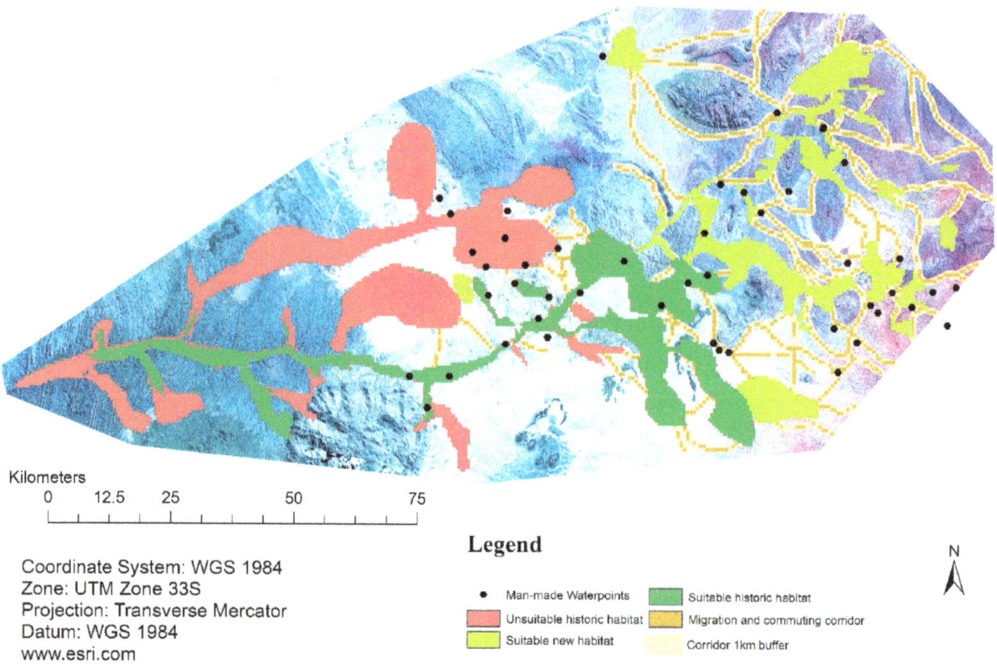

Figure 7. Unsuitable and suitable historic habitats and migration corridors for elephants.

The Ugab River lower catchment herds had 2243 km^2 of suitable historic habitat (Figure 7), which was reduced by 73.0% by 2018. However, the current Ugab River desert-dwelling elephants' overall home range has been estimated at 12,237.0 km^2, demonstrating considerable expansion. It covers more than a third of the Ugab River's catchment (about 29,175 km^2) and is larger than any historical home range for all Namib desert-dwelling elephants reported in previous studies [28,31].

The estimated historical suitable habitat started 30.0 km from the ocean to 150.0 km inland, measured in a straight line, and expanded 53.0 km east of the URUC (Figure 7). Recent expansion has increased the viable habitat by 130.7%, which covers 11.5% of the current elephant home range (Table S5).

3.4. Habitat Modification and Food Availability at the Ugab River Lower Catchment

Over the last 20 years, vegetation has decreased especially in the riverbed, where it used to be abundant and healthy (Figure 8 and Table S5), originating habitats with patches barely vegetated or dominated by dead *F. albida* and *V. erioloba*, (Supplementary Figure S3). Between 2000 and 2006, the percentage of SVeg cover increased by 4.0%, but this value reduced in 2007 and dropped sharply between 2010 and 2015 (Figure 8). Images captured in the field (Supplementary Figure S3) and matched with satellite images confirmed that many large *F. albida* trees mostly died from 2012 onwards. The PUVeg cover decreased every year between 2000 and 2020, affecting the availability of PHVeg.

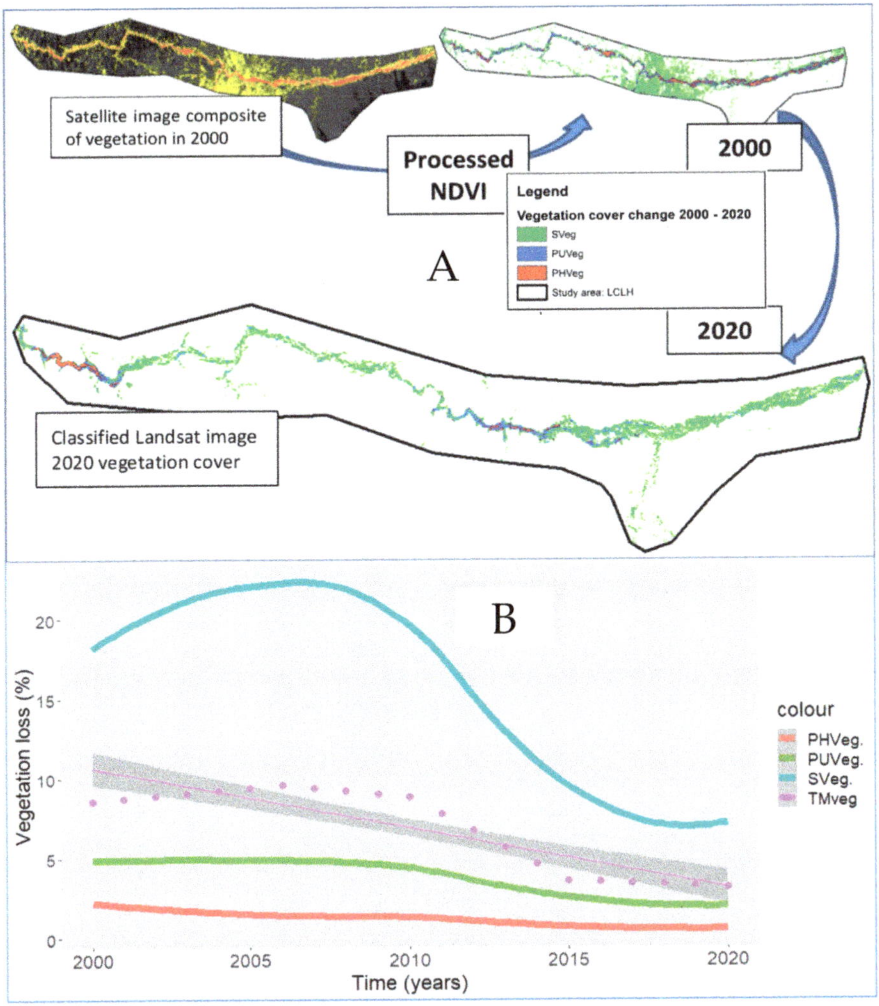

Figure 8. Vegetation changes in the Ugab River Lower Catchment. (**A**) Satellite image processed into NDVI, showing the changes between the vegetation cover in 2000 and 2020. (**B**) Percentage of vegetation cover between 2000 and 2020. The vegetation is divided into three categories: seasonal vegetation (SVeg), perennial unhealthy vegetation (PUVeg), and perennial healthy vegetation (PHVeg). (Analysed using ArcGIS Pro 3.0. https://www.esri.com, accessed on 20 March 2021 (**A**) and R 4.3.0, accessed from: https://www.r-project.org/, accessed on 20 April 2023 (**B**) [46].

Overall, it was observed that in 20 years the percentages of SVeg, PUVeg, and PHVeg decreased by approximately 11.5%, 3.0%, and 1.7%, respectively. Clearly, suitable vegetation covers for elephant habitats and palatable seasonal vegetation have declined, making the URLC habitats unsuitable for elephants, which appears to be driving them to new habitats in the URUC. Figure 8 suggests a relationship between the trends in vegetation loss and declining rainfall. These trends also coincide with those observed regarding the increase in temperature (Figure 1), drought events, and number of CDD (Table S1).

4. Discussion

4.1. Vegetation Loss, Climate Change, and Home Range Shift

Vegetation loss is directly associated with the degradation of habitats, as observed in arid and semi-arid environments, and is notably affecting biodiversity [8,100]. Desert elephants are one of many victims of this problem. Elephants are social animals that group together or form herds associated with the dynamics of their society [82]. The movement of elephant herds in the Namib Desert is influenced by the population structure and the environment [76]. Family groups of desert-dwelling elephants—composed of related adult females and their calves or juveniles—often come together or roam within the same river sections (URLC or URUC) or even subsections (LCLH, LCUH, UCLH, and UCUH) at a specific time. Their movement pattern essentially depends on water and food availability in wet and dry seasons, leading them to walk regularly upstream and downstream [30,69,72].

The desert-dwelling elephant home range averaged 12,237 km^2, showing considerable expansion compared to previously recorded data. This information is supported by emerging hot spots estimated by the OHSA model. The home range of desert-dwelling elephants was estimated to vary from 1763 to 2944 km^2 [28] to be about 2776 km^2 for elephants in South Africa's Kruger National Park and 3309 km^2 for elephants in northern Botswana [1]. The large discrepancy between the area obtained in this study and reported in others, regarding elephants from the desert and from non-desert regions, suggests that some drastic events may be affecting the desert-dwelling elephants. Climatic factors such as rainfall patterns influence the amount of food availability, quality, and abundance and can play a significant role in animal movements [28,101]. Thus, this study argues that recent home range expansions in the Namib desert toward transitional areas were due to the changes in elephant behaviours associated with food and water scarcity and historic habitat loss. These causes arise partly due to climate change, which, without giving any truce, will contribute to an uncertain future for elephants.

The drought events recorded over the last ten years in Namibia [74], with eight years declared as a state of emergency at the national or regional level (mostly in the Kunene region) (Table S1) [33], can be linked to the loss of vegetation in the Namib Desert ephemeral rivers. Figure 7 shows the decline in the vegetation cover, especially from 2012, coinciding with the year from which the number of CDD was increasing (Table S2), and the mean annual precipitation decreased (Figure 1). Therefore, the impact of reduced rainfall, high temperature, and severe drought may further pose a great risk to the desert vegetation.

The findings in this study on poor health vegetation differ from 30 years ago [29] and revealed a large-scale die-off of *A. erioloba* and *F. albida*. This vegetation is preferred by elephants, with *F. albida* playing an important role in the presence of elephants at the springs, as it is downstream that this abundant species offers them shelter. Elephants are developing an interest in consuming reeds and wild *Tamarix* sp., which are less nutritious and contain high amounts of salt [41], and this appears to be happening as a consequence of the loss of the highly nutritious seedpods of Ana trees (*Faidherbia* sp.) and other species.

Previous studies have not investigated how species that inhabit zones of highly underground water-dependent vegetation may be forced to shift their habitats. Desert elephant habitats are losing vegetation, and this event in arid environments can be associated with reduced rainfall and underground water depletion due to over abstraction [64]. Thus, the decrease in groundwater may be linked to the drying up of the natural springs (Supplementary Figure S3). This is leading desert-dwelling elephants to rely on artificial water points such as boreholes to access water (Table S3 and Supplementary Figure S2), ultimately changing seasonal movement behaviour compared to that of a similar population observed in Legget (2006) [30].

Many studies of elephant–vegetation interactions in Africa have focused more on analysing how surface water influences elephant migration [8,100,102,103] or how vegetation is impacted by human–livestock–elephant pressures [58–62,65]. Although this is important, it is argued here that there is an urgent need of analysing the impact of extreme droughts and prolonged groundwater abstraction in arid environments, especially on ephemeral rivers' groundwater flows and riparian vegetation changes. It can only be expected that frequent extreme drought events (Table S1) and poor annual rainfall will further minimise the rate of aquifer recharge. Consequently, this will reduce the water available at natural springs for elephants to dig and for humans to extract.

The recent policy on elephant conservation and management of the Ministry of Environment, Forestry and Tourism (2021) [23] presented that the elephant population growth rate is increasing between 4.2% and 6.5%, which is why the species is expanding into new areas across the country, most notably Omatjete, Kamanjab, Manketi, and Katwitwi. However, local populations such as the desert-dwelling and semi-desert-dwelling elephant populations in the Omatjete area are not growing due to low calf survival rate observed, even though the arrival of desert-dwelling elephants in the area may suggest that named-conservation actions, including the auctioning of 30 live elephants in 2021 as a conflict control measure [104,105], may also affect the population's stability and ability to sustainably provide ecological and economic services.

It should be noted that the expansion of the home range reported here does not mean population growth. Furthermore, the results suggest that the decline in the availability of important resources within historic elephant habitats is causing their home range shift (Figure 4), contributing to the establishment of new habitats (Figure 7). This, in turn, is leading to emerging conflict hot spots, further endangering this ecotype.

4.2. Farmers and Human–Elephant Conflict

For conservation to succeed, an integrated decision-making approach is needed that considers scientific methods with social values [54,106–108]. An integrated decision-making approach validates the fundamental community-based natural resources management (CBNRM) strategy on sustainable resource use in Namibia, including elephants. However, the prevailing HEC still places significant responsibility on the CBNRM program [109]. Emerging conflict hot spots, such as Omatjete (within the Ugab River upper catchment (Figure 6)), Kamanjab, Manketi, and Tsumkwe [23,105], are evolving because of limited food availability and lack of water infrastructure for wildlife.

The water infrastructure built at farms in the past rarely provided drinking dams for elephants, contributing to greater competition between wild animals and farmers in accessing these water points. The community waterpoints in communal areas of Erongo and Kunene regions are often placed in the middle of villages—to be easily accessible to the residents (humans)—yet, elephants coming to drink at the same water points are forced to walk through the villages, leading to a dangerous encounter with people that are commuting between their neighbouring homesteads. Based on these observations, building specific dams out of villages for elephants to drink could be a strategy to reduce dangerous encounters and conflict.

Increasing tourism development facilities has also led to the drilling of new boreholes for water abstraction, whereas communal farms continue to expand, accompanied by agricultural and livestock practices that require large amounts of water [17,35]. Such anthropogenic activities exert pressure on aquifers, whose recharges are increasingly limited in arid environments [64], as they happen in the region of the Ugab River. Thus, the exploitation of resources for human use, the historic habitat loss for elephants, and the complications of arid environments derived from climate change are linked to the increase in HEC observed at the upper catchments of the Ugab River.

4.3. Restoring Elephant's Historic Habitats

"Can we repair some of the damage humans have done to ecosystems and biodiversity?" [110]. Ecosystem restoration is an exciting concept, but it should be considered a secondary option behind the conservation of nature [111]. For example, desert-dwelling elephants survived the harsh conditions of the desert for many years, but the developments of unsustainable farming practices have had a severe negative impact on the megafauna habitats, suggesting that conservation strategies have not been effectively implemented.

The need to restore the URLC section is more complex than an initial restoration process. Regarding vegetation, it is important to mention that few young *F. albida* and *A. erioloba* have been recorded in ephemeral rivers [58,59], and, throughout this work, it was also shown that these species are disappearing from the Ugab River mainly due to large tree die off (Supplementary Figure S3). Therefore, planting indigenous tree species unique to this river system is critical to restoring lost vegetation. In turn, plants require more water for seeds to germinate, and limited rainfall affects the chances of new seed germination [112].

The same can be said for elephants that can no longer dig for water because the water table at the springs has fallen to 1.5 m, which is below the deepest accessible level of 1 m previously recorded in 2013 [70], after two successive drought events (Table S1). Although it is believed that habitat restoration can only occur with minimal human interference (rehabilitation), several aspects are required to achieve an entirely functional ecosystem, such as planting of trees [40], protection of seedlings, and reducing water abstraction [113]. Thus, the restoration of Namibia's ephemeral river habitat requires a combination of the two perspectives, with human intervention required for reforestation as a near-term solution [40].

5. Conclusions

The combined effects of reduced rainfall, frequent droughts, human demand for groundwater abstraction, and pressure on habitat patches in the Namib Desert have negatively impacted water availability, leading to vegetation loss and large tree die-off. As a result, elephants are expanding their home range and establishing new habitats upstream, out of the desert. The new habitats are established within commercial farms, whereas some farms are in commuting and migration corridors. Such emerging commuting patterns contribute to the HEC, and these conflicts may not only be responsible for the killing of elephants by farmers but also for endangering livestock and human life.

Reducing the number of livestock in the desert could help reduce pressure on vegetation and reduce the amount of water abstracted. The latter is particularly important, as the abstraction of water in large quantities has contributed to the reduction in underground water flow, leading to the drying up of natural springs. However, as the chances of recruiting young trees are low due to limited rainfall and severe droughts, this solution alone is not enough. Thus, assisted planting and protection of important tree species (*F. albida* and *V. erioloba*) would help to restore the riparian vegetation that supports elephants and other rare and endangered species, such as the black rhinos.

Another option is to provide financial support to local organisations in the area to sensitise the farmers to coexist with elephants and teach them about elephant behaviours to reduce retaliation killings. The killing of elephants for self-defence or fear is a reality and may increase as new conflicts emerge in areas where people are less familiar with elephants. Locating elephant drinking dams from the centre to the outskirts of the villages can considerably reduce conflict and open doors to coexistence between farmers and wildlife. Lastly, Namibia has a high tourism potential for locals to generate income. Shifting into wildlife farming would reduce pressure on vegetation, and game species in Namibia are better adapted to the dry conditions than livestock, thus paving the way for sustainable income generation through eco-tourism. For this, it would be crucial to obtain policy support and capital investment for tourism infrastructure from the central government and the private institutions. That does not only create job opportunities in the conflict's hot spot but also reduces reliance on livestock that consume more water in turn, which is negatively affecting the entire ecosystem downstream.

Supplementary Materials: The following supporting information can be downloaded at: https://www.mdpi.com/article/10.3390/su151612400/s1, Supplementary Figure S1: Estimated elephant distribution in Namibia; Supplementary Figure S2: Waterpoints frequency visits; Supplementary Figure S3: Dead vegetation; Table S1: Major climate events; Table S2: Climate observation data; Table S3: Water infrastructure and conflict events; Table S4: Elephant home range; Table S5: Vegetation cover. However, raw data with physical descriptions of the area or coordinates will not be publicly available to uphold and promote the security of free-roaming species in the region but may be requested from the corresponding author on reasonable grounds.

Author Contributions: Conceptualization, M.S. and M.J.P.; Methodology, M.S., F.S. and M.J.P.; Software, M.S. and F.S.; Validation, M.S., F.S., A.R.S. and M.J.P.; Formal analysis, M.S., F.S. and M.J.P.; Investigation, M.S., F.S. and M.J.P.; Resources, M.S. and M.J.P.; Data curation, M.S. and M.J.P.; Writing—original draft, M.S.; Writing—review and editing, M.S., F.S., A.R.S. and M.J.P.; Visualization, M.S., F.S., A.R.S. and M.J.P.; Supervision, M.J.P.; Project administration, M.S. All authors have read and agreed to the published version of the manuscript.

Funding: This research received no external funding.

Institutional Review Board Statement: Not applicable.

Informed Consent Statement: Not applicable.

Data Availability Statement: The data used can be downloaded from supplementary materials. The other data presented in this study are available on request from the corresponding author. The data that are not publicly available are due to the security of species in the region.

Acknowledgments: We thank the reviewers who have contributed to the processing of this manuscript. We also thank many Namibian conservation pioneers and elephant rangers of the Ministry of Environment, Forestry and Tourism, community conservancies, traditional authorities, and conservation organisations that joined or accommodated us for elephant tracking and community engagements at times. We look forward to collaborating with field-based institutions operating in the region, especially those that may have long-term observation data. This publication was made possible through a research permit from Namibia's National Commission on Research, Science and Technology (NCRST), with support from the Ministry of Environment, Forestry and Tourism.

Conflicts of Interest: The authors declare no conflict of interest.

References

1. Poole, J.; Granli, P. Chapter 1. Mind and Movement: Meeting the Interests of Elephants. In *An Elephant in the Room: Science and Well Being of Elephants in Captivity*; Forthman, L.D., Kane, L.F., Waldau, P., Eds.; Center for Animals and Public Policy: North Grafton, MA, USA, 2008; pp. 1–20. ISBN 9780615229843.
2. Hauenstein, S.; Kshatriya, M.; Blanc, J.; Dormann, C.F.; Beale, C.M. African Elephant Poaching Rates Correlate with Local Poverty, National Corruption and Global Ivory Price. *Nat. Commun.* **2019**, *10*, 2242. [CrossRef]
3. Schlossberg, S.; Chase, M.J.; Griffin, C.R. Poaching and Human Encroachment Reverse Recovery of African Savannah Elephants in South-East Angola despite 14 Years of Peace. *PLoS ONE* **2018**, *13*, e0193469. [CrossRef]
4. Chase, M.J.; Schlossberg, S.; Griffin, C.R.; Bouché, P.J.C.; Djene, S.W.; Elkan, P.W.; Ferreira, S.; Grossman, F.; Kohi, E.M.; Landen, K.; et al. Continent-Wide Survey Reveals Massive Decline in African Savannah Elephants. *PeerJ* **2016**, *4*, e2354. [CrossRef]
5. Redpath, S.M.; Bhatia, S.; Young, J. Tilting at Wildlife: Reconsidering Human–Wildlife Conflict. *Oryx* **2015**, *49*, 222–225. [CrossRef]
6. Nyhus, P.J. Human–Wildlife Conflict and Coexistence. *Annu. Rev. Environ. Resour.* **2016**, *41*, 143–171. [CrossRef]
7. Smith, R.J.; Muir, R.D.J.; Walpole, M.J.; Balmford, A.; Leader-Williams, N. Governance and the Loss of Biodiversity. *Nature* **2003**, *426*, 67–70. [CrossRef]
8. Laurance, W.F.; Carolina Useche, D.; Rendeiro, J.; Kalka, M.; Bradshaw, C.J.A.; Sloan, S.P.; Laurance, S.G.; Campbell, M.; Abernethy, K.; Alvarez, P.; et al. Averting Biodiversity Collapse in Tropical Forest Protected Areas. *Nature* **2012**, *489*, 290–294. [CrossRef]
9. Tilman, D.; Clark, M.; Williams, D.R.; Kimmel, K.; Polasky, S.; Packer, C. Future Threats to Biodiversity and Pathways to Their Prevention. *Nature* **2017**, *546*, 73–81. [CrossRef]
10. Hughes, A.C.; Qiao, H.; Orr, M.C. Extinction Targets Are Not SMART (Specific, Measurable, Ambitious, Realistic, and Time Bound). *Bioscience* **2021**, *71*, 115–118. [CrossRef]
11. Hodnebrog, O.; Myhre, G.; Forster, P.M.; Sillmann, J.; Samset, B.H. Local Biomass Burning Is a Dominant Cause of the Observed Precipitation Reduction in Southern Africa. *Nat. Commun.* **2016**, *7*, 11236. [CrossRef]
12. Szczys, P.; Oswald, S.A.; Arnold, J.M. Conservation Implications of Long-Distance Migration Routes: Regional Metapopu-lation Structure, Asymmetrical Dispersal, and Population Declines. *Biol. Cons.* **2017**, *209*, 263–272. [CrossRef]

13. Turpie, J.; Midgley, G.; Brown, C.; Barnes, J.I.; Pallett, J.; Desmet, P.; Tarr, J.; Tarr, P. *Climate Change Vulnerability and Adaptation Assessment for Namibia's Biodiversity and Protected Area System*; Climate Systems Analysis Group for the Ministry of Environment and Tourism: Windhoek, Namibia, 2010.
14. Bauer, H.; Tehou, A.C.; Gueye, M.; Garba, H.; Doamba, B.; Diouck, D.; Sillero-Zubiri, C. Ignoring Species Hybrids in the IUCN Red List Assessments for African Elephants May Bias Conservation Policy. *Nat. Ecol. Evol.* **2021**, *5*, 1050–1051. [CrossRef]
15. Gobush, K.S.; Edwards, C.T.T.; Balfour, D.; Wittemyer, G.; Maisels, F.; Taylor, R.D. African Savanna Elephant: Loxodonta africana. Available online: https://go.nature.com/3v9Bno6 (accessed on 18 July 2023).
16. Hart, J.; Gobush, K.; Maisels, F.; Wasser, S.; Okita-Ouma, B.; Slotow, R. African Forest and Savannah Elephants Treated as Separate Species. *Oryx* **2021**, *55*, 170–171. [CrossRef]
17. Brown, C. *Conservation and the Environment in Namibia*; Venture Media: Windhoek, Namibia, 2021.
18. Douglas-Hamilton, l. African Elephants: Population Trends and Their Causes. *Oryx* **1987**, *21*, 11–24. [CrossRef]
19. Chwalibog, A.; Ngcobo, J.N.; Nedambale, T.L.; Nephawe, K.A.; Sawosz, E. The Future Survival of African Elephants: Implications for Conservation. *Int. J. Avian Wildl. Biol.* **2018**, *3*, 379–384. [CrossRef]
20. Pinnock, D.; Bell, C. *The Last Elephants*; Penguin Random House South Africa: Cape Town, South Africa, 2019; ISBN 9781775846840.
21. di Minin, E.; Slotow, R.; Fink, C.; Bauer, H.; Packer, C. A Pan-African Spatial Assessment of Human Conflicts with Lions and Elephants. *Nat. Commun.* **2021**, *12*, 156. [CrossRef]
22. Osei-Owusu, Y.; Bakker, L. *Human-Wildlife Conflict: Elephant Technical Manual, the Wildlife Management Working Papers Report*; FAO: Rome, Italy, 2008.
23. Ministry of Environment, F. *and T. National Elephant Conservation and Management Plan 2020/2021–2029/2030*; Ministry of Environment, Forestry and Tourism: Windhoek, Namibia, 2020.
24. Leggett, K.; Godfrey, M.; Weiver, E. *Annual Report for the International Elephant Foundation*; IEF: Azle, TX, USA, 2008.
25. Brown, L.M.; Ramey, R.R.; Vinjevold, R.; Vinjevold, T.; Jauire, A. 2019 Annual Research Report Status and Distribution of Desert-Dwelling Elephants in the Hoarusib, Hoanib, and Uniab River Drainages, Kunene Region, Namibia. 2020. Available online: https://img1.wsimg.com/blobby/go/bded65a9-b30d-4b15-90f3-4d331465d33e/downloads/2019%20DEC%20Annual%20Research%20Report%20compressed.pdf?ver=1585231842554 (accessed on 27 June 2023).
26. Martin, B.R. *Transboundary Species Project—Background Study*; Environmental Information Service Namibia: Windhoek, Namibia, 2005.
27. Ishida, Y.; Van Coeverden de Groot, P.J.; Leggett, K.E.A.; Putnam, A.S.; Fox, V.E.; Lai, J.; Boag, P.T.; Georgiadis, N.J.; Roca, A.L. Genetic Connectivity across Marginal Habitats: The Elephants of the Namib Desert. *Ecol. Evol.* **2016**, *6*, 6189–6201. [CrossRef]
28. Viljoen, P.J. Spatial Distribution and Movements of Elephants (*Loxodonta africana*) in the Northern Namib Desert Region of the Kaokoveld, South West Africa/Namibia. *J. Zool.* **1989**, *219*, 1–19. [CrossRef]
29. Viljoen, P.J.; Bothma, J. du P. The Influence of Desert-Dwelling Elephants on Vegetation in the Northern Namib Desert, South West Africa/Namibia. *J. Arid. Environ.* **1990**, *18*, 85–96. [CrossRef]
30. Leggett, A.E.K. Home Range and Seasonal Movement of Elephants in the Kunene Region, Northwestern Namibia. *Afr. Zool.* **2006**, *41*, 17–36. [CrossRef]
31. Viljoen, P.J. Status and Past and Present Distribution of Elephants in the Kaokoveld, South West Africa/Namibia. *S. Afr. J. Zool.* **1987**, *22*, 247–257. [CrossRef]
32. Ministry of Environment, Forestry and Tourism. *Revised National Strategy on Wildlife Protection and Law Enforcement in the Republic of Namibia*; Ministry of Environment, Forestry and Tourism: Windhoek, Namibia, 2021.
33. Hitila, M. *An Assessment of the Capacity of the Directorate of Disaster Risk Management of Namibia*; Namibia University of Science and Technology: Windhoek, Namibia, 2019.
34. United Nations Framework Convention on Climate Change. *Republic of Namibia: Fourth National Communication to the United Nations Framework Convention; on Climate Change*; United Nations Framework Convention on Climate Change: Windhoek, Namibia, 2020.
35. Republic of Namibia Namibia. *Namibia's Updated Nationally Determined Contribution*; Ministry of Environment, Forestry and Tourism: Windhoek, Namibia, 2021.
36. Fox, J.T.; Vandewalle, M.E.; Alexander, K.A. Land Cover Change in Northern Botswana: The Influence of Climate, Fire, and Elephants on Semi-Arid Savanna Woodlands. *Land* **2017**, *6*, 73. [CrossRef]
37. Inman, E.N.; Hobbs, R.J.; Tsvuura, Z. No Safety Net in the Face of Climate Change: The Case of Pastoralists in Kunene Region, Namibia. *PLoS ONE* **2020**, *15*, e0238982. [CrossRef] [PubMed]
38. Dirkx, E.; Claus, H.; Mark, T.; Bethune, S.; Curtis, B. *Climate Change Vulnerability & Adaptation Assessment Namibia*; Climate Systems Analysis Group for the Ministry of Environment and Tourism: Windhoek, Namibia, 2008.
39. Thuiller, W.; Midgley, G.F.; Hughes, G.O.; Bomhard, B.; Drew, G.; Rutherford, M.C.; Woodward, F.I. Endemic Species and Ecosystem Sensitivity to Climate Change in Namibia. *Glob. Chang. Biol.* **2006**, *12*, 759–776. [CrossRef]
40. IPCC. *IPCC Climate Change 2023: Synthesis Report of the Intergovernmental Panel on Climate Change Sixth Assessment Report (AR6)*; IPCC: Geneva, Switzerland, 2023.
41. Mannheimer; Curtis, B. *Le Roux and Müller's Field Guide to the Trees & Shrubs of Namibia*, 2nd ed.; Mannheimer, Curtis, B., Eds.; Namibia Publishing House: Windhoek, Namibia, 2011; Volume 77, ISBN 978-99916-0-970-6.

42. Soetaert, F.; Wanke, H.; Dupuy, A.; Lusuekikio, V.; Gaucher, E.C.; Bordmann, V.; Fleury, J.M.; Franceschi, M. Toward the Sustainable Use of Groundwater Springs: A Case Study from Namibia. *Sustainability* **2022**, *14*, 3995. [CrossRef]
43. The World Bank Group, World Bank Climate Change Knowledge Portal (CCKP). *For Global Climate Data and Information!* Climate Change Knowlwdge Portal: Washington, DC, USA, 2020.
44. World Bank Group. *Climate Risk Profile: Namibia*; World Bank Group: Washington, DC, USA, 2021.
45. Gómez-Rubio, V. Ggplot2—Elegant Graphics for Data Analysis. *J. Stat. Softw.* **2017**, *77*, 1–3. [CrossRef]
46. R Core Team. *The R Project for Statistical Computing*; R Core Team: Vienna, Austria, 2020.
47. Enzerink, R.; Liefferink, D. *Co-Existing with Wildlife in Namibia's Conservancies: A Case Study on the Relationship between Human-Wildlife Conflict and Attitudes of Local Communities and the Influence of Community-Based Natural Resource Management on This Relationship*; Radboud University: Nijmegen, The Netherlands, 2017.
48. Naidoo, R.; Weaver, L.C.; de Longcamp, M.; du Plessis, P. Namibia's Community-Based Natural Resource Management Programme: An Unrecognized Payments for Ecosystem Services Scheme. *Environ. Conserv.* **2011**, *38*, 445–453. [CrossRef]
49. Klintenberg, P.; Seely, M.; Christiansson, C. Local and National Perceptions of Environmental Change in Central Northern Namibia: Do They Correspond? *J. Arid. Environ.* **2007**, *69*, 506–525. [CrossRef]
50. Seely, M.; Moser, P. Connecting Community Action and Science to Combat Desertification: Evaluation of a Process. *Environ. Monit. Assess.* **2004**, *99*, 33–55. [CrossRef]
51. Seely, M.; Wöhl, H. Connecting Research to Combating Desertification. *Environ. Monit. Assess.* **2004**, *99*, 23–32. [CrossRef] [PubMed]
52. MEFT/NACSO. *The State of Community Conservation in Namibia (Annual Report 2019)*; NACSO: Windhoek, Namibia, 2021.
53. Jones, R.; Weaver, L.C. CBNRM in Namibia: Growth, Trends, Lessons and Constraints. In *Evolution and Innovation in Wildlife Conservation*; Child, B., Suich, H., Anna, S., Eds.; Routledge: London, UK, 2012; pp. 241–260. ISBN 9781849771283.
54. Brown, J.; Bird, N. *Sustainable Natural Resource Management in Namibia: Successful Community-Based Wildlife Conservation*; Overseas Development Institute: London, UK, 2011.
55. Shaffer, L.J.; Khadka, K.K.; Van Den Hoek, J.; Naithani, K.J. Human–Elephant Conflict: A Review of Current Management Strategies and Future Directions. *Front. Ecol. Evol.* **2019**, *6*, 235. [CrossRef]
56. Trumbull, R.B.; Reid, D.L.; de Beer, C.; van Acken, D.; Romer, R.L. Magmatism and Continental Breakup at the West Margin of Southern Africa: A Geochemical Comparison of Dolerite Dikes from Northwestern Namibia and the Western Cape. *S. Afr. J. Geol.* **2007**, *110*, 477–502. [CrossRef]
57. Shikangalah, R.N.; Mapani, B.S. Ephemeral River Systems and Their Ecosystem Provisions to the Local Populations: A Review of the Huab and Ugab Rivers, Namibia. *Int. Sci. Technol. J. Namib.* **2021**, *14*, 46–62.
58. Curtis, A.B. The Status of Faidherbia Albida Trees in the Hoanib River, Namibia. *Namib. J. Environ.* **2017**, *1*, 77–91.
59. Moss, P. *Regeneration and Utilization of Faidherbia Albida and Acacia Erioloba along Ephemeral Rivers of Namibia*; Cuvilliier Verlag: Göttingen, Germany, 2006; Volume 42, ISBN 3-86537-948-6.
60. Jacobson, P.J. The Influence of Elephants on Faidherbia Albida Trees in the Northern Namib Desert: A Reappraisal. In *An Ephemeral Perspective of Fluvial Ecosystems: Viewing Ephemeral Rivers in the Context of Current Lotic Ecology*; Jacobson, P.J., Richard, W.L.D., Cherry, D.S., Angermeier, P.L., Eds.; Virginia Polytechnic Institute and State University: Blacksburg, VA, USA, 1997; pp. 111–122.
61. Douglas, C.M.S. The Distribution and Survival of Riparian Trees along a Dammed Ephemeral River. Ph.D. Thesis, King's College London, London, UK, 2015.
62. Douglas, C.M.S.; Mulligan, M.; Harrison, X.A.; Henschel, J.R.; Pettorelli, N.; Cowlishaw, G. Widespread Dieback of Riparian Trees on a Dammed Ephemeral River and Evidence of Local Mitigation by Tributary Flows. *PeerJ* **2016**, *2016*, e2622. [CrossRef]
63. Garstang, M.; Davis, R.E.; Leggett, K.; Frauenfeld, O.W.; Greco, S.; Zipser, E.; Peterson, M. Response of African Elephants (*Loxodonta africana*) to Seasonal Changes in Rainfall. *PLoS ONE* **2014**, *9*, e108736. [CrossRef]
64. Jasechko, S.; Perrone, D. Global Groundwater Wells at Risk of Running Dry. *Science* **2021**, *372*, 418–421. [CrossRef]
65. Douglas, C.M.S.; Cowlishaw, G.; Harrison, X.A.; Henschel, J.R.; Pettorelli, N.; Mulligan, M. Identifying the Determinants of Tree Distributions along a Large Ephemeral River. *Ecosphere* **2018**, *9*, e02223. [CrossRef]
66. IPCC Summary for Policymakers. *Climate Change: The Physical Science Basis. Contribution of Working Group I to the Sixth Assessment Report of the Intergovernmental Panel on Climate Change*; Masson-Delmotte, V., Zhai, P., Pirani, A., Connors, S.L., Péan, C., Berger, S., Caud, N., Chen, Y., Goldfarb, L., Gomis, I.M., Eds.; Cambridge University Press: Cambridge, UK, 2021.
67. Jacobson, J.P.; Jacobson, M.K.; Seely, K.M. Reviewed Work: Ephemeral Rivers and Their Catchments: Sustaining People and Development in Western Namibia. *J. East. Afr. Res. Dev.* **1996**, *26*, 238–241.
68. Thomson, G. Climate Change in Namibia Part 2: Current and Projected Changes. Conservation and the Environment in Namibia. *Windhoek*, 29 October 2021.
69. Viljoen, P.J. Habitat Selection and Preferred Food Plants of a Desert-Dwelling Elephant Population in the Northern Namib Desert, South West Africa/Namibia. *Afr. J. Ecol.* **1989**, *27*, 227–240. [CrossRef]
70. Ramey, E.M.; Ramey, R.R.; Brown, L.M.; Kelley, S.T. Desert-Dwelling African Elephants (*Loxodonta africana*) in Namibia Dig Wells to Purify Drinking Water. *Pachyderm* **2013**, *53*, 66–72.
71. Ministry of Environment and Tourism. *Species Management Plan: Elephants Loxodonta africana*; Ministry of Environment and Tourism: Windhoek, Namibia, 2007.

72. Viljoen, J.P.; Bothma, P.J. Daily Movements of Desert-Dwelling Elephants in the Northern Namib Desert. *S. Afr. J. Wildl. Res.* **1990**, *20*, 69–80.
73. USGS EarthExplorer: USGS. Available online: https://earthexplorer.usgs.gov/ (accessed on 21 June 2023).
74. Office of the Prime. *Minister The Annual Report 2019-2020*; Office of the Prime Minister: Windhoek, Namibia, 2021.
75. The Cooperation in International Waters in Africa. *Drought Resilience Profiles: Namibia*; The Cooperation in International Waters in Africa: Washington, DC, USA, 2021.
76. Curtis, R. Guide to Animal Tracking: Outdoor Action. Available online: https://outdooraction.princeton.edu/nature/guide-animal-tracking (accessed on 22 November 2021).
77. Nchanji, A.C.; Plumptre, A.J. Seasonality in Elephant Dung Decay and Implications for Censusing and Population Moni-toring in South-Western Cameroon. *Afr. J. Ecol.* **2001**, *39*, 24–32. [CrossRef]
78. Barnes, R.F.W.; Asamoah-Boateng, B.; Naada Majam, J.; Agyei-Ohemeng, J. Rainfall and the Population Dynamics of El-ephant Dung-Piles in the Forests of Southern Ghana. *Afr. J. Ecol.* **1997**, *35*, 39–52. [CrossRef]
79. Masunga, G.S.; Andresen, Ø.; Taylor, J.E.; Dhillion, S.S. Elephant Dung Decomposition and Coprophilous Fungi in Two Habitats of Semi-Arid Botswana. *Mycol. Res.* **2006**, *110*, 1214–1226. [CrossRef]
80. Jachmann, H. Estimating Age in African Elephants. *Afr. J. Ecol.* **1985**, *23*, 199–202. [CrossRef]
81. Shrader, A.M.; Ferreira, S.M.; McElveen, M.E.; Lee, P.C.; Moss, C.J.; van Aarde, R.J. Growth and Age Determination of African Savanna Elephants. *J. Zool.* **2006**, *270*, 40–48. [CrossRef]
82. Douglas-Hamilton, I. On the Ecology and Behaviour of the African Elephant. Ph.D. Thesis, University of Oxford, Oxford, UK, 1972.
83. Elephant Voices How to Age African Elephants. Available online: https://www.elephantvoices.org/features-guide/139-elephantvoices/education/808-how-to-age-african-elephants.html (accessed on 4 August 2022).
84. Harris, R.; Winter, C.; Pitot, C.; Shiweda, M. *EHRA 2019 Annual Report*; Elephant-Human Relations Aid: Swakopmund, Namibia, 2020.
85. Nelson, T.A.; Boots, B. Detecting Spatial Hot Spots in Landscape Ecology. *Ecography* **2008**, *31*, 556–566. [CrossRef]
86. Zerbe, K.; Polit, C.; McClain, S.; Cook, T. Optimized Hot Spot and Directional Distribution Analyses Characterize the Spatiotemporal Variation of Large Wildfires in Washington, USA, 1970−2020. *Int. J. Disaster Risk Sci.* **2022**, *13*, 139–150. [CrossRef]
87. Getis, A.; Ord, J.K. The Analysis of Spatial Association by Use of Distance Statistics. *Geogr. Anal.* **1992**, *24*, 189–206. [CrossRef]
88. Hazen, E.L.; Abrahms, B.; Brodie, S.; Carroll, G.; Welch, H.; Bograd, S.J. Where Did They Not Go? Considerations for Generating Pseudo-Absences for Telemetry-Based Habitat Models. *Mov. Ecol.* **2021**, *9*, 5. [CrossRef] [PubMed]
89. Evans, L.J.; Goossens, B.; Davies, A.B.; Reynolds, G.; Asner, G.P. Natural and Anthropogenic Drivers of Bornean Elephant Movement Strategies. *Glob. Ecol. Conserv.* **2020**, *22*, e00906. [CrossRef]
90. Loarie, S.R.; Van Aarde, R.J.; Pimm, S.L. Fences and Artificial Water Affect African Savannah Elephant Movement Patterns. *Biol. Conserv.* **2009**, *142*, 3086–3098. [CrossRef]
91. Ibrahim, M.; Al-Mashagbah, A. Change Detection of Vegetation Cover Using Remote Sensing Data as a Case Study: Ajloun Area. *Civ. Environ. Res.* **2016**, *8*, 1–5.
92. Ajibade, F. Fidelis.; Adewumi, J.R. Application of GIS and Remote Sensing Technique to Change Detection in Land Use/Land Cover Mapping of Igbokoda, Ondo State, Nigeria. *J. Appl. Sci. Process Eng.* **2016**, *3*.
93. Choate, M.J.; Rengarajan, R.; Storey, J.C.; Beckmann, T. Landsat 8 Thermal Infrared Sensor Scene Select Mechanism Open Loop Operations. *Remote Sens.* **2021**, *13*, 617. [CrossRef]
94. Opedes, H.; Mücher, S.; Baartman, J.E.M.; Nedala, S.; Mugagga, F. Land Cover Change Detection and Subsistence Farming Dynamics in the Fringes of Mount Elgon National Park, Uganda from 1978–2020. *Remote Sens.* **2022**, *14*, 2423. [CrossRef]
95. Al-Kindi, K.M.; Alqurashi, A.F.; Al-Ghafri, A.; Power, D. Assessing the Impact of Land Use and Land Cover Changes on Aflaj Systems over a 36-Year Period. *Remote Sens.* **2023**, *15*, 1787. [CrossRef]
96. Roy, P.S.; Behera, M.D.; Murthy, M.S.R.; Roy, A.; Singh, S.; Kushwaha, S.P.S.; Jha, C.S.; Sudhakar, S.; Joshi, P.K.; Reddy, C.S.; et al. New Vegetation Type Map of India Prepared Using Satellite Remote Sensing: Comparison with Global Vegeta-tion Maps and Utilities. *Int. J. Appl. Earth Obs. Geoinf.* **2015**, *39*, 142–159. [CrossRef]
97. Weier, J.; Herring, D. Measuring Vegetation (NDVI & EVI). Available online: https://www.earthobservatory.nasa.gov/features/MeasuringVegetation (accessed on 2 September 2022).
98. Islam, K.; Jashimuddin, M.; Nath, B.; Nath, T.K. Land Use Classification and Change Detection by Using Multi-Temporal Remotely Sensed Imagery: The Case of Chunati Wildlife Sanctuary, Bangladesh. *Egypt. J. Remote Sens. Space Sci.* **2018**, *21*, 37–47. [CrossRef]
99. Rudnick, D.; Ryan, S.; Beier, P.; Cushman, S.; Dieffenbach, F.; Epps, C.; Gerber, L.; Hartter, J.; Jenness, J.; Kintsch, J.; et al. The Role of Landscape Connectivity in Planning and Implementing Conservation and Restoration Priorities. *Issues Ecology. Ecol. Soc. Am.* **2012**, *16*, 1–23. Available online: https://www.esa.org/esa/wp-content/uploads/2013/03/issuesinecology16.pdf (accessed on 2 September 2022).
100. Druce, D.J.; Shannon, G.; Page, B.R.; Grant, R.; Slotow, R. Ecological Thresholds in the Savanna Landscape: Developing a Protocol for Monitoring the Change in Composition and Utilisation of Large Trees. *PLoS ONE* **2008**, *3*, e3979. [CrossRef] [PubMed]
101. Beirne, C.; Meier, A.C.; Brumagin, G.; Jasperse-Sjolander, L.; Lewis, M.; Masseloux, J.; Myers, K.; Fay, M.; Okouyi, J.; White, L.J.T.; et al. Climatic and Resource Determinants of Forest Elephant Movements. *Front. Ecol. Evol.* **2020**, *8*, 96. [CrossRef]

102. Purdon, A.; Mole, M.A.; Chase, M.J.; van Aarde, R.J. Partial Migration in Savanna Elephant Populations Distributed across Southern Africa. *Sci. Rep.* **2018**, *8*, 11331. [CrossRef]
103. Birkett, P.J.; Vanak, A.T.; Muggeo, V.M.R.; Ferreira, S.M.; Slotow, R. Animal Perception of Seasonal Thresholds: Changes in Elephant Movement in Relation to Rainfall Patterns. *PLoS ONE* **2012**, *7*, e38363. [CrossRef]
104. Schwartz, M. Namibia to sell off wild elephants in controversial auction. *Mongabay*, 29 January 2021.
105. Thomson, G.C. The Story Behind the Namibian Elephant Auction. Conservation Namibia. *Windhoek*, 23 February 2021.
106. Brown, K. Integrating Conservation and Development: A Case of Institutional Misfit. *Front. Ecol. Environ.* **2003**, *1*, 479. [CrossRef]
107. Hobbs, J.S. Community Participation in Biodiversity Monitoring. Ph.D. Thesis, University of York, York, UK, 2012.
108. Johnson, C.M.; Poulin, M.; Graham, M. Towards an Integrated Approach to the Conservation and Sustainable Use of Bio-diversity: Lessons Learned from the Rideau River Biodiversity Project on JSTOR. *Hum. Ecol. Rev.* **2003**, *10*, 40–55.
109. Boudreaux, K. *Community Conservation in Namibia: Devolution as a Tool for the Legal Empowerment of the Poor. Mercatus Policy Paper*; Environmental Information Service Namibia: Windhoek, Namibia, 2010; pp. 1–25.
110. Vaughn, K.J.; Porensky, L.M.; Wilkerson, M.L.; Balachowski, J.; Peffer, E.; Riginos, C.; Young, T.P. Restoration Ecology. *Nat. Educ. Knowl.* **2010**, *3*, 66.
111. Young, T.P. Restoration Ecology and Conservation Biology. *Biol. Conserv.* **2000**, *92*, 73–83. [CrossRef]
112. Favier, C.; Aleman, J.; Bremond, L.; Dubois, M.A.; Freycon, V.; Yangakola, J.M. Abrupt Shifts in African Savanna Tree Cover along a Climatic Gradient. *Glob. Ecol. Biogeogr.* **2012**, *21*, 787–797. [CrossRef]
113. Perring, M.P.; Standish, R.J.; Price, J.N.; Craig, M.D.; Erickson, T.E.; Ruthrof, K.X.; Whiteley, A.S.; Valentine, L.E.; Hobbs, R.J.; Perring, M.P.; et al. Advances in Restoration Ecology: Rising to the Challenges of the Coming Decades. *Ecosphere* **2015**, *6*, 1–25. [CrossRef]

Disclaimer/Publisher's Note: The statements, opinions and data contained in all publications are solely those of the individual author(s) and contributor(s) and not of MDPI and/or the editor(s). MDPI and/or the editor(s) disclaim responsibility for any injury to people or property resulting from any ideas, methods, instructions or products referred to in the content.

sustainability

Article

Habitat Selection: Autumn and Winter Behavioral Preferences of Water Deer (*Hydropotes inermis*) in Northeast China

Yue Sun [1], Zongzhi Li [1], Junda Chen [1], Romaan Hayat Khattak [1,2], Zhensheng Liu [1,3,*] and Liwei Teng [1,3,*]

[1] College of Wildlife and Protected Areas, Northeast Forestry University, Harbin 150040, China; sunyue@nefu.edu.cn (Y.S.); lizz@nefu.edu.cn (Z.L.); chenjunda1998@nefu.edu.cn (J.C.); romaanhayatkhattak@nefu.edu.cn (R.H.K.)
[2] Institute of Zoology, Guangdong Academy of Science, Guangzhou 510000, China
[3] Key Laboratory of Conservation Biology, National Forestry and Grassland Administration, Harbin 150040, China
* Correspondence: zhenshengliu@163.com (Z.L.); tenglw1975@163.com (L.T.)

Abstract: The wild water deer (*Hydropotes inermis*) population has declined rapidly over recent decades and has reached an endangered status in China. Therefore, it is important to understand their habitat selection to effectively protect both existing and emerging populations. This paper used the data of 11 habitat factors in Baishan Musk Deer National Nature Reserve in the autumn and winter from 2018 to 2019 to conduct a habitat selection study of water deer by resource selection function analysis. The results indicated that in both the autumn and winter, water deer preferred grasslands at sunny and middle slopes, dominated by *Artemisia carvifolia* and *A. argyi*, respectively. In addition, the resource selection function showed that the height of dominant herbage, hiding cover, distance from water, and distance to human settlements greatly contribute to the habitat selection of water deer in the cold season. The correct prediction rate of the resource selection function model exceeded 80%, highlighting its suitability for predicting the habitat selection of water deer. The outcomes of this study provide an effective scientific basis for the conservation and restoration of water deer, and valuable enlightenment for implementing a sustainable development strategy in northeast China.

Keywords: water deer; habitat selection; resource selection function; redistribution

Citation: Sun, Y.; Li, Z.; Chen, J.; Khattak, R.H.; Liu, Z.; Teng, L. Habitat Selection: Autumn and Winter Behavioral Preferences of Water Deer (*Hydropotes inermis*) in Northeast China. *Sustainability* **2023**, *15*, 12181. https://doi.org/10.3390/su151612181

Academic Editor: Vasilios Liordos

Received: 30 June 2023
Revised: 6 August 2023
Accepted: 8 August 2023
Published: 9 August 2023

Copyright: © 2023 by the authors. Licensee MDPI, Basel, Switzerland. This article is an open access article distributed under the terms and conditions of the Creative Commons Attribution (CC BY) license (https://creativecommons.org/licenses/by/4.0/).

1. Introduction

The relationship between animals and their habitats is a central component of wildlife ecology [1]. A habitat is defined as the space in which an individual, population, or community can complete their whole life process, as it provides the basic environmental conditions for wildlife, such as water, food, and shelter [2]. Hence, to protect wildlife populations, it is centrally important to understand their habitat selection. Habitat selection directly affects the survival rate of wildlife, thus also ultimately affecting population persistence [3]. However, global climate change and other factors impose direct or indirect impacts on the wildlife habitat environment, causing changes in population size [4] and distribution ranges [5–7] and even lead to extinction [8]. The average global temperature has increased by approximately 0.74 °C over the last century [9], and the fifth assessment report of the United Nations Intergovernmental Panel on Climate Change proposed that the global climate will continue to be warm [10]. Many studies have suggested that in response to this warming, species will migrate toward higher elevations or higher latitudes [11–14]. Furthermore, the choice of habitats by wild animals also represents a balance between physiological needs and capture risk [15,16]. Fast-growing human populations and rising global temperatures have primarily destroyed vast portions of wildlife habitats, thus forcing wildlife to live near human settlements [17,18]. The human shield hypothesis argues that humans exert different top-down pressures on the apex (hereafter dominant) predator in a system, thus indirectly helping subordinate competitors and facilitating greater spatial overlap between humans and subordinate species [19,20]. The crop attraction hypothesis argues that many large grazing animals (such as ungulates) frequently feed on agricultural crops close to human settlements [21]. Both hypotheses suggest that even species that would otherwise prefer to avoid humans may choose to approach human settlements to ensure their survival. Thus, understanding habitat selection of large grazing or subordinated animals requires assessing the importance of human-created habitats upon their decisions of habitat use.

This study addresses the habitat selection of the endangered water deer (*Hydropotes inermis*), a small-sized ruminant that belongs to the genus of *Hydropotes* and the family of Cervidae. It is endemic to East Asia with a natural distribution throughout China (e.g., Liaodong Peninsula, North China Plains, and both sides of the Yangtze River) and the Korean Peninsula [22]. In the early 1990s, China had a wild water deer population of about 10,000–30,000 individuals [23]. A study conducted in 2013 showed that because of habitat loss caused by excessive human disturbance, the population size and distribution of water deer shrunk dramatically. This has resulted in an isolated "island" distribution of less than 10,000 individuals [24]. The water deer is currently classified as a national key protected class II species in China, and according to the red list of the International Union for Conservation of Nature, its status is vulnerable [25].

In China, studies on the habitat selection of water deer have mainly focused on the southern parts of their distribution, while studies on the northeastern parts are rare. Zhang conducted a study in southern China (i.e., the Poyang Lake national nature reserve in Jiangxi Province), and found that water deer preferred *Miscanthus sacchairflorus*, *Phragmitas*, low slopes, and vegetation coverage ranging from 50% to 69%, while avoiding habitats with strong human disturbance in 2019 [26]. In 2022, Han et al. also conducted a study in southern China (Nanjing, Jiangsu Province). The authors found that water deer preferred evergreen deciduous broad-leaved forests and dominant shrubs with a height of no less than 120 cm; however, habitats close to intersections and those with a high disturbance index were avoided [27].

There has been no report on the water deer distribution in northeast China since the 1950s, and it is reasonable to assume that the species had become locally extinct, until its rediscovery in Jilin Province in 2019 [28]. Also in 2019, water deer were photographed in the Primorskiy region of Russia, which is currently the northernmost water deer distribution [29]. According to Teng (2007), the coastal tidal flats in Yancheng, Jiangsu Province, China, were once the northernmost ends of the distribution area of water deer in China [30]. Today, the rediscovery of water deer in the northeast has updated the northernmost end of the distribution of this species in China. This new finding has led to considerable speculations regarding the habitat selected by water deer in recent years in different regions. The authors suggest that for the robust conservation of small and fragmented populations of water deer in northeast China, it is vital to study their habitat selection.

In previous studies, water deer have been shown to avoid anthropogenic habitats in southern China, thus suggesting that the human shield hypothesis was not supported. Individuals of the northern population could respond differently to habitat disturbances, because northern habitat conditions may be more stressful (larger predation pressure or less foraging resources in natural habitats). Since the living environment of water deers will likely be different from that of southern China, their response to habitat conditions may be consistent with the human shield hypothesis or the crop attraction hypothesis. In the current study, the importance of different habitat factors on the habitat selection of water deer is assessed in northeast China. The implications derived from this study may help to provide reliable information for the protection and restoration of water deer populations in northeast China.

2. Materials and Methods

2.1. Study Area

This study was conducted from September 2018 to December 2019 in Baishan Musk Deer National Nature Reserve (126°29′50″–126°45′27″ E, 41°36′43″–41°49′54″ N), located in the southeast of Baishan City, Jilin Province, China (Figure 1). The reserve covers an area of 219.95 km^2 [31]. The study area is adjacent to Sandaogou Forest Farm in Baishan City in the west and Weishahe Forest Farm Shiye District in Linjiang City in the north and east, and faces North Korea across the Yalu River to the south. The water system mainly includes the Yalu River, a length of 35 km of which flows through the nature reserve. The main differences between their southern study site and the northeast site examined in the present study are the daily temperature, longitude, latitude, duration of snow cover, and vegetation type.

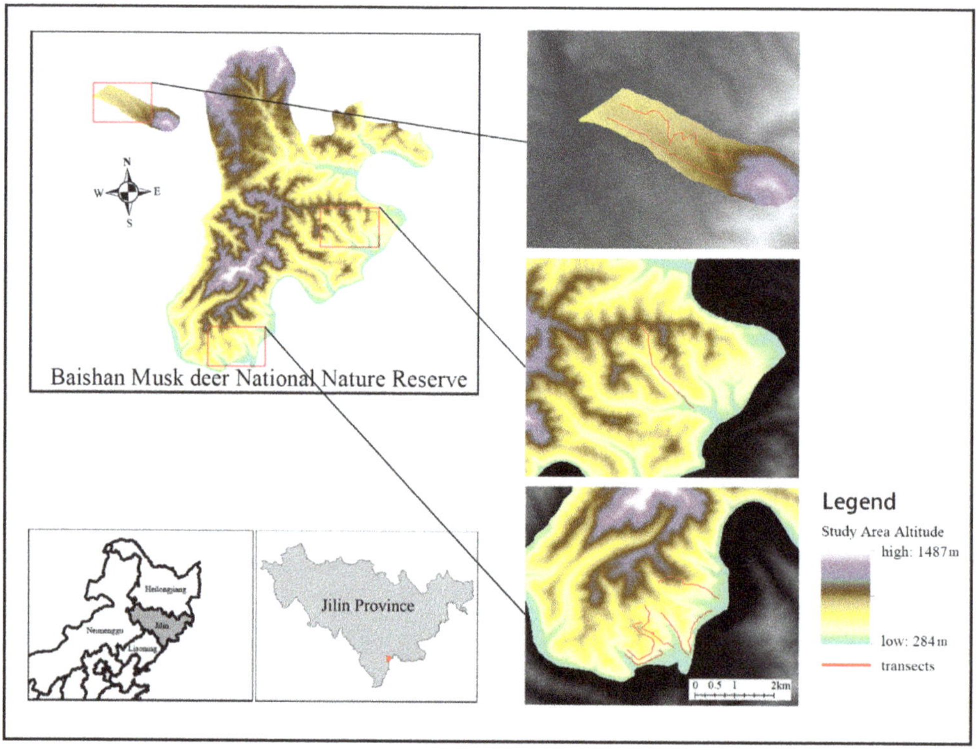

Figure 1. The left part is a location map showing the Baishan Musk Deer National Nature Reserve in Baishan City, Jilin Province, China. The right part is the survey transects.

2.2. Climate

The study area is the coldest region in Jilin Province. The terrain in the reserve is complex, and the temperature difference between day and night is quite large. The area is part of the temperate continental East Asian monsoon climate zone. The annual average temperature is 3–5 °C with a frost-free period of 115–140 days. The annual sunshine hours are 2232.6 h, and the annual average precipitation is 800–1000 mm [32]. The freezing period starts in the reserve in late October, and the thaw usually begins in early April. The maximum depth of the frozen layer is 1.36 m, the ice thickness of the water surface is 0.8–1.0 m, and the maximum thickness of the snow in average years is 44 cm [33].

2.3. Flora and Fauna

The forest vegetation of the reserve belongs to the typical flora of Changbai Mountains. The zonal vegetation is broad-leaved Korean pine forest and contains a large area of broad-leaved mixed forests and broad-leaved forests. The wild animal resource in the reserve is abundant.

2.4. Data Collection

We conducted a random transect line survey in the known and potential distribution areas of water deer in the reserve during autumn and winter, with 9 random transect lines per season. The observer walked along the transect line and when the activity trace of water deer was found, a 10 m × 10 m experimental plot centered at the trace was established to estimate habitat features used by the deer. In addition, a 10 m × 10 m control plot was established at an appropriate place in a random direction and was 500–1000 m away from the experimental plot and the transect line. The habitat factors of experimental plots and control plots recorded included vegetation types, dominant plants, elevation, distance to human settlements, distance from water, herbage coverage, height of dominant herbage, hiding cover, slope degree, slope position, and slope direction (Table 1). In the current study, a total of 49 experimental plots and 47 control plots were recorded in autumn, and a total of 45 experimental plots and 44 control plots were recorded in winter.

Table 1. Detailed description of various habitat factors.

Habitat Factors	Abbreviation	Measure Methods
Vegetation type	VT	Including grassland, coniferous forest, broad-leaved forest, shrub, cropland, and forest edge.
Dominant plant	DP	The plant with more than 70%, including *Artemisia caruifolia, Amphicarpaea edgeworthii, A. argyi, Betula platyphylla, Quercus mongolica, Senna nomame.*
Slope positions	SP	Divide the hillside into three equal parts, uphill position: top part, mid-slope position: middle part, downhill position: bottom part.
Slope direction	SD	Measured in the counterclockwise direction by the military compass Type 65 with the true north direction at 0°. Sunny slope (157.5°–337.5°), Shady slope (337.5°–157.5°).
Herbage coverage (%)	HEC	The average of 5 small plots (1 m × 1 m) sampled at center and 4 corners of the experimental plots or control plots
Height of dominant herbage (cm)	HDH	The average of 5 small plots (1 m × 1 m) sampled at center and 4 corners of the experimental plots or control plots
Slope degree (°)	S	Measured by the military compass Type 65.
Elevation (m)	E	Measured by global positioning instrument.
Distance from water (m)	DW	The distance between the center of plots and water resource
Distance to human settlements (m)	DHS	The distance between the center of plots and human settlement.
Hiding cover (%)	HIC	Erect a 1 m wooden pole at the center of the plot, measure the visibility of the pole at 20 m away from the center in the east, south, west, and north directions, that is, the percentage of the length of the pole in the total length can be seen, and then calculate the average value.

In addition to water deer, other ungulate species such as musk deer (*Moschus moschiferus*) and roe deer (*Capreolus pygargus*) also inhabit the protection area. One possibility to determine which species the feces come from is to relate the feces to the habitat they were found in. Feces from water deer are often found in habitats at lower elevations and lower slopes, while feces from musk deer and roe deer are usually found in habitats at higher elevations and larger slopes. According to "A Guide to the Mammals of China", activity (Figure 2a,b,d) in combination with habitat can be used to identify water deer [34]. Apart from differences in habitat, the feces can be distinguished using differences in their appearance. Water deer feces are smaller particles and are slenderer than those of roe deer (Figure 2c), and the feces pile is more concentrated than that of roe deer; moreover, the distance between feces piles is smaller. Therefore, by combining habitat and fecal morphology, activity traces can be identified as belonging to water deer.

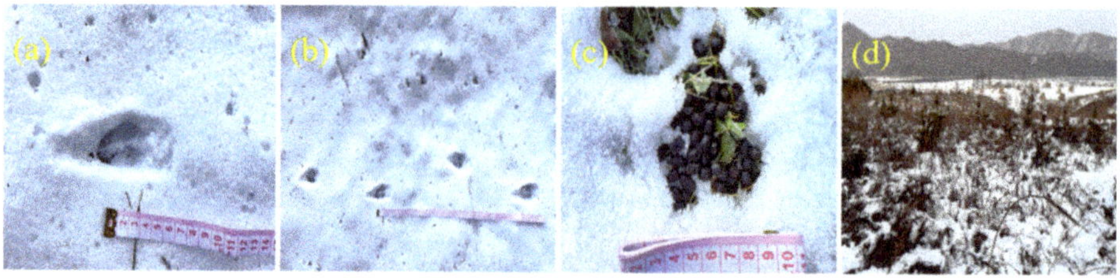

Figure 2. (**a**) Water deer tracks with a length of about 0.05 m; (**b**) water deer track spacing of about 0.3 m; (**c**) water deer feces featuring small and slender granules, concentrated in a feces pile; (**d**) a typical habitat of water deer where their tracks and feces were found in the habitat.

2.5. Data Analysis

Chi-square tests were used to analyze the differences in non-numerical habitat factors between experimental plots and control plots [35]. The single-sample K-S test was used to examine whether the numerical habitat factors conform to a normal distribution [36]. The independent *t*-test and Mann–Whitney U test were used to analyze habitat factors with normal distribution and non-normal distribution, respectively (*: $p < 0.05$, **: $p < 0.01$, ***: $p < 0.001$), indicating the differences in numerical habitat factors between experimental plots and control plots.

The contribution of these 11 factors to habitat selection by water deer was tested using principal component analysis. Those that have greater biological significance habitat factors were selected for logistic regression, and the "Forward-LR" method (based on assumptions for the probability of the likelihood ratio test and forward stepwise selection variables) was used to screen habitat factors.

The resource selection function model was used to analyze the influence of different habitat factors on the habitat selection of water deer [37]. The resource selection function model is a linear logarithmic model that includes multiple independent habitat variables: $\omega(x) = \exp(\beta_0 + \beta_1 x_1 + \beta_2 x_2 + \ldots + \beta_k x_k)$ [38], where x represents different independent habitat variables and β represents the selection coefficient. Then, the probability of species choosing habitats is calculated as $T(x) = \exp(\beta_0 + \beta_1 x_1 + \beta_2 x_2 + \ldots + \beta_k x_k)/[1 + \exp(\beta_0 + \beta_1 x_1 + \beta_2 x_2 + \ldots + \beta_k x_k)]$. The selection coefficient β can be estimated by the logistic regression coefficient [39]. The function model of resource selection based on selected habitat factor variables is $P = e^z/(1 + e^z)$, $z = \beta_0 + \beta_1 x_1 + \beta_2 x_2 + \ldots + \beta_k x_k$, where e is the natural number and P is the probability of habitat selection. The receiver operating characteristic (ROC) curve can be used to gauge the accuracy of the resource selection function model with the following evaluation criteria: if the value of the area under the curve (AUC) is 0.5–0.6, failure; 0.6–0.7, poor; 0.7–0.8, general; 0.8–0.9, good; 0.9–1.0, excellent [40,41]. All data processing was performed using Origin 2021 (OriginLab Corporation in Northampton, MA, USA) and SPSS 20.0 (International Business Machines Corporation in Armonk, NY, USA).

3. Results

3.1. Habitat Selection of Water Deer in Autumn

The chi-square test showed that vegetation types ($\chi^2 = 74.730$, $df = 5$, $p < 0.001$), dominant plants ($\chi^2 = 80.106$, $df = 18$, $p < 0.001$), slope positions ($\chi^2 = 7.898$, $df = 3$, $p < 0.05$), and slope direction ($\chi^2 = 51.184$, $df = 2$, $p < 0.001$) differed between experimental plots and control plots (Table S1). Compared with control plots, in autumn, the water deer prefers grasslands on middle and sunny slopes, with *A. carvifolia* as the dominant plant (Figure 3).

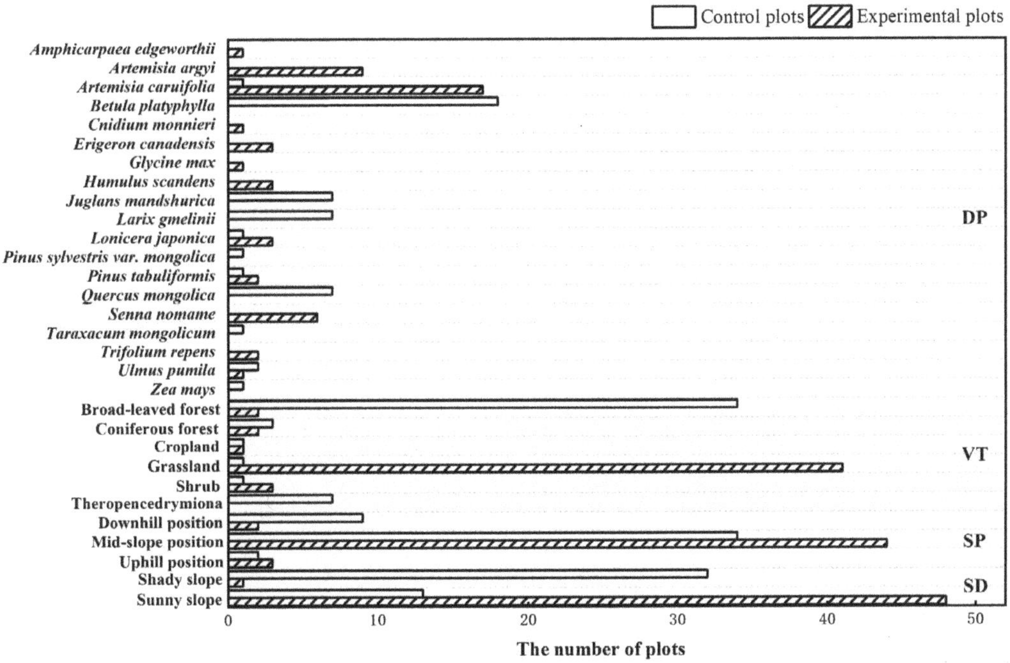

Figure 3. Selection of water deer regarding non-numeric habitat factors in autumn for both experimental and control plots (DP: dominant plant, VT: vegetation type, SP: slope positions, SD: slope direction).

The mann–Whitney U test showed that experimental plots with traces of water deer were found about 70.89 m father away from human settlements than control plots ($Z = -5.594$, $p < 0.001$), and they were also covered to a higher degree by herbage (23.99%, $Z = -5.032$, $p < 0.001$) and hiding cover (26.06%, $Z = -5.341$, $p < 0.001$). Furthermore, in experimental plots, the dominant herbage was about 28.29 cm higher ($Z = -5.779$, $p < 0.001$) and at an elevation that was about 51.76 m lower ($Z = -2.661$, $p < 0.01$). The slope degree ($Z = -5.267$, $p < 0.001$) was 12.82° lower in experimental plots than in control plots. Both plot types were located within the same distances from water sources ($Z = -1.130$, $p > 0.05$) (Figure 4, Table S2). Compared with control plots, in autumn, the water deer prefer habitats farther away from human settlements, with higher herbage coverage, higher hiding cover, lower elevation, higher height of dominant herbage, and gentler slope degree.

The results showed that the height of dominant herbage (HDH), elevation (E), and slope position (SP) have greater biological significance. Hence, these three habitat factors (HDH, E, and SP) were selected for logistic regression. Finally, the habitat factor that was entered into the resource selection function model was HDH. The function model of resource selection based on selected habitat factor variables is $P = e^z/(1 + e^z)$, $z = 2.864 - 0.041 \times HDH$ (Table 2). The correct prediction rate of the model was 85.4%. ROC curve analysis showed that AUC = 0.840, indicating that the prediction result of the resource selection function was good. The results showed that the largest contribution rate of habitat factors was the height of dominant herbage.

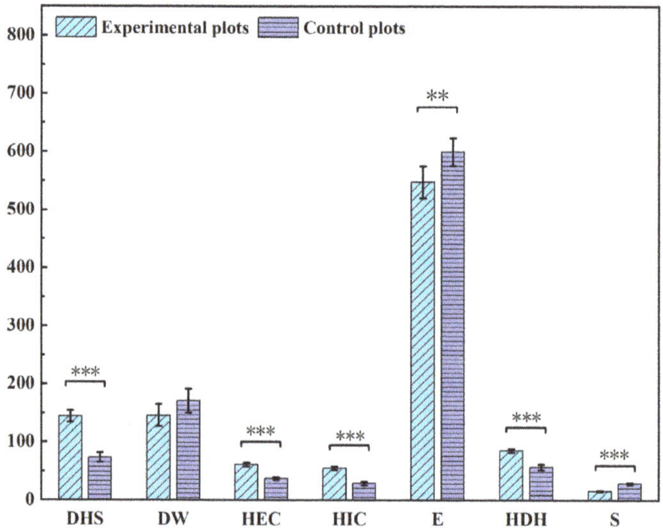

Figure 4. Comparison of numerical habitat factors between experimental plots and control plots in autumn (DHS: distance to human settlements (m), DW: distance from water (m), HEC: herbage coverage (%), HIC: hiding cover (%), E: elevation (m), HDH: height of dominant herbage (cm), S: slope degree (°). **: $p < 0.01$, ***: $p < 0.001$.

Table 2. Selection coefficient of water deer resource selection function in autumn.

Habitat Factors	Regression Coefficient	Wald Chi-Square	*p*-Value
HDH (height of dominant herbage/cm)	−0.041	17.575	0.000 ***
Constant	2.864	15.610	0.000 ***

***: $p < 0.001$.

3.2. Habitat Selection of Water Deer in Winter

The chi-square test showed that vegetation types ($\chi^2 = 49.906$, $df = 6$, $p < 0.001$) and dominant plants ($\chi^2 = 72.223$, $df = 16$, $p < 0.001$) were significantly different between experimental plots and control plots in winter; however, both slope positions ($\chi^2 = 5.596$, $df = 2$, $p > 0.05$) and slope direction ($\chi^2 = 2.885$, $df = 1$, $p > 0.05$) were not significantly different between experimental plots and control plots (Table S3). Compared with control plots, in winter, the water deer prefers grasslands in middle and sunny slopes, with *A. argyi* as the dominant plant (Figure 5).

The results of the *t*-test and Mann–Whitney U test showed that experimental plots were found to be around 218.61 m closer to water sources than the control plots ($Z = -3.929$, $p < 0.001$); experimental plots were also covered to a higher degree by herbage (28.49%, $Z = -5.143$, $p < 0.001$) and had higher hiding cover (18.16%, $Z = -4.119$, $p < 0.001$). Furthermore, the experimental plots showed a higher dominant herbage by about 37.81 cm ($Z = -5.238$, $p < 0.001$). Slope degree ($Z = -6.177$, $p < 0.001$) was 14.34° lower in experimental plots than in control plots. Both plot types were located within the same distance to human settlements ($t = -0.336$, $p > 0.05$) and had the same elevation ($Z = -1.350$, $p > 0.05$) (Figure 6, Table S4). Compared with control plots, in winter, the water deer prefers habitats with a shorter distance from the water, higher herbage coverage, higher hiding cover, higher height of dominant herbage, and gentler slope degree.

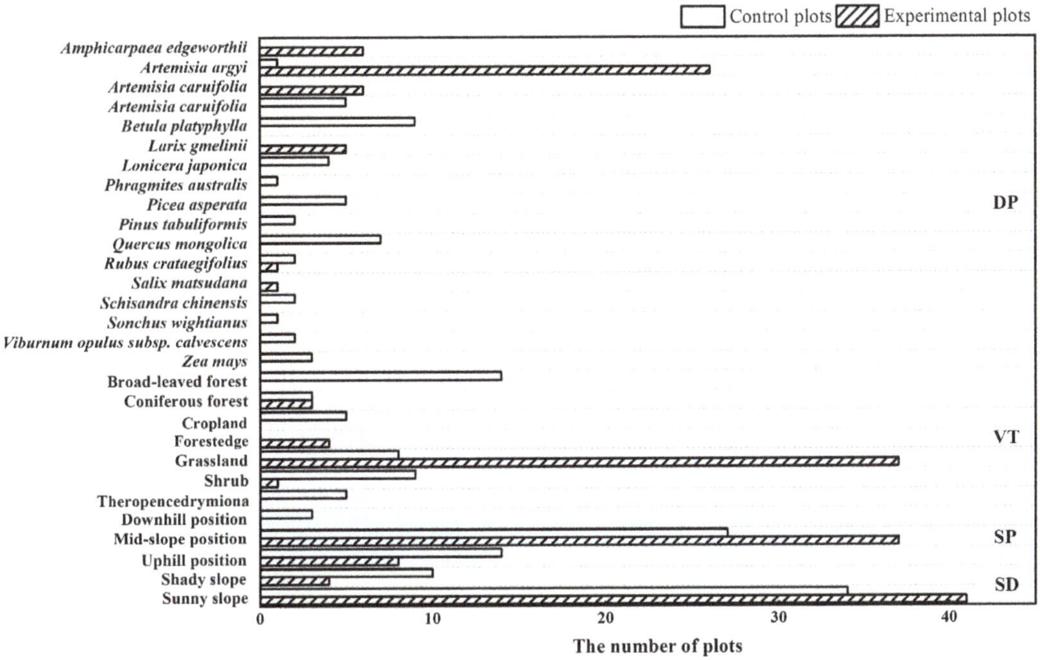

Figure 5. Selection of water deer regarding non-numeric habitat factors in winter for both experimental plots and control plots (DP: dominant plant, VT: vegetation type, SP: slope positions, SD: slope direction).

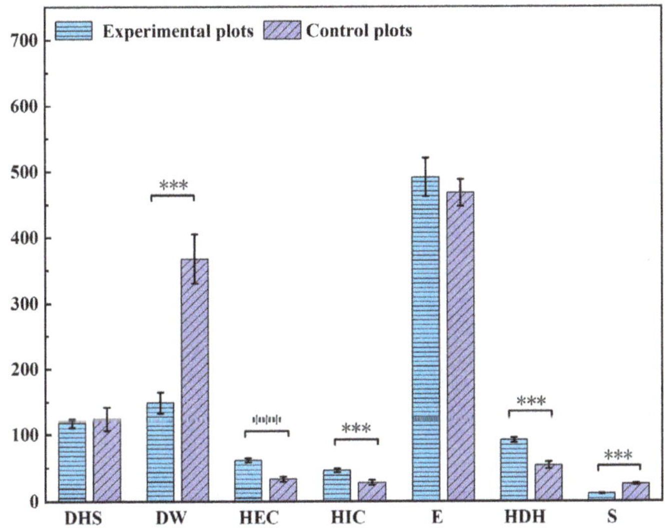

Figure 6. Comparison of numerical habitat factors between experimental plots and control plots in winter (DHS: distance to human settlements (m), DW: distance from water (m), HEC: herbage coverage (%), HIC: hiding cover (%), E: elevation (m), HDH: height of dominant herbage (cm), S: slope degree (°)). ***: $p < 0.001$.

Hiding cover (HIC), distance from water (DW), distance to human settlements (DHS), and slope position (SP) have greater biological significance. The four habitat factors of HIC, DW, DHS, and SP were selected for logistic regression. Finally, the following habitat factors were entered into the resource selection function model: DHS, DW, and HIC. The function model of resource selection based on selected habitat factor variables is $P = e^z/(1 + e^z)$, $z = 0.533 - 0.013 \times DHS + 0.009 \times DW - 0.032 \times HIC$ (Table 3). The total correct prediction rate of the model was 79.8%. ROC curve analysis showed that AUC = 0.867, indicating that the prediction results of the resource selection function was good. The results show that the contribution rate of these factors can be ordered from large to small as hiding cover > distance to human settlements > distance from water.

Table 3. Selection coefficient of water deer resource selection function in winter.

Habitat Factors	Regression Coefficient	Wald Chi-Square	*p*-Value
DHS (distance to human settlements/m)	−0.013	6.804	0.009 **
DW (distance from water/m)	0.009	16.007	0.000 ***
HIC (hiding cover/%)	−0.032	7.805	0.005 **
Constant	0.533	0.669	0.413

**: $p < 0.01$, ***: $p < 0.001$.

4. Discussion

Habitat selection of wildlife is one of the most important issues for researchers and conservationists, as it guides the formulation of conservation strategies for endangered species [42]. In the current study, habitat selection research was conducted on rediscovered water deer in cold regions of China. The obtained results showed that there are many similarities between water deer habitat selection in northeast China and southern China. Both prefer grasslands with higher herbage coverage, higher hiding cover, and smaller slope degree. However, there are also many differences between the southern study site and the northeastern study site, which are reflected in aspects of daily temperature, longitude, latitude, duration of snow cover, and vegetation types.

The vegetation type comprehensively reflects the characteristics of food composition, temperature, light levels, terrain, and landform of the habitat. It meets the physiological and ecological needs of animals to the greatest extent, and is an important habitat factor in the habitat selection of water deer [43]. The vegetation type water deer prefer in autumn and winter is grassland. A similar preference of these deer was also reported in other regions of China, such as their habitat selection in spring, summer, and autumn at Yancheng Nature Reserve and in winter at Dafeng Milu Natural Nature Reserve in Jiangsu Province [44,45]. A year-round habitat selection study of water deer in Shanghai Binjiang Park also found a similar preference for grasses with higher herbage height [46]. Other small ruminants had similar preferences, such as the Indian muntjac (*Muntiacus muntjak*), the preferred habitat of which was shrub grassland and reclaimed grassland; further, the height of trees in the habitat was the main influencing factor for the habitat selection of the Indian muntjac [47].

Furthermore, *A. carvifolia* and *A. argyi* are dominant plants for water deer in northeast China. In 2021, Li et al. conducted a study in Jilin Province, China, which was similar to the present study, and they also found that water deer preferred *A. carvifolia* as the dominant plant habitat [48]. *A. carvifolia* and *A. argyi* both belong to the Compositae family. Many previous studies on the diet of water deer have found that Compositae occupy a certain proportion in China, for example, up to 11.34% of their diet in Jilin Province, and up to 4% of their diet in Jiangxi Province [33,49]. Kim et al. studied the diet of Korean water deer (*Hydropotes inermis argyropus*) in Gyeonggi Province, Korea, and found that up to 28.4% of their diet is composed of Compositae [50]. In conclusion, it is clear that water deer habitat selection is closely related to its feeding habits. In addition, the height of dominant herbage of the experimental plots (autumn: 85.31 ± 3.00, winter: 91.33 ± 3.50) was significantly higher than that of control plots (autumn: 57.02 ± 5.00, winter: 53.52 ± 5.22). This result suggests that the water deer prefers a habitat where the herb height exceeds its own shoulder height (the shoulder height of adult water deer is 50–57 cm) [24]. Furthermore, the results of the resource selection function model show that the dominant herb height and hiding cover are habitat factors that contribute more to water deer habitat selection in the cold season. On the one hand, water deer are timid and alert, and the high herbage could supply good hiding areas for them [51]. On the other hand, winter is the season when water deer mate; hence, their higher activity would attract predators and a good hiding area is essential [52]. To survive and avoid natural enemies, habitats with high hiding cover and higher dominant herbage are their preferred activity areas.

In addition to the vegetation factor, other factors also affected the habitat selection of water deer. In northeast China, prolonged snow accumulation increases the difficulty of survival for water deer, as in winter, obtaining water and food resources is more difficult in the northern part than in the southern part. First, the obtained results show that in both autumn and winter, water deer chose a habitat closer to water. This factor is also the habitat factor that contributes most to the results of the resource selection function. The results of water deer selection of suitable distribution areas in Korea similarly showed that water deer prefer areas that are relatively close to water [53]. Second, ruminants are generally considered to migrate to lower elevations in late autumn and early winter as a strategy to find areas with shallow snow cover [54]. This is highly consistent with the results of the present study, which showed that the ecological habitat of water deer tends to be at relatively low elevation (autumn: experimental plots 547.91 $\pm$ 27.56 m, control plots: 599.70 $\pm$ 23.93 m). Snow-covered vegetation makes it more difficult to obtain food, and a habitat at lower elevation has higher temperatures and may possess relatively abundant food resources. Third, in Poyang Lake Wetland of Jiangxi Province, China, water deer prefer a distance to human settlements of 500–999 m [26]. However, the results of this study indicated that the preferred distance to human settlements was 100–200 m in both autumn and winter, which is far shorter than the results obtained for Poyang Lake Wetland. Similar to the crop attraction hypothesis, the timid water deer may be closer to human settlements because the human activity area offers domestic waste or farmland residues such as pea seedlings, soybean leaves, peanut leaves, and sweet potato leaves, which provide food sources for water deer. Some Western studies have found that other deer species also support the crop attraction hypothesis; for example, red deer spending time on the pasture increased with increasing availability, but not in a proportional manner, resulting in the strength of the trade-off varying with habitat availability driven by landscape-level variability. The seasonal variation in the trade-off may be due to the seasonally varying abundance of forage and cover in the different habitats [55]. Even in some parts of the West, large numbers of fallow, red, and roe deer have caused some damage to crops and forestry [56,57].

5. Conclusions and Suggestions

In the current study, during the cold season of northeast China, water deer activities mainly focused on grasslands in the mid-slope position and on sunny slopes, dominated by *A. carvifolia* and *A. argyi*. In addition, water deer preferred a higher dominant herbage, higher hiding cover, and closer distance to water and human settlements. The correct prediction probability of the resource selection function model exceeds 80%, indicating that the model can predict the cold season habitat selection of water deer in Baishan Musk Deer Natural Reserve. This result has positive significance for further research on water deer and the development of scientific conservation and management plans in this reserve. Water deer were rediscovered in northeast China, which is the result of the effective restoration of the ecological environment that provides more habitats for wildlife. For protected areas, the construction of wildlife nature reserves needs to be further strengthened to protect both the existing and potential habitats of water deer. In the future, reserves should pay close attention to water sources, the activities of surrounding human settlements, and grassland protection; further, patrols should be strengthened to eliminate outside interference, which will further protect water deer. Based on these efforts, water deer could better adapt to habitats in cold regions.

6. Limitations and Future Research

Due to limited data availability and imperfect knowledge, this study has some limitations. For example, global climate change is indirectly or directly affecting the distribution of wild animals, and whether Jilin Province is the northernmost edge of the future distribution of water deer in China will continue to be monitored. For future research, the methodology of habitat selection studies must be further optimized. We will focus on the combination of changes in climatic factors in north and south China in the past decades and the assessment of habitat suitability of water deer in China. With data availability and in-depth research, these issues will be further explored in our future work.

Supplementary Materials: The following supporting information can be downloaded at: https://www.mdpi.com/article/10.3390/su151612181/s1, Table S1: Characteristics of habitat selection in autumn; Table S2: Mann Whitney U-test of autumn quadrat; Table S3: Characteristics of habitat selection in winter; Table S4: Independent sample *t*-test and Mann Whitney U-test of winter quadrat.

Author Contributions: Conceptualization, Y.S., Z.L. (Zongzhi Li), R.H.K., J.C., Z.L. (Zhensheng Liu) and L.T.; methodology, Y.S., Z.L. (Zongzhi Li) and J.C.; validation, Y.S. and J.C.; formal analysis, Z.L. (Zhensheng Liu) and L.T.; investigation, Z.L. (Zongzhi Li); data curation, Y.S. and Z.L. (Zongzhi Li); writing—original draft preparation, Y.S., Z.L. (Zongzhi Li), R.H.K. and J.C.; supervision Z.L. (Zhensheng Liu) and L.T.; writing—review and editing, Y.S., Z.L. (Zhensheng Liu), R.H.K., J.C., Z.L. (Zongzhi Li) and L.T. All authors have read and agreed to the published version of the manuscript.

Funding: This research was funded by the Key Laboratory of Conservation Biology, National Forestry and Grassland Administration, China; the Heilongjiang Touyan Innovation Team Program for Forest Ecology and Conservation; and National Natural Science Foundation of China (32070519, 32071649).

Institutional Review Board Statement: Not applicable.

Informed Consent Statement: Not applicable.

Data Availability Statement: The data presented in this study are openly available in figshare at https://doi.org/10.6084/m9.figshare.23963316.

Acknowledgments: We thank the staff of the Baishan Musk Deer National Nature Reserve for their support in the field work of our study.

Conflicts of Interest: The authors declare no conflict of interest.

References

1. Morrison, M.L.; Marcot, B.G.; Mannan, W. Wildlife-Habitat Relationship: Concepts and Applications. *J. Range Manag.* **2007**, *57*, 980–981.
2. Grinnell, J. Field Tests of Theories Concerning Distributional Control. *Am. Nat.* **1917**, *51*, 115–128. [CrossRef]
3. Gaillard, J.-M.; Hebblewhite, M.; Loison, A.; Fuller, M.; Powell, R.; Basille, M.; Van Moorter, B. Habitat–performance relationships: Finding the right metric at a given spatial scale. *Philos. Trans. R. Soc. B Biol. Sci.* **2010**, *365*, 2255–2265. [CrossRef]
4. Furstenburg, D.; Scholtz, M. Global Climate Change and Animal Production in Southern Africa: A Short Review. Submitted: Livestock Science. 2009 (Supplement–10th Wrld. Cong. Anim. Prod.). Available online: https://www.researchgate.net/publication/306055750_Global_climate_change_and_animal_production_in_southern_Africa_a_short_review (accessed on 29 June 2023).
5. Ehrlén, J.; Morris, W.F. Predicting changes in the distribution and abundance of species under environmental change. *Ecol. Lett.* **2015**, *18*, 303–314. [CrossRef]
6. Singh, B.; Mainali, K.; Jiang, Z.; Thapa, A.; Subedi, N.; Awan, M.N.; Ilyas, O.; Luitel, H.; Zhou, Z.; Hu, H. Projected distribution and climate refugia of endangered Kashmir musk deer Moschus cupreus in greater Himalaya, South Asia. *Sci. Rep.* **2020**, *10*, 1511. [CrossRef]
7. Zhu, G.; Papeş, M.; Armsworth, P.R.; Giam, X. Climate change vulnerability of terrestrial vertebrates in a major refuge and dispersal corridor in North America. *Divers. Distrib.* **2022**, *28*, 1227–1241. [CrossRef]
8. Campos, F.A.; Morris, W.F.; Alberts, S.C.; Altmann, J.; Brockman, D.K.; Cords, M.; Pusey, A.; Stoinski, T.S.; Strier, K.B.; Fedigan, L.M. Does climate variability influence the demography of wild primates? Evidence from long-term life-history data in seven species. *Glob. Chang. Biol.* **2017**, *23*, 4907–4921. [CrossRef]

9. Pacifici, M.; Foden, W.B.; Visconti, P.; Watson, J.E.M.; Butchart, S.H.M.; Kovacs, K.M.; Scheffers, B.R.; Hole, D.G.; Martin, T.G.; Akçakaya, H.R.; et al. Assessing species vulnerability to climate change. *Nat. Clim. Chang.* **2015**, *5*, 215–224. [CrossRef]
10. IPCC. Summary for Policymakers. In *Global Warming of 1.5 °C. An IPCC Special Report on the Impacts of Global Warming of 1.5 °C above Pre-Industrial Levels and Related Global Greenhouse Gas Emission Pathways, in the Context of Strengthening the Global Response to the Threat of Climate Change, Sustainable Development, and Efforts to Eradicate Poverty;* Masson-Delmotte, V., Zhai, P., Pörtner, H.-O., Roberts, D., Skea, J., Shukla, P.R., Pirani, A., Moufouma-Okia, W., Péan, C., Pidcock, R., et al., Eds.; Cambridge University Press: Cambridge, UK; New York, NY, USA, 2022; pp. 1–24. [CrossRef]
11. Parmesan, C.; Yohe, G. A globally coherent fingerprint of climate change impacts across natural systems. *Nature* **2003**, *421*, 37–42. [CrossRef]
12. Kelly, A.E.; Goulden, M.L. Rapid shifts in plant distribution with recent climate change. *Proc. Natl. Acad. Sci. USA* **2008**, *105*, 11823–11826. [CrossRef]
13. Chen, I.-C.; Hill, J.K.; Ohlemüller, R.; Roy, D.B.; Thomas, C.D. Rapid Range Shifts of Species Associated with High Levels of Climate Warming. *Science* **2011**, *333*, 1024–1026. [CrossRef]
14. Prato, T. Adaptively Managing Wildlife for Climate Change: A Fuzzy Logic Approach. *Environ. Manag.* **2011**, *48*, 142–149. [CrossRef]
15. Leslie, J.D.M.; Bowyer, R.T.; Kie, J.G. Life-History Strategies of Ungulates. *J. Mammal.* **1999**, *80*, 1067–1069. [CrossRef]
16. Underwood, H.B. Deer of the World: Their Evolution, Behaviour, and Ecology. *Val. Geist* **2000**, *75*, 196.
17. Boydston, E.E.; Kapheim, K.M.; Watts, H.E.; Szykman, M.; Holekamp, K.E. Altered behaviour in spotted hyenas associated with increased human activity. *Anim. Conserv.* **2003**, *6*, 207–219. [CrossRef]
18. Inskip, C.; Zimmermann, A. Human-felid conflict: A review of patterns and priorities worldwide. *Oryx* **2009**, *43*, 18. [CrossRef]
19. Shannon, G.; Cordes, L.S.; Hardy, A.R.; Angeloni, L.M.; Crooks, K.R. Behavioral Responses Associated with a Human-Mediated Predator Shelter. *PLoS ONE* **2014**, *9*, e94630. [CrossRef]
20. Moll, R.J.; Cepek, J.D.; Lorch, P.D.; Dennis, P.M.; Robison, T.; Millspaugh, J.J.; Montgomery, R.A. Humans and urban development mediate the sympatry of competing carnivores. *Urban Ecosyst.* **2018**, *21*, 765–778. [CrossRef]
21. Atickem, A.; Loe, L.E.; Stenseth, N.C. Individual Heterogeneity in Use of Human Shields by Mountain Nyala. *Ethology* **2014**, *120*, 715–725. [CrossRef]
22. Won, C.; Smith, K.G. History and current status of mammals of the Korean Peninsula. *Mammal Rev.* **2002**, *29*, 3–36. [CrossRef]
23. Ohtaishi, N. *Deer of China: Biology and Management*; Elsevier: Amsterdam, The Netherlands, 1993; pp. 4–7.
24. Li, Y.K.; Shan, J.H.; Li, J.; Yuan, F.K.; Miao, L.J.; Xie, G.Y. Research Advances on the Ecology of Chinese Water Deer (*Hydropotes inermis*). *Chin. J. Wildl.* **2013**, *34*, 270–273.
25. Harris, R.B.; Duckworth, J.W. *Hydropotes inermis*. The IUCN Red List of Threatened Species 2015: e.T10329A22163569. Available online: https://www.iucnredlist.org/species/10329/22163569 (accessed on 15 June 2023).
26. Zhang, N. Home Range and Habitat Selection of Chinese Water Deer (*Hydropotes inermis*) after Release in Poyang Lake during the Dry Season. Master's Thesis, Jiangxi Normal University, Nanchang, China, 2019.
27. Han, Q.; Liang, T.; Zhang, M.Y.; Liu, M.M.; Wang, Z.Q.; Lu, C.H. Analysis of activity rhythm and habitat selection of water deer based on the infrared-triggered camera technology in autumn and winter in urban forest park. *Chin. J. Ecol.* **2022**, *41*, 963.
28. Li, Z.Z.; Wu, J.P.; Teng, L.W.; Liu, Z.S.; Wang, B.K.; Liu, Y.C.; Xu, T. The Rediscovery of Water Deer (*Hydropotes inermis*) in Jilin Province. *Chin. J. Zool.* **2019**, *54*, 108–112.
29. Belyaev, D.A.; Jo, Y.-S. Northernmost finding and further information on water deer *Hydropotes inermis* in Primorskiy Krai, Russia. *Mammalia* **2021**, *85*, 71–73. [CrossRef]
30. Teng, L.W. *Population of Chinese Water Deer (Hydropotes inermis) in Yancheng National Nature Reserve and Feasibility of Reintroduction Chinse Water Deer to Shanghai*; Research Report of Postdoctoral Fellow; East China Normal University: Shanghai, China, 2007.
31. Chen, L.C. A brief discussion on the causes of problems in the former musk Deer National Nature Reserve of Baishan in Jilin Province and the results of rectification. *Agric. Dev. Equip.* **2021**, 230–231. Available online: https://kns-cnki-net-443.webvpn.nefu.edu.cn/kcms/detail/detail.aspx?FileName=NJJY202103112&DbName=CJFQ2021 (accessed on 29 June 2023).
32. Li, H.J. Investigation and Analysis of Wild Animals in Jilin Baishan Original Musk Deer National Nature Reserve. *For. Prospect. Des.* **2021**, *50*, 52–55.
33. Ding, Y.X. Study on Diets Analysis of Water Deer (*Hydropotes inermis*) in the Baishan Musk Deer National Nature Reserve in Jilin Province. Master's Thesis, Northeast Forestry University, Harbin, China, 2022.
34. Smith, A.T.; Xie, Y. *A Guide to the Mammals of China. Changsha, Hunan Province, China*; Hunan Education Publishing House: Changsha, China, 2009; pp. 481–482.
35. Chen, Y.-T.; Chen, M.C. Using chi-square statistics to measure similarities for text categorization. *Expert Syst. Appl.* **2011**, *38*, 3085–3090. [CrossRef]
36. Rosenthal, R. An Application of the Kolmogorov-Smirnov Test for Normality with Estimated Mean and Variance. *Psychol. Rep.* **1968**, *22*, 570. [CrossRef]
37. Lennon, J.J. Resource selection functions: Taking space seriously? *Trends Ecol. Evol.* **1999**, *14*, 399–400. [CrossRef]
38. Murtaugh, P.A. The Statistical Evaluation of Ecological Indicators. *Ecol. Appl.* **1996**, *6*, 132–139. [CrossRef]
39. David, W.H.J.; Stanley, L.; Rodney, X.S. *Applied Logistic Regression*; John Wiley and Sons: New York, NY, USA, 2013.
40. Swets, J.A. Measuring the accuracy of diagnostic systems. *Science* **1988**, *240*, 1285–1293. [CrossRef]

41. Boyce, M.S.; Vernier, P.R.; Nielsen, S.E.; Schmiegelow, F.K. Evaluating resource selection functions. *Ecol. Model.* **2002**, *157*, 281–300. [CrossRef]
42. Xia, X.; Ren, J.; Li, L.; Wang, H.; Song, Y.; Yang, D.; Jiang, Z. Autumn-winter habitat selection by the re-wild Milu (*Elaphurus davidianus*) at the early stage after release in Dongting Lake Wetland, China. *Biodivers. Sci.* **2021**, *29*, 1087–1096. [CrossRef]
43. Zheng, X.; Bao, Y.X.; Ge, B.M.; Zheng, R.Q. Seasonal changes in habitat use of black muntjac (*Muntiacus crinifrons*) in Zhejiang. *Acta Theriol. Sin.* **2006**, *26*, 201–205.
44. Zhang, E.D.; Teng, L.W.; Wu, Y.B. Habitat selection of the Chinese water deer (*Hydropotes inermis*) in Yancheng Reserve, Jiangsu Province. *Acta Theriol. Sin.* **2006**, *26*, 49–53.
45. Zhang, X.L.; Zhang, E.D. Distribution Pattern of *Hydropotes inermis* in various habitats in Jiangsu Dafeng Peredavids deer state nature reserve. *Sichuan J. Zool.* **2002**, *21*, 19–22.
46. He, X. Spatial Behavioural Ecology of the Chinese Water Deer. Ph.D. Thesis, East China Normal University, Shanghai, China, 2013.
47. Teng, L.; Liu, Z.; Song, Y.-L.; Zeng, Z. Forage and bed sites characteristics of Indian muntjac (*Muntiacus muntjak*) in Hainan Island, China. *Ecol. Res.* **2004**, *19*, 675–681. [CrossRef]
48. Li, Z.Z.; Liu, Z.S.; Mi, S.H.; Wu, J.P.; Teng, L.W. Habitat selection of the Chinese water deer at Baishan Musk Deer Natural Reserve in spring and summer. *Acta Ecol. Sin.* **2021**, *41*, 1625–1633.
49. Huang, Y. Food-Habits of Chinese Water Deer (*Hydropotes inermis*) in Poyang Lake Area. Master's Thesis, Jiangxi Normal University, Nanchang, China, 2016.
50. Kim, B.J.; Lee, N.S.; Lee, S.D. Feeding diets of the Korean water deer (*Hydropotes inermis argyropus*) based on a 202 bp rbcL sequence analysis. *Conserv. Genet.* **2011**, *12*, 851–856. [CrossRef]
51. Cederlund, G.; Lindström, E. Effects of severe winters and fox predation on roe deer mortality. *Acta Theriol.* **1983**, *28*, 129–145. [CrossRef]
52. Miao, L.J. The Population Viability Analysis, Distribution and Habitat Selection of Chinese Water Deer (*Hydropotes inermis*) in Poyang Lake. Master's Thesis, Jiangxi Normal University, Nanchang, China, 2015.
53. Choi, J.; Lee, S. Application of Habitat Evaluation Procedure with Quantifying the Eco-Corridor in the Process of Environmental Impact Assessment. *Int. J. Environ. Res. Public Health* **2019**, *16*, 1437. [CrossRef] [PubMed]
54. Nelson, M.E. Winter range arrival and departure of white-tailed deer in northeastern Minnesota. *Can. J. Zool.* **2011**, *73*, 1069–1076. [CrossRef]
55. Godvik, I.M.R.; Loe, L.E.; Vik, J.O.; Veiberg, V.; Langvatn, R.; Mysterud, A. Temporal scales, trade-offs, and functional responses in red deer habitat selection. *Ecology* **2009**, *90*, 699–710. [CrossRef]
56. Putman, R.J.; Moore, N.P. Impact of deer in lowland Britain on agriculture, forestry and conservation habitats. *Mammal Rev.* **2002**, *28*, 141–164. [CrossRef]
57. Gordon, I.J.; Hester, A.J.; Festa-Bianchet, M. REVIEW: The management of wild large herbivores to meet economic, conservation and environmental objectives. *J. Appl. Ecol.* **2004**, *41*, 1021–1031. [CrossRef]

Disclaimer/Publisher's Note: The statements, opinions and data contained in all publications are solely those of the individual author(s) and contributor(s) and not of MDPI and/or the editor(s). MDPI and/or the editor(s) disclaim responsibility for any injury to people or property resulting from any ideas, methods, instructions or products referred to in the content.

sustainability

MDPI

Article

Estimation of Fishery Losses from Great Cormorants during the Wintering Period in Greek Lagoons (Ionian Sea, W. Greece)

George Katselis [1], Spyridon Konstas [2] and Dimitrios K. Moutopoulos [1,*]

[1] Department of Fisheries & Aquaculture, University of Patras, 30200 Mesolongi, Greece; gkatselis@upatras.gr
[2] Management Unit of Acheloos Valley and Amvrakikos Gulf Protected Areas, Natural Environment & Climate Change Agency, 47150 Aneza Artas, Greece; spiroskonstas@hotmail.com
* Correspondence: dmoutopo@upatras.gr

Abstract: The present study aims through a modeling approach to quantify fishing losses from the impact of great cormorants (*Phalacrocorax carbo sinensis*) during their wintering period in Greek lagoons. A number of assumptions were incorporated into the model regarding fish population growth, species distribution, age (or size) of fish caught, and the different fishing strategies that could be applied in the studied lagoons. The results indicated that the mean value of daily economic losses ranged from 0.614 to 1.075 €·bird^{-1}·day^{-1}, whereas the ratios of biomass losses to landings biomass and of economic losses to economic profit ranged from 0.18:1 to 3.80:1 and from 0.14:1 to 4.18:1, respectively, depending on the lagoon. The results supported a strong competitive relationship between great cormorants and fisheries in lagoons of the Amvrakikos Gulf.

Keywords: fisheries–marine bird interactions; fisheries management; economic losses; Amvrakikos Gulf; Lefkada island

Citation: Katselis, G.; Konstas, S.; Moutopoulos, D.K. Estimation of Fishery Losses from Great Cormorants during the Wintering Period in Greek Lagoons (Ionian Sea, W. Greece). *Sustainability* **2023**, *15*, 12007. https://doi.org/10.3390/su151512007

Academic Editor: Vasilios Liordos

Received: 3 July 2023
Revised: 2 August 2023
Accepted: 3 August 2023
Published: 4 August 2023

Copyright: © 2023 by the authors. Licensee MDPI, Basel, Switzerland. This article is an open access article distributed under the terms and conditions of the Creative Commons Attribution (CC BY) license (https://creativecommons.org/licenses/by/4.0/).

1. Introduction

Coastal lagoons are crucial components of local people's cultural heritage, coastal environment, and economic vitality [1]. They serve as transitional areas between the open ocean and inland waters and are characterized by environmental fluctuations (for a review, see [2]). Given that these systems are habitats with an increasing concentration of fishing activities, because several fish species use them as part of their life histories for spawning, feeding, and refuge [3], the majority of them are protected by international conventions for the conservation of biodiversity (e.g., Natura 2000). Around the Mediterranean, an area of at least 6500 km^2 of coastal lagoons [4] is exploited as fishing grounds by local fisher associations [5]. The Greek lagoons cover a total surface of about 350 km^2 [6] and fishery exploitation is a common extensive culture, based on seasonal ongoing migration movements of fry and adult euryhaline fish species between the sea and lagoons (e.g., mugilids: flathead grey mullets *Mugil cephalus* Linnaeus, 1758, thicklip grey mullet *Chelon labrosus* (Risso, 1827), leaping mullet *Chelon saliens* (Risso, 1810), golden grey mullet *Chelon auratus* (Risso, 1810), and thinlip grey mullet *Chelon ramada* (Risso, 1827); gilthead seabream *Sparus aurata* Linnaeus, 1758; European seabass *Dicentrarchus labrax* (Linnaeus, 1758); and European eel *Anguilla anguilla* (Linnaeus, 1758)). Fishery production mainly comprises mugilids (about 56%) [6].

Aquatic birds can be seen as an integral part of these transitional aquatic systems, where an ongoing conflict with human activities exists. In fact, bycatch, a global issue, is one of the most significant factors contributing to the decline of aquatic bird populations (e.g., [7–9]). It is estimated that 400,000 birds per year are incidentally caught in gillnets globally [10], and in some cases, this taxon shares common resources with the fishers [11–18]. In contrast, in the lagoons of the southern Baltic Sea in Poland, the dominant piscivorous bird species, the Goosander, does not negatively impact fishing activities because it primarily preys on small fish species that are not the target of fishers [19].

The great cormorant *Phalacrocorax carbo sinensis* (Linnaeus, 1758), hereafter cormorant, is a species of piscivorous aquatic bird whose population has considerably increased throughout Europe in recent decades. This increase is due to protection (EU Directive 79/409, Directive 2009/147) and to increased food availability leading to eutrophication of aquatic habitats and commercial fish stocking [11,13,20–22]. The European population is estimated at 401,000–512,000 pairs, which equates to 803,000–1,020,000 mature individuals [23]. In the southern wetlands of Europe, the cormorant appears year-round but in greater numbers during the winter period [24–27]. In Greece, the species' breeding population has increased over time [27], which is accompanied by an increase in its population on wintering grounds [28]. There are six cormorant colonies in Greece, with a total breeding population of 5600 pairs, while in winter, their population amounts to approximately 22,000 individuals [29].

Studies on the cormorant's diet, energy requirements, impact on fish populations and fisheries, ethology [13,30–33], and management issues [13,28,34,35] have been extensively carried out worldwide. On the other hand, the fish-eating habits of the species and its increased numbers have caused severe conflicts with fisheries activities in many countries [21,30]. Cormorants' diet is described as opportunistic, because they do not select specific species or sizes of fish [24] but rather focus on fish species available in the ecosystem in which they settle. The European cormorant population is estimated to consume about 1000 t of fish per day [11] and severe economic losses are recorded in freshwater fish farms ranging from 150 to 1500 €$\cdot 10^{-1}$ ha^{-1} [11,12,17]. In the Czech Republic, the annual losses of carp pond aquaculture for 2019 are estimated to be EUR 4 million [36], while in a few countries/regions (e.g., Belgium (Waloon), Finland, Romania, the Czech Republic, Slovakia, Germany (Saxony), and Latvia), recognizing the conflict between cormorants and aquaculture has taken the form of financial compensation or subsidies programs [12]. In the northern Baltic Sea, there is an ongoing debate about the role of cormorants in the coastal ecosystem, where some studies suggest that cormorants can regulate fish populations [13,14,16] or that the effects have been considered to be site dependent [15] or have no effect [37].

In Greece, the effects of cormorants on fisheries range from minor (in lakes) [25,38] to relatively significant (in lagoons) [24,26]. Cormorants often prey on fish species with low commercial value, but this pattern appears to be reversed when colonies of fish-eating birds are concentrated in habitats with high fish concentrations, such as lagoons [26]. In Greek lagoons, Mugilidae comprise 60–70% of daily preyed biomass from the species [24,26], revealing a possible strong conflict between fisheries. However, the above assessments must be considered under-valued, because they estimate the direct losses by birds which are composed of fish individuals with sizes smaller than the commercial size [13,16]. If cormorants prey upon substantial number of fish at younger age classes, the yields of the fishers might be affected after a time delay [39]. This is anticipated to be more noticeable in a large culture, such as lagoons [11,12,17], than in coastal fisheries [13–16,37], justifying the fisher's perception of the negative impact of cormorants in lagoon fishery production [24].

In the present study, we estimate the catch losses due to cormorants' predation on fish resources during their wintering period in the lagoons of the Amvrakikos Gulf, using modeling techniques that estimate the loss in biomass and income based on individuals preyed upon by the cormorant population when the fish enters the exploited phase. The model used is a common population dynamics model [40] incorporating an economic component, which has been estimated for grey mullets. The basic idea of the model is that the fish size of grey mullets preyed on by cormorants in lagoons of western Greece [24,26] is smaller than the minimum length size of capture [41], and therefore, the effect of fish prey will be obvious as losses in fishery production after a time lag [39]. Moreover, in order to cover most of the uncertainties of input variables, the model was run for 60,000 iterations where the input variables, randomly, varied between a range (from available information), delineating the cormorant effect on lagoon fisheries as an expected range of values (biomass or income).

2. Materials and Methods

2.1. Site and Data Scources

The Amvrakikos Gulf (Western Greece, Ionian Sea) is a fjord-like hydrological regime (400 km^2) [42] around which 15 lagoons are located (Figure 1), covering a total area of about 96.2 km^2 that are protected under the Ramsar convention (www.ramsar.org (accessed on 27 May 2023)) and are part of the Natura 2000 network (http://ec.europa.eu/environment/nature/natura2000/index_en.htm (accessed on 27 May 2023)). Most of the lagoons are traditional fishery fields, exploited as common extensive cultured systems based on the seasonal entrance of young fish into the lagoons and the autumn-to-winter offshore fish migration. In six lagoons, for each lagoon, the proportions of the annual fisheries landings to the total (during 1980–2008) landings per species ranged from 0.1 to 0.56 (mostly between 0.45 and 0.56) for mugilids, from 0.08 to 0.38 for European eels, from 0.09 to 0.58 (mostly between 0.09 and 0.15) for gilthead seabream, and from 0.02 to 0.15 for gobies (mainly *Zosteriosessor ophiocephalus*) [42].

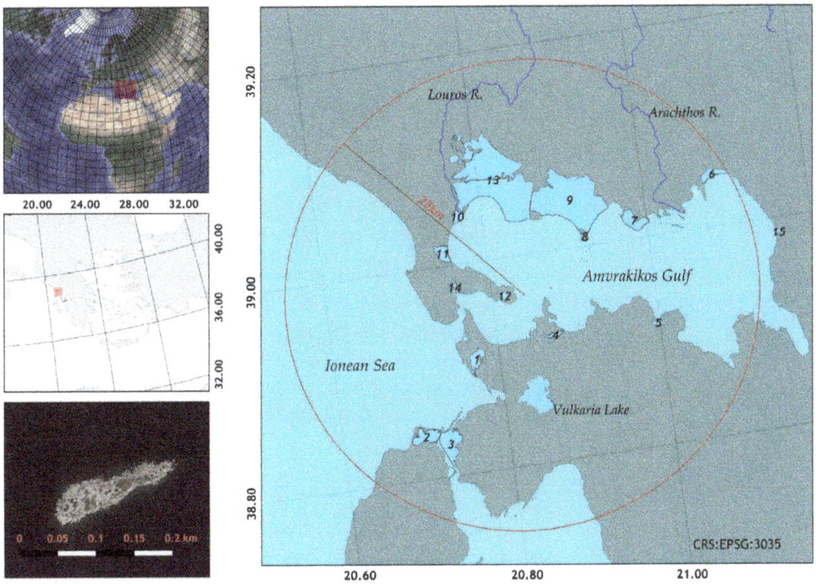

Figure 1. Map of study area. The red cycle indicated the daily flight zone of the great cormorants that roosted at Gaidaros (center of cycle and in the left photo at the bottom). Numbers indicate the lagoons studied, 1: Saltini, 2: Palaio, 3: Avlemonas, 4: Myrtari, 5: Rouga, 6: Agrilos, 7: Koftra-Palaiompouka, 8: Sakolesi, 9: Logarou, 10: Tsopeli, 11: Mazoma, 12: Pogonitsa, 13: Tsoukalio-Rouga, 14: Vathi.

In the Amvrakikos Gulf, the recorded annual wintering population of cormorants ranged from 1800 to 16,000 individuals during 2002–2022 (data provided by the Management Unit of Acheloos Valley and Amvrakikos Gulf Protected Areas and Hellenic Ornithological Society) using the midwinter census method [43]. Cormorants mainly roost on the small rocky island called Gaidaros (surface 5.6×10^{-3} km^{-2}), located in the western part of the gulf (Figure 1) and their daily excursion for feeding extends up to a radius of 25 km (daily flight zone: DFZ) [31,32] (Figure 1).

Fisheries data consisted of the annual landings per species and lagoon during 1977–2020, provided by the Fishery Department of Preveza, Arta, and Lefkada island. Fisheries data for mugilids were disaggregated at three commercial categories (Cat), namely, "mugilids" Cat A (above 400 g), Cat B (between 150 and 400 g), and Cat C (between 80 and 150 g).

2.2. Modeling Approach

The basic concept of the model based on the fish size of mugilids that is consumed by the cormorants is smaller compared to the minimum length size of capture by fishers. Thus, the daily biomass (B_{cr}) of a fish species consumed by one individual of cormorant per day was estimated by

$$B_{cr} = DFI = N_{cr} \cdot W_{cr}$$

where the *DFI* is the daily food intake of a cormorant, N_{cr} is the number of individuals of prey fish per cormorant, and W_{cr} is the mean weight of the fish species. The N_{cr} individuals are converted at the same length (L_{cr}) using the weight–length relationship, $W = a \cdot L^b$ (*a* and *b* the parameters of the length–weight relationships of fish species), and at W_{cr} and the age of preyed individuals (t_{cr}) using the length–age estimated by the von Bertalanffy equation, $L_t = L\infty \cdot (1 - e^{-k(t-t_o)})$ ($L\infty$, k, t_o are the parameters of the equation for each fish species).

When only the natural mortality existed (M), the number of individuals (N_c) at age t_c of exploitation for fish species was estimated by the following equation [42]:

$$N_c = N_{cr} \cdot e^{-(t_c - t_{cr}) \cdot M}$$

The fishing biomass (B_c) which corresponds to the fish species individuals that consumed by the cormorant per day is

$$B_c = catchR \cdot N_c \cdot W_c$$

where W_c is the mean weight of species at the fishing age (t_c), which is estimated by the equation $W_c = a \cdot \left(L\infty \cdot \left(1 - e^{-k(t_c - t_o)}\right)\right)^b$, and *catchR* is the catch rate.

The economic losses (E_c) are $E_c = B_c \cdot v_j$, where *v* is the economic value (€) on each weight category (Cat) *j* (mean weight> 400 g: Cat A, 150 ≤ mean weight ≤ 400 g: Cat B, and 80 ≤ mean weight ≤ 150 g: Cat C).

In the case of the inclusion of *i* fish species in the predator's trophic spectrum, the number of individuals of species *i* consumed by one cormorant (N_{cri}) is

$$N_{cr_i} = \frac{B_{cr}}{W_{cr_i}} \cdot q_i \quad (1)$$

where *q* is the biomass proportion of species *i* in daily consumed biomass.

On a daily basis, the losses in fishing production (tB_c), the proportion of tB_c per weight category j (P_{rj}), the economic losses (tE_c), the average time that the losses will appear after the consumption of a fish (aT), and the average length of a consumed fish (aL_{cr}) per each individual cormorant were estimated, respectively:

$$tB_c = \sum_{j=1}^{3} \sum_{i=1}^{5} \left(B_{c_{t_{j,i}}}\right)$$

$$Pr_j = \sum_{i=1}^{5} \left(\frac{B_{c_{i,j}}}{tB_c}\right)$$

$$tE_c = \sum_{j=1}^{3} \sum_{i=1}^{5} \left(B_{c_{t_{j,i}}} \cdot v_j\right)$$

$$aT = \sum_{i=1}^{5} \frac{(N_{c_i} \cdot (t_{c_i} - t_{cr_i}))}{\sum_{i=1}^{5} (N_{c_i})}$$

$$aL_{cr} = \sum_{i=1}^{5} \frac{(N_{cr_i} \cdot L_{cr_i})}{\sum_{i=1}^{5} (N_{cr_i})}$$

2.3. Model Calibration and Sensitivity Analysis

The model was calibrated using the estimated biological parameters found in the literature (Table A1, Appendix A) and received random values around the M_i (Table A1) from a normal distribution (average value = 1; SD = 0.1) (Mcg), for the q_i from a uniform distribution at range [0.08–0.40], for the Lcr_i from a normal distribution (average value = 19.3 cm; SD = 1.3 cm; Table A1), for the DFI from a normal distribution (average value = 0.18 kg: SD = 0.04 kg) and for the *catchR* from a uniform distribution at range [0.2–1.0]. At the time of capture (t_c), the model received random values according to three cases: (1) from a beta distribution ($\alpha = 1$, $\beta = 3$) at a range of integer values from 1 to 6, (2) from a uniform distribution at a range of integer values from 1 to 6; and (3) from a symmetrical distribution of the case (1) at a range of integer values from 1 to 6. Three different fishing strategies were followed based on targeted fish size: all fish sizes (1), non-targeted fishing (2), and targeted fishing of the larger fish sizes (3).

For each fishing strategy, the model was iterated 20,000 times, and in each iteration, the variables and the estimates of the model were recorded (Table A2, Appendix B). In each iteration, the Mcg was kept the same for all species, whereas the other random values were independent. For each fishing strategy, the distribution of the time that the losses would appear was also estimated.

A Kruskal–Wallis test (test-statistic; $p = 0.05$) was applied to check the significant differences found in tB_c and tE_c among the different fishing strategies, and a Mann–Whitney test (U; $p = 0.05$) was also used to identify the differences found in the mean values of tB_c and tE_c among the different fishing strategies [44].

Linear relationships among the log-transformed tB_c and tE_c with *Mcg*, *aLcr*, *catchR*, and Wcr_i were also applied based on a multi-regression model (MREG) for each fishing strategy:

$$Log(Vb) = c + b_1 \cdot Mcg + b_2 \cdot aLcr + b_3 \cdot catchR + \sum_{i=1}^{5} (k_i \cdot W_{cr_i}) + SE$$

where V_b is tB_c or tE_c, the c, b's, and k_i are coefficients estimated by the least squares regression techniques and SE is the standard error of the estimate. Significant variables used in the final model were selected using the backward stepwise variable selection method (F-to-remove; $p \leq 0.05$) [44].

A sensitivity analysis provides an estimate of changes in the tB_c and tE_c values produced by the fluctuation of the above-stated parameters. This determines the parameters with a major influence on the tB_c, tE_c predictions. The analysis was conducted by means of successive simulations involving all parameters included in the model, varying by 20% above or below their initial baseline values (mean values of *Mcg*, a*Lcr*, *catchR*, and Wcr_i).

2.4. Fishing Patterns in the Lagoons

A multivariate hierarchical cluster analysis based on the Bray–Curtis similarity index was also used to define similarities/dissimilarities of fisheries patterns among the studied lagoons using for each lagoon the mean (for the period 2002–2020) proportion of mugilid landings per each category and the mean proportion of simulated weight categories for each fishing strategy.

2.5. Distribution of Losses at Lagoons and Time Appearance of Losses

The total losses X (X = tB_c or TE_c) per lagoon *s* at a given year was estimated by

$$X_s = \frac{A_s}{\sum_{s=1}^{S}(A_s)} \times WP \times N_{bird} \times X_l$$

where A is the area of each lagoon, WP the wintering period (65 days) of cormorants in the study lagoons, N_{bird} the number of cormorants at a given year, and *l* the different fishing strategies.

The time distribution of appearance of losses after the predation time in the lagoon *s* is the same as the *aT* distribution of the fishing pattern that is used to classify the lagoon.

The quantity Xs per year (Xs_y) is calculated as $Xs_y = \sum_{d=1}^{6} \left(Fr_{y+d} \cdot Xs_{y-d-1} \right)$ where the y is the year of Xs estimation, Fr the proportion of X at years d after fishes preyed on birds.

The above model was developed in Excel (Microsoft Corporation, 2018), and the maps were created in QGIS (QGIS Development Team, 2009).

3. Results

The descriptive statistic of model variables and results are given in Table A2. The mean value of tE_c ranged from 0.614 (FP1) to 1.075 € bird^{-1} day^{-1} (FP3) and the mean value of tW_c ranged from 0.196 (FP1) to 0.203 kg bird^{-1} day^{-1} (FP2). Both variables (tE_c and tW_c) showed statistically significant differences among the fishing patterns (tE_c: FP3 > FP2 > FP1 and tW_c: FP3 = FP1 < FP2) (Kruskal–Wallis test; $p < 0.05$; Mann–Whitney test: U = 93.10^8; $p < 0.05$) (Table 1). The 95% of tE_c values ranged from 0.151 to 1.538 € bird^{-1} day^{-1} (FP1), from 0.221 to 2.064 € bird^{-1} day^{-1} (FP2), and from 0.306 to 2.351 € bird^{-1} day^{-1} (FP3). The 95% of tW_c values ranged from 0.061 to 0.399 kg bird^{-1} day^{-1} (FP1), from 0.059 to 0.424 kg bird^{-1} day^{-1} (FP2), and from 0.057 to 0.429 kg bird^{-1} day^{-1} (FP3) (Table 1).

The weight category composition differed among the different fishing strategies (χ^2 = 3214.9; df 4; $p < 0.05$). FP1 was mostly characterized by Cat B (58%), and FP2 by both Cat A (52%) and Cat B (39%), whereas FP3 was mostly characterized by a high percentage of Cat A (81%) (Figure 2).

Table 1. Mean value and standard deviation (SD) of tE_c and tW_c and range of their values that correspond at the range of cumulative frequency (CF) from 0.025 to 0.975, per fishing pattern (FP). n is the number of model iterations, and the same letter marks the non-statistically significant mean values of groups (Mann–Whitney test: U; $p > 0.05$).

	tE_c (€·bird^{-1}·day^{-1})		tW_c (kg·bird^{-1}·day^{-1})		
FP	Mean (SD)	CF Range (0.025–0.975)	Mean (SD)	CF Range (0.025–0.975)	n
1	0.614 ^{a} (0.376)	0.151–1.538	0.196 ^{a} (0.093)	0.061–0.399	20,000
2	0.886 ^{b} (0.499)	0.221–2.064	0.203 ^{b} (0.099)	0.059–0.424	20,000
3	1.075 ^{c} (0.554)	0.306–2.351	0.198 ^{a} (0.099)	0.057–0.429	20,000
Total	0.857 (0.518)	0.186–2.112	0.199 (0.097)	0.058–0.416	60,000

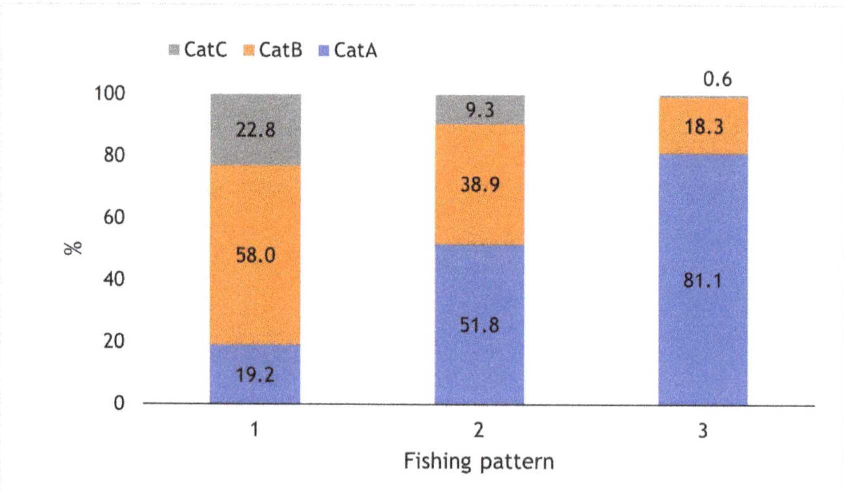

Figure 2. Weight category composition of tWc according to the fishing pattern.

The time appearance of economic and biomass losses after the preyed time differed among the fishing strategies ($\chi^2 > 55{,}167$; Df = 10; $p < 0.05$) for tE_c and tW_c. For fishing strategy 1, 92% of economic and 94% of biomass losses appeared during the 1st and 2nd year after the impact. For fishing strategy 2, 76% of economic and 78% of biomass losses appeared during the 2nd and 3rd year after the impact, whereas for fishing strategy 3, 80% of economic and 79% of biomass losses appeared during the 4th and 5th year after the impact (Table 2). The tE_c and tW_c exhibited a strong linear relationship ($\log(tE_c) = 1.476 + 1.043 \times \log(tW_c) \pm 0.321$; $R^2 = 0.739$; df:1.599; $p < 0.05$), and therefore, the multi-regression and sensitivity analysis applied only to the tE_c. At all fishing strategies the multi-regression analysis was significant ($R^2 > 0.651$; df:1,200; $p < 0.05$). The *catchP* and the fish species weight during consumption by a bird (Wcr_i) exhibited positive association with the tE_c, whereas the M_{cg} and L_{cr} showed a negative association with the tE_c.

Table 2. Losses of tE_c (€·bird^{-1}·day^{-1}) and tW_c (kg·bird^{-1}·day^{-1}) at years (aT) after the impact per each different fishing strategy (FS). Brackets indicate the ratio and bold numbers indicate the highest values.

	FS	aT (Years)						Mean Value
		1	2	3	4	5	6	
tE_c	1	0.160 (**0.26**)	0.407 (**0.66**)	0.046 (0.07)	0.001 (0.00)	0.000 (0.00)	0.000 (0.00)	0.615
	2	0.017 (0.01)	0.317 (**0.36**)	0.350 (**0.39**)	0.155 (0.17)	0.034 (0.03)	0.002 (0.00)	0.879
	3	2.857 (2.65)	0.006 (0.00)	0.088 (0.08)	0.358 (**0.33**)	0.509 (**0.47**)	0.112 (0.10)	1.075
	total	0.059 (0.06)	0.243 (0.28)	0.162 (0.18)	0.171 (0.20)	0.181 (0.21)	0.038 (0.04)	0.856
tW_c	1	0.064 (0.32)	0.121 (0.61)	0.010 (0.05)	0.000 (0.00)	0.001 (0.00)	0.000 (0.00)	0.197
	2	0.006 (0.03)	0.082 (0.41)	0.075 (0.37)	0.029 (0.14)	0.006 (0.03)	0.000 (0.00)	0.201
	3	9.922 (4.99)	0.001 (0.00)	0.018 (0.09)	0.068 (0.34)	0.090 (0.45)	0.019 (0.09)	0.198
	total	0.023 (0.11)	0.068 (0.34)	0.035 (0.17)	0.032 (0.16)	0.032 (0.16)	0.006 (0.03)	0.199

Sensitivity analysis indicated that changes by ±20% in the tE_c exhibited changes from 0.48% (FP3: Wcr_{LS}) to 54.49% (FP3: Lcr). Between fish species, the highest change in the tE_c was shown by the *M. cephalus* (from 7.99% to 9.69%) and by the *C. aurata* (from 5% to 6.65%). Changes in *catchP* by ±20% exhibited changes in tE_c ranging from 19.9% to 24.9%. Changes in *Mcg* by ±20% exhibited changes in tE_c ranging from 15.89% to 18.89% for FP1, from 23.13% to 30.09% for FP2, and from 28.94% to 40.73% for FP3. Changes in a*Lcr* by ±20% exhibited changes in tE_c ranging from 6.03% to 6.42% for FP1, from 27.04% to 37.07% for FP2, and from 35.27% to 54.49% for FP3 (Table 3).

The mean total yield of lagoons during 2002–2020 ranged from 1.40 (±1.01) t·km^{-2} (No10, Tsopeli) to 4.40 (±3.59) t·km^{-2} (#14, Vathi), whereas the mean yield of mugilids ranged from 0.11 (±0.09) t·km^{-2} (No12, Pogonitsa) to 1.61 (±1.58) t·km^{-2} (No14, Vathi). The mean proportion of mugilids to total lagoon landings was 0.44 and ranged from 0.05 (No12, Pogonitsa) to 0.55 (No3, Avlemonas). The corresponding estimates for Cat A ranged from 0.21 (#9, Logarou) to 1 (No6, Agrilos); for Cat B, they ranged from 0 (No6, AGR) to 0.59 (No9, Logarou); and for Cat C, they ranged from 0 (No6, Agrilos) to 0.35 (No3, Avlemonas).

Hierarchical clustering in the mean (for the period 2002–2020) proportion of mugilids landings per commercial category showed that (Figure 3 and Table 4), in a similarity index higher than 0.75, Logarou, Tsoukalio, and Vathi lagoons followed fishing strategy 1; Palaio, Avlemonas, Koftra-Palaiompouka, Pogonitsa, Saltini, and Myrtari lagoons followed fishing strategy 2; and Agrilos, Tsopeli, and Mazoma lagoons followed fishing strategy 3. In the Rouga and Sakolesi lagoons, no losses by birds were estimated, because we are not aware on the applied fishing strategy (Figure 3 and Table 4).

Table 3. Coefficients and standard error (in brackets) of multi-regression analysis among the natural log-transformed tEc and the independent variables (Int.Var), and sensitivity analysis per fishing strategy. N is the number of model iterations. Wcr is the mean weight of CS: *Chelon saliens*, CA: *Chelon aurata*, MC: *Mugil cephalus*, CR: *Chelon ramada*, CL: *Chelon labrosus*.

	Coefficients (Standard Error)			Sensitivity Analysis		
	Fishing Strategy					
	1	2	3	1	2	3
Constant (c)	−1.62 (0.0823)	0.449 (0.0704)	1.660 (0.0358)			
Int. Var				%changes in tEc when Int.Var change ±20%		
catchP	1.851 (0.0109)	1.863 (0.0093)	1.853 (0.0047)	19.92–24.88	20.04–25.06	19.93–24.90
Mcg	−0.860 (0.0254)	−1.31 (0.0213)	−1.700 (0.0109)	15.89–18.89	23.13–30.09	28.94–40.73
aLcr	−0.010 (0.0040)	−0.08 (0.0034)	−0.110 (0.0017)	6.03–6.42	27.04–37.07	35.27–54.49
Wcr_{CS}	0.001 (0.0002)	0.000 (0.0001)	0.000 (9.5223)	0.92–0.93	0.51–0.51	0.48–0.48
Wcr_{CA}	0.007 (0.0002)	0.007 (0.0001)	0.008 (0.0000)	5.00–5.26	5.57–5.90	6.24–6.65
Wcr_{MC}	0.011 (0.0002)	0.012 (0.0001)	0.012 (9.5908)	7.99–8.68	8.46–9.24	8.84–9.69
Wcr_{CR}	0.005 (0.0002)	0.005 (0.0001)	0.004 (9.6104)	4.25–4.44	4.04–4.21	3.47–3.60
Wcr_{CL}	0.002 (0.0002)	0.002 (0.0001)	0.001 (9.5399)	1.90–1.94	1.46–1.48	1.22–1.24
R^2	0.651	0.734	0.918			
n	20,000	20,000	20,000			

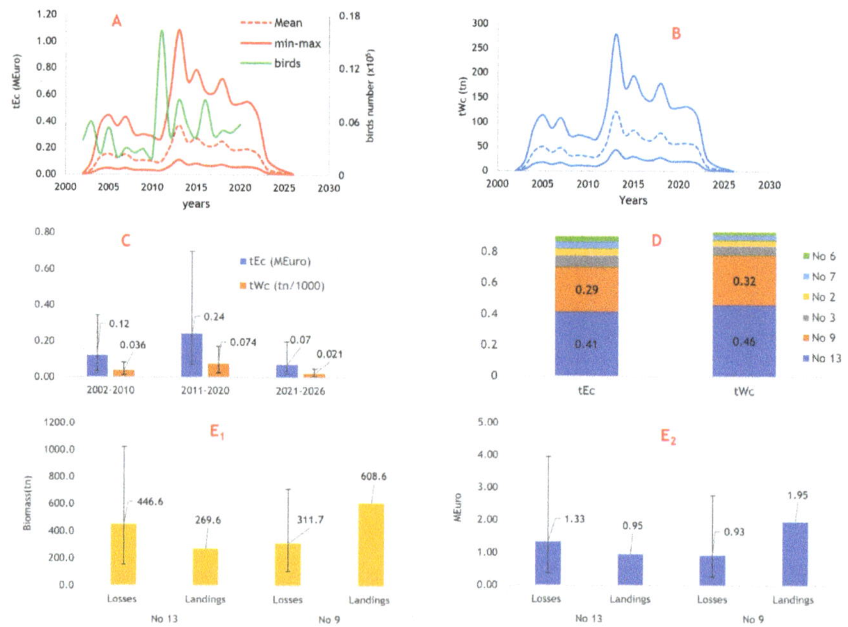

Figure 3. Time series of economic losses and cormorant number of (**A**), biomass losses (**B**), losses (economic and biomass) per period (**C**), distribution of losses to lagoons (**D**), sum of biomass losses and mugilid landings for lagoons Nos 13 and 9 (E_1), and economic losses and economic of mugilid landings for lagoons Nos 13 and 9, for the period 2002–2020 (E_2). The bars indicate the 95% confidence limits. Codes for lagoons are shown in Figure 1.

Table 4. Mean total yield (Yield T), mean yield of mugilids (Yield M), weight proportion of mugilids to total landings (PrpM), n the years by available data, mean proportion of weight categories of mugilids (Cat A, Cat B, and Cat C) and fishing strategy (FS) and estimated fishing strategy (FSe) by cluster analysis, for 2002–2020. No indicates the codes of the studied lagoons according to Figure 1. * indicates lagoons Nos 1 and 4, for which fishing strategy 2 was applied.

No	Area (km^2)	Yield T (t·km^{-2})	Yield M (t·km^{-2})	PrpM	n	Cat A	Cat B	Cat C	FSe
1	2.26	Unknown fishing activity and fishing data during 2002–2020							2 *
2	3.29	1.30 (0.75)	0.48 (0.22)	0.37	19	0.42	0.29	0.29	2
3	5.25	1.36 (0.80)	0.76 (0.39)	0.55	14	0.37	0.28	0.35	2
4	0.66	Unknown fishing activity and fishing data during 2002–2020							2 *
5	0.56	Unknown fishing activity and fishing data							
6	1.94	0.29 (0.15)	0.13 (0.10)	0.45	5	1.00	0.00	0.00	3
7	3.22	1.28 (0.78)	0.19 (0.16)	0.14	6	0.54	0.32	0.15	2
8	0.39	Unknown fishing activity and fishing data							
9	30.64	2.65 (0.83)	1.24 (0.56)	0.46	16	0.21	0.59	0.20	1
10	1.12	1.40 (1.01)	0.65 (0.43)	0.46	19	0.75	0.20	0.05	3
11	1.90	2.29 (1.84)	0.60 (0.52)	0.26	18	0.74	0.16	0.10	3
12	0.47	2.39 (2.40)	0.11 (0.09)	0.05	12	0.58	0.42	0.00	2
13	43.90	0.73 (0.27)	0.36 (0.14)	0.48	17	0.31	0.49	0.20	1
14	0.29	4.40 (3.59)	1.61 (1.58)	0.36	18	0.41	0.47	0.13	1
					1	0.18	0.59	0.24	1
				FS	2	0.50	0.40	0.10	2
					3	0.81	0.19	0.01	3

The time series of economic and biomass losses caused by the cormorants in the lagoons of the Amvrakikos Gulf from 2002 to 2020 are forecast up to 2026. Their picks follow the time series of bird numbers with a delay of about 2 years (Figure 3A,B). The mean annual number of cormorants was significantly (Mann–Whitney test: $U = 12$; $p < 0.05$) increased in 2002–2010 and 2011–2020. More specifically, the number of cormorants ranged from 1848 to 5823 bird·year^{-1} (mean value of 3239.33 bird·year^{-1}) during the period 2002–2010 and from 4181 to 8375 bird·year^{-1} (mean value of 5591.77 bird·year^{-1}, excluding an extreme value of 16,236 birds) in 2011–2020.

The estimated economic losses ranged from 0.04 to 0.35 MEuro·year^{-1} (mean value of 0.12 MEuro·year^{-1}) during 2002–2010, from 0.07 to 0.70 MEuro·year^{-1} (mean value of 0.24 MEuro·year^{-1}) during 2011–2020, and from 0.02 to 0.20 MEuro·year^{-1} (mean value of 0.07 MEuro·year^{-1}) during 2021–2026 (Figure 3C). The estimated landing losses ranged from 12.49 to 82.71 t·year^{-1} (mean value of 35.78 t·year^{-1}) during 2002–2010, from 25.88 to 172.04 t·year^{-1} (mean value of 74.31 t·year^{-1}) during 2011–2020, and from 7.22 to 48.69 t·year^{-1} (mean value of 20.89 t·year^{-1}) during 2021–2026 (Figure 3C). Six lagoons (Figure 1: No13, Tsoukalio-Rodia; No9, Logarou; No3, Avlemonas; No2, Palaio; No7, Koftra-Palaiompouka; and No6, Agrilos) cumulatively contributed to the 89.7% and the 92.8% of the total tE_c and tW_c, respectively, with the first two lagoons (Tsoukalio-Rodia and Logarou) exhibiting 70.2% and 78.1% of the total tE_c and tW_c, respectively (Figure 3D).

In the Tsoukalio-Rodia lagoon (Figure 1: No13), the sum of the biomass losses during 2002–2020 was 446.6 t (ranging from 157.87 to 1024.82 t), and in Logarou (Figure 1: No9), it was 311.7 t (ranging from 110.19 t to 715.33 t). The sum of landings was 269.6 t and 608.6 t for Tsoukalio-Rodia and Logarou (Figure 1: Nos 13 and 9), respectively (Figure 3E_1), and the ratio of biomass losses to landings biomass ranged from 0.58:1 to 3.80:1 and from 0.18:1 to 1.17:1, respectively. The ratio of economic losses to income from landings ranged from 0.41:1 to 4.18:1 and from 0.14:1 to 1.42:1, respectively (Figure 3E_1,E_2).

4. Discussion

In the present study, an attempt was made through modeling to estimate the losses of lagoon fisheries from predation of cormorants during the wintering period. A substantial number of uncertainties have been incorporated into the model regarding the fish population growth, fish size and species distribution, age (or size) of fish that might be caught, and catchability as well as the different fishing strategies followed by the lagoon fishers. On the other hand, the large number of the model's iterations reduced the uncertainties of a large number of the aforementioned parameters.

Mugilids are the dominant fish group in the Greek lagoon fisheries [6] (>50%), and they also comprise the dominant prey of cormorants during their wintering period (74% of DFI in Mesolongi-Aitoliko lagoons: [45]; 65% of DFI in Amvrakikos Gulf lagoons: [24]). The range of DFI for mugilids in the model agrees with previous work in the study area [24]. The length of preyed fish (except for *C. saliens*) ranged from 15 to 24 cm [45], supporting the basic assumption that the losses induced by the impact of cormorants would appear in future [41].

The model assumed (a) that natural mortality was constant during the years, (b) that preyed fish for each species were of the same size, (c) spatial homogeneity of species composition among the lagoons, (d) that mean preyed fish species composition was constant day to day as well as during the years, (e) spatial distribution of the cormorants, and (f) that the wintering period of the cormorants was defined as 65 days.

(a) The natural mortality M was ±42% in each species, which might include the interannual variability and the variability driven by the von Bertalanffy estimates (L_∞ and k participate in the M estimation; Table A1).
(b) The maximum number of preyed individuals was approximately 7 (Table A2). Given that mugilids exhibit a schooling behavior according to the individuals' size [46], the likelihood that individuals of the same species in a feeding area for cormorants are the same size was high.
(c) The studied lagoons exhibited differences in physicochemical variables [42], which explains the spatial expansion of mugilids according to their preferences [46]. Model iterations recorded a series of cases that simulated variable species composition. In each case, this affects the average value of the model's estimates but not the range of distributions.
(d) The analysis indicated that most of the losses are predicted to be in the two largest lagoons (No 13, Tsoukalio-Rodia and No 9, Logarou; Figure 1). In these lagoons, the high representation of mugilids is maintained at the same levels [42]. According to fishers, most of the losses in the above-mentioned lagoons were on the gilthead seabream. However, the gilthead seabream (except for a small lagoon: No 12, Pogonitsa; Figure 1) consists of a relatively small proportion to landings (<15%), and in some cases, the production was supported by enrichment programs [42]. It seems that during periods of low temperatures, gilthead seabream searches for favorable sites and is concentrated near to the communication, with the sea channels being an easy target for cormorants. In each case, this impact is temporarily limited to a few days, and it might not be sufficient to alter the prey species' composition in the study lagoons.
(e) The bird-days establishment of the cormorants in each lagoon ranged from 12 to 110 bird·days·year^{-1}·ha^{-1}, which is in agreement with the estimates reported by a previous study [13] (inland waters/lagoons: 20 to 100 bird·days·year^{-1}·ha^{-1}). Certain wetlands have been excluded from our estimations (e.g., the Amvrakikos Gulf, rivers Louros and Arachthos, Lake Voulkaria). The area of the gulf is apparently used by cormorants in limited situations for feeding, such as preying on small pelagics, which consists of 2.8% of DFI. Given also that freshwater fish species are not included in the feeding spectra of cormorants [24], the adjacent lakes and the upper system of rivers have not been considered in this study.
(f) The wintering population of cormorants in the studied area referred to the maximum number of midwinter estimates. However, the first appearance of the cormorants in the Mediterranean is observed from the mid October and the last ones in mid April, with a progressive increase during the winter and a gradual decrease after

the maximum appearance [47]. Through the use of a normal distribution, the timing of the appearance was simulated (mean = 90 days, SD = 25.7 days) and indicated that the recorded bird value (maximum value of distribution) multiplied by 64.39 days was equal to the bird-days estimated from the simulated appearance of cormorants. Thus, the 65 days estimated as the wintering period could be considered as a reliable estimate.

The model's sensitivity analysis revealed that the Mcg and the a*Lcr* mostly affected the losses, according to fishing strategy. The most sensitive scenario was FP3 (fishing targeting greater-size fishes). In the case of high natural mortality, some of the individual fish were not caught in this scenario, so low values of M indicate greater losses than high values of M, which indicate smaller losses. The *aLcr* also affected the losses. Greater losses than the preyed *aLcr* are suggested when the preyed occurs at low length. This is also expected due to the fact that when the *aLcr* is low, a higher preyed mortality (consumed more individuals) was estimated rather than when the *aLcr* is high (consumed fewer individuals, and thus lower preyed mortality). In relation to fish species, sensitivity analysis indicated that the losses are more sensitive to the preyed flathead grey mullets, but the effect on losses was rather low (≈8 to 9.7%). It should be noted that for all fish species, the impact of cormorants referred to the prospective fishing biomass. From flathead grey mullet, a boutarga-type product called *"avgotaracho"* (dried, salted, and waxed ripe ovaries) is made, which corresponds to 7% of fishing biomass of the species [48,49]. Thus, the economic losses incurred by Cat A for flathead grey mullet were underestimated by 130%, revealing an important effect on losses for this species.

The ratio of biomass losses to landing biomass and the ratio of economic losses to incomes from landings differed between the two largest lagoons, supporting different measures against the impact of cormorants. The ratios of biomass losses to landings biomass and the ratio of economic losses to income from landings that ranged from 0.58:1 to 3.80:1 and from 0.18:1 to 1.17:1 (biomass) and from 0.41:1 to 4.18:1 and from 0.14:1 to 1.42:1 (economic) support a strong competitive relationship between cormorants and fisheries in the lagoons of the Amvrakikos Gulf. It seems that the yearly losses in biomass by cormorants is sufficient to justify this detrend of production. Finally, it is worth noting that the estimates in this study applied to the direct losses by cormorants and were underestimated due to the fact that, during the feeding, significant numbers of fish are injured [13,38] (mainly of commercial size), most of which die shortly after or survive and are later caught by fishers and sold at low value or discarded.

Author Contributions: Conceptualization, G.K.; methodology, G.K.; validation, G.K. and D.K.M.; formal analysis, G.K. and S.K.; resources, G.K. and S.K.; data curation, G.K. and S.K.; writing—review and editing, G.K., S.K., and D.K.M.; supervision, G.K. All authors have read and agreed to the published version of the manuscript.

Funding: This research received no external funding.

Institutional Review Board Statement: Not applicable.

Informed Consent Statement: Not applicable.

Data Availability Statement: Data supporting results of the study can be provided upon request to the first author.

Conflicts of Interest: The authors declare no conflict of interest.

Appendix A

Table A1. von Bertalanffy's equation coefficients, length–weight equation coefficients, natural mortality (M), marker values (v) per weight category, daily food intake of birds (DFI), length of fish consumed (Lcr).

		Mugilids Species					
		Chelon saliens	*Chelon aurata*	*Mugil cephalus*	*Chelon ramada*	*Chelon labrosus*	
Total length	$L\infty$ (cm)	32.99	69.30	79.10	56.33	47.20	
Weight	$W\infty$ (gr)	299.77	2656.31	4960.19	1873.70	1128.28	
	k	0.261	0.136	0.151	0.179	0.200	
	t_0	−0.470	−1.140	−0.100	−0.856	−0.400	
$W = a \times TL^b$	a	0.00784	0.0056	0.0072	0.0055	0.008	
	b	3.018	3.130	3.041	3.160	3.010	
	Reference			[43]		[50]	
	Tm (°C)	18					Reference
	M #	0.587	0.312	0.322	0.396	0.448	[51]
	weight category	Cat A	Cat B	Cat C			
	Weight range	>400 gr	150–400 gr	80–150 gr			[48]
	v ($€.kg^{-1}$)	[24]	[51]	[43]			market values 2015–2020
(DFI) Daily food intake of Birds,	DFI for mugilids (gr)	169.6					[24]
(Lcr) Length of fish consumed	Lcr for mugilids (cm)	15–24					[24]
	q	0.134	0.089	0.147	0.341	0.293	[24]

$\text{Log}(M) = -0.0066 - 0.279 \log(L_\infty) + 0.6543 \log(K) + 0.4634 \log(Tm) \pm 0.245$ [51].

Appendix B

Table A2. Descriptive statistics of input and output of model data for 60,000 iterations.

No	Species	Situation	Variable	abr	Mean	SD	Min	Max	
Input data									
1	*Chelon saliens*	consumed	W (gr)	CS Wcr	36.30	12.03	4.02	100.22	estimated by #17 and #6
2	*Chelon aurata*		W (gr)	CA Wcr	36.41	12.07	4.74	98.40	estimated by #17 and #7
3	*Mugil cephalus*		W (gr)	MC Wcr	36.38	12.05	3.41	106.27	estimated by #17 and #8
4	*Chelon ramada*		W (gr)	CR Wcr	36.40	12.02	4.73	96.57	estimated by #17 and #9
5	*Chelon labrosus*		W (gr)	CL Wcr	36.40	12.04	5.31	98.09	estimated by #17 and #10
6	*Chelon saliens*		W Proportion	CS q	0.20	0.05	0.08	0.40	random distribution
7	*Chelon aurata*		W Proportion	CA q	0.20	0.05	0.08	0.40	random distribution
8	*Mugil cephalus*		W Proportion	MC q	0.20	0.05	0.08	0.40	random distribution
9	*Chelon ramada*		W Proportion	CR q	0.20	0.05	0.08	0.40	random distribution
10	*Chelon labrosus*		W Proportion	CL q	0.20	0.05	0.08	0.40	random distribution
11	*Chelon saliens*		L (cm)	CS Lcr	19.30	1.30	14.25	25.10	normal distribution
12	*Chelon aurata*		L (cm)	CA Lcr	19.30	1.31	13.79	25.07	normal distribution
13	*Mugil cephalus*		L (cm)	MC Lcr	19.30	1.30	13.94	24.93	normal distribution
14	*Chelon ramada*		L (cm)	CR Lcr	19.31	1.31	13.77	24.51	normal distribution
15	*Chelon labrosus*		L (cm)	CL Lcr	19.30	1.30	12.84	24.79	normal distribution
16			mean length (cm)	Lcr	19.09	0.63	16.10	21.69	estimated by #11 to #15
17			DFI (kg)	DFI	0.18	0.04	0.03	0.32	normal distribution
18			individuals	Ncr	3.12	0.68	0.60	6.80	estimated by #1 to #15
19	*Chelon saliens*		M range	CS M	0.59	0.06	0.34	0.83	estimated by Mi and #24
20	*Chelon aurata*		M range	CA M	0.31	0.03	0.18	0.44	estimated by Mi and #24
21	*Mugil cephalus*		M range	MC M	0.32	0.03	0.19	0.46	estimated by Mi and #24
22	*Chelon ramada*		M range	CR M	0.40	0.04	0.23	0.56	estimated by Mi and #24
23	*Chelon labrosus*		M range	CL M	0.45	0.04	0.26	0.64	estimated by Mi and #24
24			M fluctuation	Mcg	1.00	0.10	0.58	1.42	normal distribution
25			catchability (proportion)	catchP	0.60	0.23	0.20	1.00	random distribution

Table A2. *Cont.*

No	Species	Situation	Variable	abr	Mean	SD	Min	Max	
26	*Chelon saliens*		Lc (cm)	CS Lc	26.87	2.87	18.66	31.31	beta distribution and random distribution
27	*Chelon aurata*		Lc (cm)	CA Lc	37.25	7.85	21.47	49.20	beta distribution and random distribution
28	*Mugil cephalus*	captured	Lc (cm)	MC Lc	42.55	9.89	23.68	56.95	beta distribution and random distribution
29	*Chelon ramada*		Lc (cm)	CR Lc	35.52	6.68	21.09	45.27	beta distribution and random distribution
30	*Chelon labrosus*		Lc (cm)	CL Lc	32.31	5.25	19.97	40.00	beta distribution and random distribution
Output model									
31	*Chelon saliens*		W (gr)	CS Wc	9.85	8.93	0.20	78.55	estimated
32	*Chelon aurata*		W (gr)	CA Wc	55.55	33.37	4.07	341.29	estimated
33	*Mugil cephalus*		W (gr)	MC Wc	75.00	46.60	3.91	433.98	estimated
34	*Chelon ramada*		W (gr)	CR Wc	36.68	21.99	2.36	240.28	estimated
35	*Chelon labrosus*		W (gr)	CL Wc	22.31	14.48	0.88	135.13	estimated
36	*Chelon saliens*		E (Euro)	CS Ec	0.02	0.01	0.00	0.12	estimated
37	*Chelon aurata*		E (Euro)	CA Ec	0.25	0.20	0.00	2.05	estimated
38	*Mugil cephalus*	captured	E (Euro)	MC Ec	0.39	0.30	0.01	2.60	estimated
39	*Chelon ramada*		E (Euro)	CR Ec	0.15	0.11	0.00	1.44	estimated
40	*Chelon labrosus*		E (Euro)	CL Ec	0.06	0.04	0.00	0.41	estimated
41			aT (year)	aT (year)	3.46	1.43	1.00	6.00	estimated
42			fishery biomass (kg)	Wc	0.20	0.10	0.02	0.76	estimated
43			E (Euro)	Ec	0.86	0.52	0.05	4.45	estimated
44			mean proportion	Cat A	0.50	0.33	0.00	1.00	estimated
45			mean proportion	Cat B	0.39	0.28	0.00	1.00	estimated
46			mean proportion	Cat C	0.11	0.16	0.00	1.00	estimated

References

1. Perez-Ruzafa, A.; Mompeán, C.; Marcos, C. Hydrographic, geomorphologic and fish assemblage relationships in coastal lagoons. *Hydrobiologia* **2007**, *577*, 107–125. [CrossRef]
2. GFCM. *Mediterranean Coastal Lagoons: Sustainable Management and Interactions among Aquaculture, Capture Fisheries and the Environment*; FAO Fish. Technical Paper; FAO: Rome, Italy, 2015; p. 95.
3. Mateus, M.; Almeida, D.; Simonson, W.; Felgueiras, M.; Banza, P.; Batty, L. Conflictive uses of coastal areas: A case study in a southern European coastal lagoon (Ria de Alvor, Portugal). *Ocean Coast. Manag.* **2016**, *132*, 90–100. [CrossRef]
4. Pearce, F.; Crivelli, A.J. *Characteristics of Mediterranean Wetlands*; MedWet Publications: Arles, France, 1994.
5. Kapetsky, J.M. Coastal Lagoon Fisheries around the World: Some Perspectives On Fishery Yields, and Other Comparative Fishery Characteristics. In *Studies and Reviews-General Fisheries Council for the Mediterranean (FAO)*; FAO: Rome, Italy, 1984; Volume 61.
6. Koutrakis, E.; Conides, A.; Parpoura, A.C.; Van Ham, E.H.; Katselis, G.; Koutsikopoulos, C. Lagoon fisheries' resources in Hellas. In *State of Hellenic Fisheries*; Papaconstantinou, C., Zenetos, A., Vassilopoulou, V., Tserpes, G., Eds.; Hellenic Centre for Marine Research: Athens, Greece, 2007; pp. 223–232.
7. Marchowski, D.; Leitner, M. Conservation implications of extraordinary Greater Scaup (*Aythya marila*) concentrations in the Odra Estuary, Poland. *Condor* **2019**, *121*, duz013. [CrossRef]
8. Marchowski, D.; Jankowiak, Ł.; Ławicki, Ł.; Wysocki, D.; Chylarecki, P.W. Fishery bycatch is among the most important threats to the European population of Greater Scaup *Aythya marila*. *Bird Conserv. Int.* **2020**, *30*, 176–193. [CrossRef]
9. Marchowski, D. Bycatch of Seabirds in the Polish Part of the Southern Baltic Sea in 1970–2018: A Review. *Acta Ornithol.* **2022**, *56*, 139–158. [CrossRef]
10. Zydelis, R.; Small, C.; French, G. The incidental catch of seabirds in gillnet fisheries: A global review. *Biol. Conserv.* **2013**, *162*, 76–88. [CrossRef]
11. Steffens, W. Great Cormorant *Phalacrocorax carbo* is threatening fish populations and sustainable fishing in Europe. *Am. Fish. Soc. Symp.* **2011**, *75*, 189–200.
12. Seiche, K.; Gerdeaux, D.; Gwiazda, R.; Lévai, F.; Musil, P.; Nemenonoks, O.; Strod, T.; Carss, D. Cormorant-fisheries conflicts in Carp pond areas in Europe and Israel: An INTERCAFE overview. In *Action 635 Final Report V NERC Centre for Ecology & Hydrology on Behalf of COST*; European Cooperation of Science and Technology: Brussels, Belgium, 2012; p. 129. ISBN 978-1-906698-10-2. Available online: http://www.intercafeproject.net/pdf/Carp_Ponds_Manual_FOR_WEB.pdf (accessed on 27 May 2023).
13. Ostman, O.; Bostrom, M.K.; Bergstrom, U.; Andersson, J.; Lunneryd, S.-G. Estimating Competition between Wildlife and Humans-A Case of Cormorants and Coastal Fisheries in the Baltic Sea. *PLoS ONE* **2013**, *8*, e83763. [CrossRef] [PubMed]

14. Hansson, S.; Bergström, U.; Bonsdorff, E.; Vetemaa, M. Competition for the fish-fish extraction from the Baltic Sea by humans, aquatic mammals, and birds. *ICES J. Mar. Sci.* **2018**, *75*, 999–1008. [CrossRef]
15. Ostman, O.; Bergenius, M.; Bostrom, M.K.; Lunneryd, S.G. Do cormorant colonies affect local fish communities in the Baltic Sea? *Can. J. Fish. Aquat. Sci.* **2012**, *69*, 1047–1055. [CrossRef]
16. Veneranta, L.; Heikinheimo, O.; Marjomäki, T.J. Cormorant (Phalacrocorax carbo) predation on a coastal perch (*Perca fluviatilis*) population: Estimated effects based on PIT tag mark-recapture experiment. *ICES J. Mar. Sci.* **2020**, *77*, 2611–2622. [CrossRef]
17. Adámek, Z.; Kajgrová, L. Great cormorant (*Phalacrocorax carbo sinensis*) occurrence in carp aquacultural ponds: A case study from the South Bohemia (Czech Republic) pond region. *Aquacult. Int.* **2022**, *30*, 2541–2556. [CrossRef]
18. Russell, R.W. Comparative demography and life history tactics of seabirds: Implications for conservation and marine monitoring. *Am. Fish. Soc. Symp.* **1999**, *23*, 51–76.
19. Marchowski, D.; Mohr, A.; Ławicki, Ł.; Jankowiak, Ł. Warmer winters increase the breeding success of the Goosander: The case of the Pomeranian Lake District in Poland. *Ardea* **2022**, *110*, 31–40. [CrossRef]
20. Van Eerden, M.R.; Koffijberg, K.; Platteeuw, M. Riding on the crest of the wave: Possibilities and limitations for a thriving population of migratory Cormorants Phalacrocorax carbo in man-dominated wetlands. *Ardea* **1995**, *83*, 1–9. [CrossRef]
21. Russell, I.C.; Dare, P.J.; Eaton, D.R.; Armstrong, J.D. *Assessment of the Problem of Fish-Eating Birds in Inland Fisheries in England and Wales*; Report to the Ministry of Agriculture, Fisheries and Food; Project Number VC0104; MAFF: London, UK, 1996; 130p.
22. Herrmann, C.; Bregnballe, T.; Larsson, K.; Leivits, M.; Rusanen, P. Population Development of Baltic Bird Species: Great Cormorant (*Phalacrocorax carbo sinensis*). HELCOM Baltic Sea Environment Fact Sheets. 2019. Available online: https://helcom.fi/baltic-sea-trends/environment-fact-sheets/biodiversity/population-development-of-great-cormorant/ (accessed on 31 December 2019).
23. BirdLife International. Species Factsheet: Phalacrocorax carbo, 2023. Available online: http://datazone.birdlife.org/species/factsheet/great-cormorant-phalacrocorax-carbo (accessed on 24 July 2023).
24. Liordos, V.; Zogaris, S.; Papandropoulos, D. Great Cormorant Phalacrocorax carbo food and human perceptions at the Amvrakikos Gulf, western Greece. In Proceedings of the 7th International Conference on Cormorants, Villeneuve, Switzerland, 23–26 November 2005; Van Eerden, M.R., van Rijn, S., Keller, V., Eds.; Wetlands International-IUCN Cormorant Research Group: Lelystad, The Netherlands, 2005; pp. 102–111.
25. Goutner, V.; Papakostas, G.; Economidis, P.S. Diet and growth of Great cormorant (Phalacrocorax carbo) nestlings in a Mediterranean estuarine environment (Axios Delta, Greece). *Israel J. Zool.* **1997**, *43*, 133–148. [CrossRef]
26. Liordos, V.; Goutner, V. Spatial Patterns of Winter Diet of the Great Cormorant in Coastal Wetlands of Greece. *Waterbirds* **2007**, *30*, 103–111. [CrossRef]
27. Kazantzidis, S. Status of the breeding population of Great Cormorants in Greece in 2012. In *National Reports from the 2012 Breeding Census of Great Cormorants Phalacrocorax carbo in Parts of the Western Palearctic*; Bregnballe, T., Lynch, J., Parz-Gollner, R., Marion, L., Volponi, S., Paquet, J.-Y., Van Eerden, M.R., Eds.; IUCN-Wetlands International Cormorant Research Group Report; Technical Report from DCE—Danish Centre for Environment and Energy, Aarhus University; DCE: Aarhus, Denmark, 2013; Volume 22, pp. 51–54. Available online: http://dce2.au.dk/pub/TR22.pdf (accessed on 27 May 2023).
28. Liordos, V.; Pergantis, F.; Perganti, I.; Roussopoulos, Y. Long-term population trends reveal increasing importance of a Mediterranean wetland complex (Messolonghi lagoons, Greece) for wintering waterbirds. *Zool. Stud.* **2014**, *53*, 12. [CrossRef]
29. Carss, D.N.; Marzano, M. Reducing the Conflict between Cormorants and Fisheries on a Pan-European Scale. Summary and National Overviews. Final Report Concerted Action Q5CA-2000 2005, 1387. Available online: http://www.intercafeproject.net/pdf/REDCAFESummaryandNationalOverview.pdf (accessed on 27 May 2023).
30. Cowx, I. Policy Department Structural and Cohesion Policies. Directorate-General for Internal Policies. 2013. Available online: http://ec.europa.eu/environment/nature/cormorants/files/Cowx_Report_for_Parliament.pdf (accessed on 27 May 2023).
31. Platteeuw, M.; van Eerden, M.R. Time and energy constraints of fishing behavior in breeding Cormorants Phalacrocorax carbo sinensis at lake Ijsselmeer, The Netherlands. *Ardea* **1995**, *83*, 223–234.
32. Paillisson, J.M.; Carpentier, A.; Le Gentil, J.; Marion, L. Space utilization by a cormorant (*Phalacrocorax carbo* L.) colony in a multi-wetland complex in relation to feeding strategies. *C. R. Biol.* **2004**, *327*, 493–500. [CrossRef]
33. van Dam, C.; Asbirk, S. Cormorants and Human Interests. In Proceedings of the Workshop towards an International Conservation and Management Plan for the Great Cormorant (Phalacrocorax carbo), Lelystad, The Netherlands, 4 October 1996.
34. Bildsøe, M.; Jensen, I.B.; Vestergaard, K.S. Foraging behaviour of cormorants Phalacrocorax carbo in pound nets in Denmark: The use of barrel nets to reduce predation. *Wildl. Biol.* **1998**, *4*, 129–136. [CrossRef]
35. Keller, T.M.; Lanz, U. Great Cormorant Phalacrocorax carbo sinensis management in Bavaria, southern Germany—What can we learn from seven winters with intensive shooting? *Vogewelt* **2003**, *124*, 189–200.
36. Polensky, J.; Regenda, J.; Adamek, Z.; Cisar, P. Prospects for the monitoring of the great cormorant (*Phalacrocorax carbo sinensis*) using a drone and stationary cameras. *Ecol. Inform.* **2022**, *70*, 101726. [CrossRef]
37. Lehikoinen, A.; Heikinheimo, O.; Lehtonen, H.; Rusanen, P. The role of cormorants, fishing effort and temperature on the catches per unit effort of fisheries in Finnish coastal areas. *Fish. Res.* **2017**, *190*, 175–182. [CrossRef]
38. Dimitriou, E.; Kapareliotis, A.; Akovitiotis, K.; Pergantis, F.; Perifanos, M.; Tsagarakis, E. Recording of injured fish from attacks by the species *Phalacrocorax carbo* Linnaeus 1758, Great cormorant, in the closed lagoon "Buka" at the mouth of the Achelos river. In Proceedings of the 11th Panhellenic Conference of Ichthyologists, Preveza, Greece, 10–13 April 2003.

39. Salmi, J.A.; Auvinen, H.; Raitaniemi, J.; Kurkilahti, M.; Lilja, J.; Maikola, R. Perch (*Perca fluviatilis*) and pikeperch (*Sander lucioperca*) in the diet of the great cormorant (*Phalacrocorax carbo*) and effects on catches in the Archipelago Sea, Southwest coast of Finland. *Fish. Res.* **2015**, *164*, 26–34. [CrossRef]
40. King, M. Fisheries Biology. In *Assessment and Management*; Oxford Press: London, UK, 1995; p. 342.
41. Katselis, G.; Koukou, K.; Moutopoulos, D. Yield per recruit and spawning stock biomass models for the management of four Mugilidae species in Mesolonghi—Aitoliko lagoon (W. Greece). *Int. Aquat. Res.* **2010**, *2*, 155–162.
42. Katselis, G.; Moutopoulos, D.K.; Dimitriou, E.; Koutsikopoulos, C. Long-term changes of fisheries landings in enclosed gulf lagoons (Amvrakikos gulf, W Greece): Influences of fishing and other human impacts. *Estuar. Coast. Shelf Sci.* **2013**, *131*, 31–40. [CrossRef]
43. HOS. Hellenic Ornithological Society. 2021. Available online: https://www.ornithologiki.gr/el/oi-draseis-mas/diatirisi-erevna/parakoloythisi-katagrafi-poulion/115-mesoxeimoniatikes-katametriseis-ydrovion-poulion (accessed on 27 May 2023).
44. Zar, J.H. *Biostatistical Analysis*, 5th ed.; Prentice-Hall/Pearson: Upper Saddle River, NJ, USA, 2010.
45. Liordos, V. *Biology and Ecology of Great Cormorant (Phalacrocorax carbo L. 1758) Populations Breeding and Wintering in Greek Wetlands*; Aristotle University of Thessaloniki: Serres, Greece, 2004; 308p.
46. Oren, O.H. *Aquaculture of Grey Mullets*; CUP Archive: Cambridge, UK, 1981; 507p, ISBN 052122926X-9780521229265.
47. Bregnballe, T.; Frederiksen, M.; Gregersen, J. Seasonal distribution and timing of migration of Cormorants Phalacrocorax carbo sinensis breeding in Denmark. *Bird Study* **1997**, *44*, 257–276. [CrossRef]
48. Katselis, G.; Koutsikopoulos, C.; Rogdakis, Y.; Lachanas, T.; Dimitriou, E.; Vidalis, K. A model to estimate the annual production of roes (avgotaracho) of flathead mullet (Mugil cephalus) based on the spawning migration of species. *Fish. Res.* **2005**, *75*, 138–148. [CrossRef]
49. Dimitriou, E.; Katselis, G.; Moutopoulos, D.K.; Milios, K.; Malamis, A.; Koutsikopoulos, C. Description of the processing stages of a protected designation of origin fish product: The Greek caviar avgotaracho messolongiou. *Agric. Econ. Rev.* **2016**, *17*, 50–62.
50. Froese, R.; Pauly, D. (Eds.) FishBase. World Wide Web Electronic Publication. 2023. Available online: www.fishbase.org (accessed on 1 February 2023).
51. Pauly, D. On the interrelationships between natural mortality, growth parameters and mean environmental temperature in 175 fish stocks. *J. Cons. Explor. Mer.* **1981**, *39*, 175–192. [CrossRef]

Disclaimer/Publisher's Note: The statements, opinions and data contained in all publications are solely those of the individual author(s) and contributor(s) and not of MDPI and/or the editor(s). MDPI and/or the editor(s) disclaim responsibility for any injury to people or property resulting from any ideas, methods, instructions or products referred to in the content.

sustainability

MDPI

Article

Birds' Flight Initiation Distance in Residential Areas of Beijing Are Lower than in Pristine Environments: Implications for the Conservation of Urban Bird Diversity

Luqin Yin [1,2], Cheng Wang [1,2,*], Wenjing Han [1,2] and Chang Zhang [1,2]

[1] Research Institute of Forestry, Chinese Academy of Forestry, Beijing 100091, China; yinluqin@caf.ac.cn (L.Y.); hanwenjing@caf.ac.cn (W.H.); song.yan.zheng@163.com (C.Z.)

[2] Key Laboratory of Tree Breeding and Cultivation and Urban Forest Research Centre, National Forestry and Grassland Administration, Beijing 100091, China

* Correspondence: wch8361@163.com; Tel.: +86-10-6288-8361

Abstract: (1) Background: With rapid urbanization, birds are facing a variety of challenges. Evaluating bird behaviour changes in response to urbanization can help us understand how to make them coexist sustainably with humans. We aimed to investigate whether birds inhabiting residential areas differed in their escape behaviour and their influencing factors. (2) Methods: We used the flight initiation distance (FID), the horizontal distance between the observer and the target bird when it escapes, to measure the escape behaviour of birds. We chose 40 urban residential areas within the 5th ring road in Beijing and conducted surveys each month for one year. We applied Generalized Linear Mixed Models (GLMMs) separately to evaluate the response variable of FID for the total species, the most common species, and the other species. (3) Results: Birds that appear more frequently in residential areas or as ground foragers, insectivores, and omnivores are better adapted to human interference and have shorter FIDs. Individual initial conditions affect bird FID, and environmental characteristics can be used as predictors for the most common birds. Tree canopy coverage was found to positively affect FID, while floor area ratio (FAR) is negatively correlated with FID. (4) Conclusions: Our results demonstrated that birds in residential areas have been adapting to the human environment, and urban tree canopies can provide refuge for birds to avoid human interference. Our study focused on the response of bird FIDs to human interference and urban trees under high urbanization, which has substantial practical implications for urban managers to improve habitat quality to ensure that birds coexist with human beings.

Keywords: flight initiation distance; adaptation; residential area; tree canopy coverage; floor area ratio; conservation of urban bird diversity; friendly to bird; Beijing

Citation: Yin, L.; Wang, C.; Han, W.; Zhang, C. Birds' Flight Initiation Distance in Residential Areas of Beijing Are Lower than in Pristine Environments: Implications for the Conservation of Urban Bird Diversity. *Sustainability* **2023**, *15*, 4994. https://doi.org/10.3390/su15064994

Academic Editor: Vasilios Liordos

Received: 11 February 2023
Revised: 5 March 2023
Accepted: 8 March 2023
Published: 10 March 2023

Copyright: © 2023 by the authors. Licensee MDPI, Basel, Switzerland. This article is an open access article distributed under the terms and conditions of the Creative Commons Attribution (CC BY) license (https://creativecommons.org/licenses/by/4.0/).

1. Introduction

Urban green spaces, with refuges, stepping stones, and abundant food sources, provide habitats for birds; and birds represent the primary type of wildlife that people engage with on a daily basis [1]. With rapid urbanization, an increasing number of people live in cities, and birds face a variety of challenges. Urban environments are heterogeneous, therefore, birds living in proximity to humans have to adapt to local conditions by altering their behavioural response to human interference [2]. Evaluating how bird behaviour changes in response to urbanization informs our understanding of how species respond to human-induced environmental changes [3]. Birds can be categorized into three types based on their response to urbanization: urban avoiders, urban adapters, and urban exploiters [4]. Adapters and exploiters adapt to highly urbanized areas by interacting with humans [5,6]. A critical trait of such birds is the ability to tolerate high levels of disturbing stimulation by humans [4]. Successful urban species modify their escape behaviour in response to new conditions, as evolutionary mismatches between fear responses and environmental conditions are likely to have negative fitness consequences [3].

Escape behaviour is a typical adaptive behaviour that follows vigilance, and many response distances can be used as a quantitative measure of a bird's tolerance to human-caused disturbances [7–11]. Different response distances of birds facing risks [8,12,13], including start distance (SD), detection distance (DD), physiological initiation distance (PID), alarm distance (AD), flight initiation distance (FID), and flight distance (FD), can partly represent their ability to perceive risks and escape behaviour. These parameters reflect the level of dependence and adaptability of birds to humans [14]. FID—the distance at which the bird moves away in response to the approach of a perceived threat—has been used to define a minimal setback distance and provide a measurement of risk-taking behaviour [3,9,10,15,16]. FID is a balance of optimization of benefits associated with escape weighed against disturbances (such as loss of feeding opportunity and energy cost of flight), which may vary with ambient conditions [17,18]. FID indicates how well a species or population has adapted to chronic environmental stress [19].

Among all response distances, SD, AD, and FID are the most frequently used for measuring escape behaviour. Some studies have suggested that FID is positively correlated with SD and AD [20,21]. Nevertheless, the relationship between SD and FID is not present under low perceived risk of predation, and birds are more relaxed in urban habitats with friendly people [22–25]; this may generate a spurious positive relationship between FID and SD [20]. Compared with AD, FID is the most visible and directly measurable behavioural response that can be observed plainly through the behaviour change of birds taking off suddenly. Therefore, we only chose FID as the most reliable response variable in this study.

Different stimulations faced by individuals influence their behaviour. Different stimulus types include vehicles (walk, bicycle, vehicle, motorcycle) [26–29], approach speeds or direction [30–32], colour of clothes [25], and noise [1]. Bicycles evoke longer FIDs than pedestrians [26], and photography was associated with longer FIDs [33]. In general, intense stimulation led to lower bird FIDs and led to complete escape. Human feeding behaviour and abundance of food in cities result in lower bird FIDs [14,22]. Birds may become habituated to certain continuous stimulations or disturbances, depending on their experience shaped by behavioural flexibility and phenotypic plasticity [4,6,19,22,34]. Considering flexible behaviour and gradual adaptation to the habitat, it is necessary to explore the bird behaviour response to human disturbance in residential areas: adapting or escaping.

Among these external factors, many studies focus on macro differences between urban and rural areas [5,19,24,25,35,36], which illustrates that birds adapt to urbanization by changing escape behaviour when facing people or potential predators [4]. Species living in closed habitats are likely to have smaller FIDs, and those residing in open habitats should have greater opportunities for trade-offs. Habitat was classified as open grassland, shrubs, trees, forests [37–39]. There is a lack of detailed quantitative habitat variables to study FID responses to different habitats. Urban trees provide habitat and shelter for birds from predators and human interference. We added quantitative variables using tree canopy coverage to reflect habitat characteristics.

Beijing is an important node for bird migration, with a large number of birds. There are 503 species (List of terrestrial wildlife in Beijing (2021) (http://yllhj.beijing.gov.cn/ztxx/ysdw/ml/202110/t20211027_2522201.shtml, accessed on 1 March 2023)) in Beijing and 448 species (http://www.birdreport.cn/, accessed on 1 March 2023) in the downtown area, of which about 100 species (https://www.fx361.com/page/2016/0307/3276550.shtml, accessed on 1 March 2023) are common species. Although FID has been relatively well studied in birds, there is a lack of experimental studies clarifying birds FID in residential areas. We aimed to test whether birds inhabiting residential areas differed in their escape behaviour (measured as FID), and whether birds modified and optimized their FID when facing human residential areas with different external environment and tree canopy coverage. By increasing the individual initial conditions, we also generated an effect related to the initial conditions. We hypothesized that differences in individual and external environmental characteristics can affect the responses of birds. We applied a mixed model approach and compared FID between bird species, focusing on the relative importance of factors, including frequency, initial behaviour, height, visibility, floor area ratio, adjacent road ratio, and canopy coverage, on the escape behaviour of birds. We focused on the implications of urban human settlements with different tree canopy coverage that alter bird behaviour in the context of human wildlife interactions. Data were collected from residential areas in the main unban area of Beijing, which is a relatively closed and independent residential area with complete infrastructure.

2. Materials and Methods

2.1. Study Area and Sampling Sites

Beijing (39°28′–41°05′ N, 115°20′–117°30′ E) is the capital of China, spanning 16,410.54 km^2 and with a population of 21.89 million people in 2020 (http://www.stats.gov.cn/tjsj/, accessed on 25 January 2023). Beijing has the sub-humid and semi-arid continental monsoon climate of the northern temperate zone, with four distinct seasons (average annual temperature of 12.06°and average annual rainfall of 545.3 mm from 1978; http://tjj.beijing.gov.cn/, accessed on 25 January 2023). The study area contained the main urban area within the 5th ring road in Beijing (667.35 km^2). We chose 40 urban residential areas distributed evenly along eight radiation directions within the 5th ring road as sample sites (Figure 1, Table S1). All studied residential areas are relatively closed and independent settlements with similar urban infrastructure (multi-story houses, concrete buildings, central squares, green space, gardens, etc.) and are surrounded by urban roads, green spaces, and buildings.

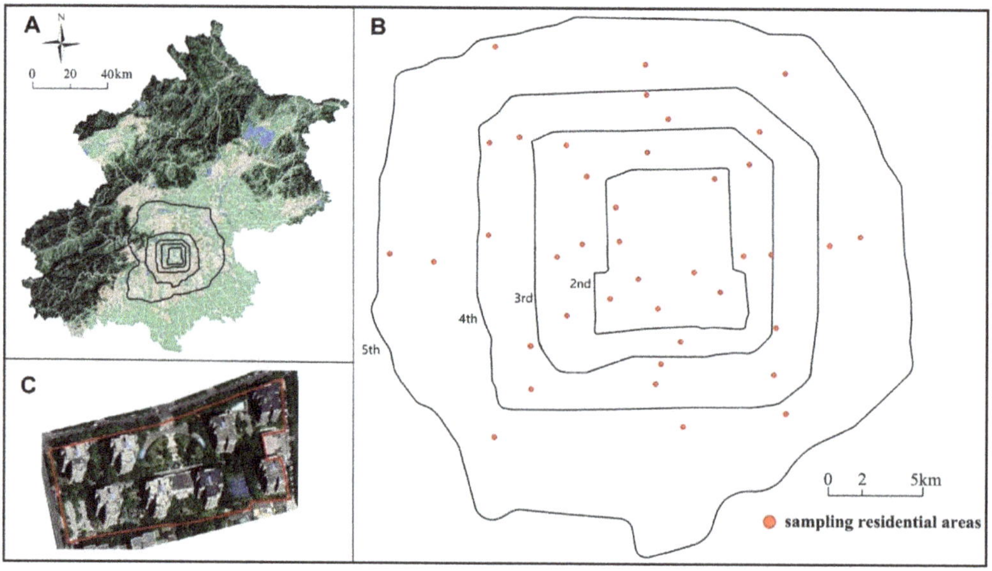

Figure 1. (**A**) The topographic map of Beijing. The light brown areas are building areas, blue areas are waters, light green areas are green spaces in the plain area, and dark green areas are green spaces in the mountain area. (**B**) The right figure shows 40 sampled residential areas located within the 5th ring road of Beijing. (**C**) An example of residential areas. The red line is the boundary of residential areas.

2.2. Field Surveys

During the entire year from June 2020 to May 2021, our survey was conducted on weekdays to reduce the fluctuation of pedestrian volume and from 7:00 to 10:00 on sunny days, when birds were most active. Each residential area was sampled once per month. We focused on birds that were foraging or engaged in "relaxed behaviour," such as preening or roosting. Neither highly vigilant and obviously alarmed birds, nor nesting birds were approached [40]. Any flight due to obvious relocation for foraging or other disturbances was not recorded [9]. The distance to the closest individual was determined when a group of birds was encountered [19]. We chose transects mainly along the walkway at each site covering the entire area. When a target bird was located with a pair of binoculars, the observer approached the bird by walking toward it in a straight line at a constant speed of 0.5 m/s [41]. Before data collection, the observer was trained to maintain a consistent stride length and a constant pace. FID was recorded as the horizontal distance between the observer and the target bird when it flew, ran, or hopped in response to being approached [34], using a laser rangefinder (Trupulse 200). During each survey, we avoided re-sampling individuals by focusing on birds in different geographic locations and not re-sampling the same location repeatedly. In addition, observers wore similar drab clothing across all surveys to control for the confounding effects of observer appearance on FID data. We eliminated the Eurasian Tree Sparrow (*Passer montanus*) to avoid influencing the results which have already adapted to the urban artificial environment.

2.3. Variables Collection

A total of 7 variables were used to characterize each individual or environmental characteristic of the 40 sampling sites. They were classified into three different groups: species frequency (1 variable), individual initial condition (3 variables), and external environment characteristics (3 variables) (Table 1).

Table 1. Variables of birds FID in residential areas.

Type	Variables	Description
Species frequency	Frequency	Total individuals of each species recorded during the survey reflecting species frequency
Individual initial condition	Initial behaviour (IB) Height(m) Visibility level (VL)	Initial behaviour of the target bird. Initial height of the target bird off the ground. Initial visibility of the target bird (Figure 2).
External environment characteristic	Floor area ratio (FAR) Adjacent roads ratio (%) (ARR) Tree canopy coverage (%) (CANOPY)	The total floor area is divided by the area of each site. The length of urban main and secondary roads is divided by the perimeter of each site. The area covered by the tree canopy is divided by the area of each site.

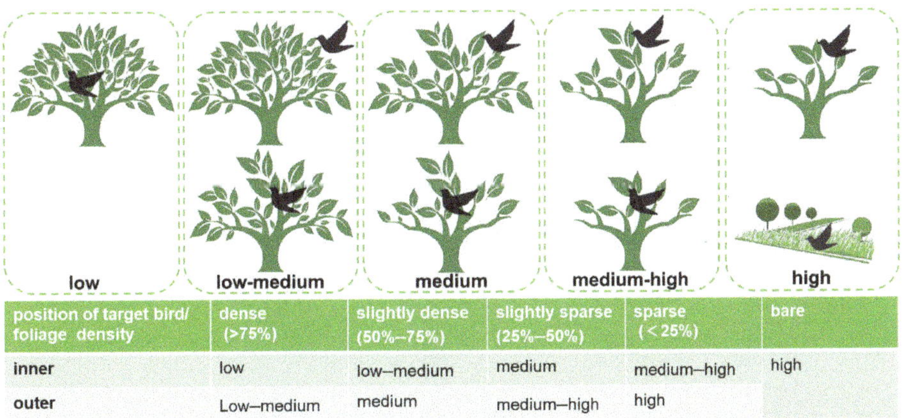

Figure 2. Diagram of visibility level according to the foliage density and the position of target bird.

2.3.1. Species Frequency

Species frequency was recorded as the total number of individuals of each species during the survey at all sites. In order to facilitate the analysis, we also classified species frequency into four levels: the most common (MCS, $n \geq 200$), common (CS, $50 \leq n < 200$), rare (RS, $10 \leq n < 50$), and extremely rare (ERS, $n < 10$).

2.3.2. Individual Initial Conditions

Unlike previous studies that only focused on individuals on the ground or in an open space to avoid the effects of height and vegetation cover [21,24], we also recorded the height of the ground where the target bird was on the ground or in a tree at the start of the approach [16], and classified birds' initial behaviours into forage, feed, drink, perch, walk, and play [3]. Similar to habitat openness [40,41], we defined the visibility level for the individuals who were approached. Visibility level was classified based on the foliage density of a tree or shrub estimated by the eye where the target bird was perched and the position of the target bird in the vegetation (Figure 2): low, low-medium, medium, medium-high, and high (bare vegetation, open lawn, or ground without any shelter). All information was collected by one person to avoid inter-observer variation [8].

2.3.3. External Environment Characteristics

External environment characteristics were calculated based on the visual interpretation method using satellite imagery from Google Earth (September 2019) and combined with field investigation for each site. The floor area ratio of buildings was calculated by dividing the total floor area by the area of each site, which reflected the overall environment and population density. Considering the impact of the surrounding environment on the residential district, the adjacent road ratio was defined as the proportion of urban main and secondary roads adjacent to the residential district, which was equal to the road length divided by the perimeter of each site. The tree canopy coverage ratio was increased to reflect the environment for each site and was equal to the area covered by the tree canopy divided by the area of each site, which could also provide refuges for birds.

2.4. Data Analysis

All statistical analyses were performed using the R version 4.0.3. Firstly, we found the data did not conform to the normal distribution with the Shapiro–Wilk test and still presented an abnormal distribution after log transformation. Therefore, we chose the Kruskal–Wallis non-parametric test method to compare bird FID among different groups, including residence type, frequency level, diet [42], foraging site [1], visibility level, and initial behaviour. We also conducted multiple comparisons between each level using Dunn post hoc tests (p-values adjusted with the Benjamini–Hochberg method). A mixed model procedure was applied to study escape behaviour in relation to frequency, individual initial behaviour (height and visibility level), and environmental characteristics (floor area ratio, adjacent road ratio, and canopy coverage). Considering there were significant differences among residence types, we applied generalized linear mixed models (GLMMs) to evaluate the response variables for species with different residence types, but the models did not apply to migratory species in which the best-supported model (the lowest AICc) was without any variables (Table S3). Hence, we applied generalized linear mixed models (GLMMs) separately to evaluate the response variables of bird FID for the total species, the most common species, and the other species. To control for temporal and interspecific variation, we included census months and species as random factors.

All explanatory variables were standardized before the analyses for the comparison of the model parameter estimates. The effects of these explanatory variables on bird FIDs were tested using the model selection procedure. For each model, a multi-model inference procedure was applied using the R MuMIn package. This method allowed us to perform model selection by creating a set of models with all possible combinations of the initial variables and sorting them according to the Akaike information criterion corrected (AICc) fitted with maximum likelihood. We selected all models with ΔAICc < 2 and used the model-averaging approach to estimate the parameters. To evaluate the relative importance of the explanatory variables of bird FIDs, we calculated the relative effect of the parameter estimates for each of the variables among the three models. We also fitted the curves for different variables with 95% confidence intervals. All analyses were run by using the package "lme4" [43], "MuMIn" [44], " sjPlot " [45] and "forestplot" [46].

3. Results

3.1. Birds FID in Residential Area

Over the 12 months of the survey, we collected 1360 valid FIDs of 31 species recorded within 40 residential areas (Table S2). Spotted Dove (*Streptopelia chinensis*) and Azure-winged Magpie (*Cyanopica cyanus*) were the most abundant bird species recorded, accounting for 24.93% and 22.21% of all individuals, respectively. Light-vented bulbul (*Pycnonotus sinensis*) and common magpie (*Pica pica*) were the two other common species.

For the pooled data, the estimated mean FID was 8.58 m. A total of 20 species had FIDs shorter than the mean FID, while 11 species had FIDs longer than the mean FID (Figure 3). Some species (e.g., human commensals), such as Spotted Dove, Common Magpie, Azure-winged Magpie, Light-vented Bulbul, and Common Hoopoe (*Upupa epops*), had FIDs significantly shorter than the mean FID. While some species hardly appeared in the residential areas, such as Grey-capped Greenfinch (*Chloris sinica*), Brown Shrike (*Lanius cristatus*) and Daurian Redstart (*Phoenicurus auroreus*), had longer FIDs (Figure 3). For different residence types, the estimated status mean FID of a resident bird was 7.86 m whereas, for a migratory bird, it was 9.60 m. Similarly, the estimated mean FID of the four most common birds was 6.83 m, whereas for the other species, it was 9.00 m.

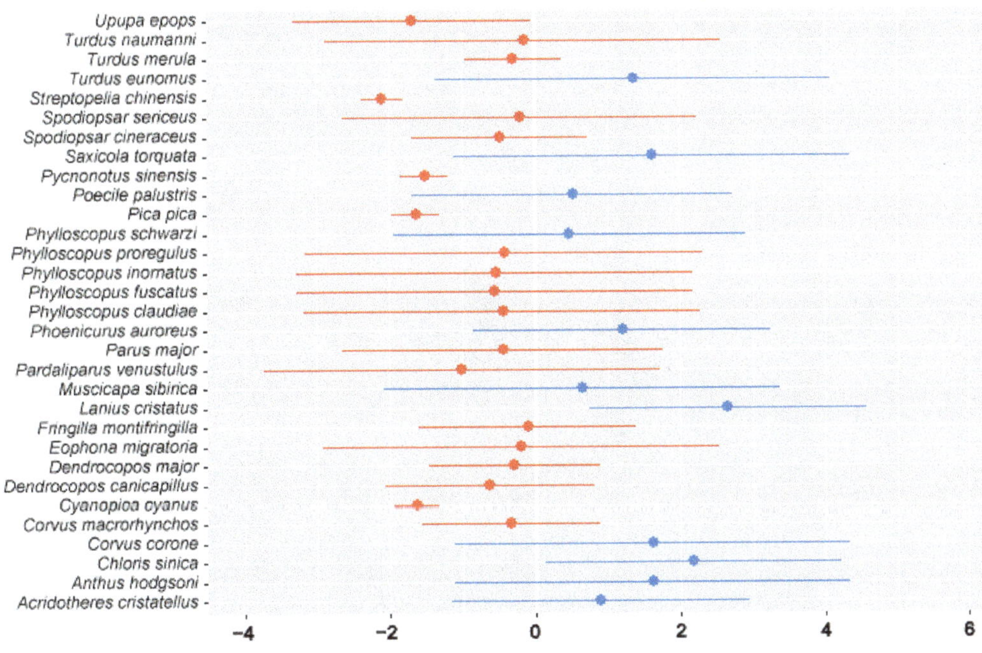

Figure 3. Comparing estimated FIDs of different species. Red shows lower than the estimated mean FID, while blue shows longer than the estimated mean FID.

3.2. Birds FID of Different Groups

For different bird groups, there were significant differences between the FIDs (Figure 4). The FID of resident birds was significantly shorter than that of migratory birds (including summer breeders, winter visitors, and passengers) (Figure 4A). Similarly, the species that appeared more frequently in residential areas had shorter FIDs. The most common species had FIDs significantly shorter than those of the common and rare species (Figure 4B). For the foraging guilds, birds with different diets had significantly different FIDs. Granivore species had FIDs shorter than those of insectivores and omnivores, and carnivores had the longest (Figure 4C). The FIDs of ground foragers and canopy gleaners were similar, but significantly shorter than that of the hawking flycatcher (Figure 4D).

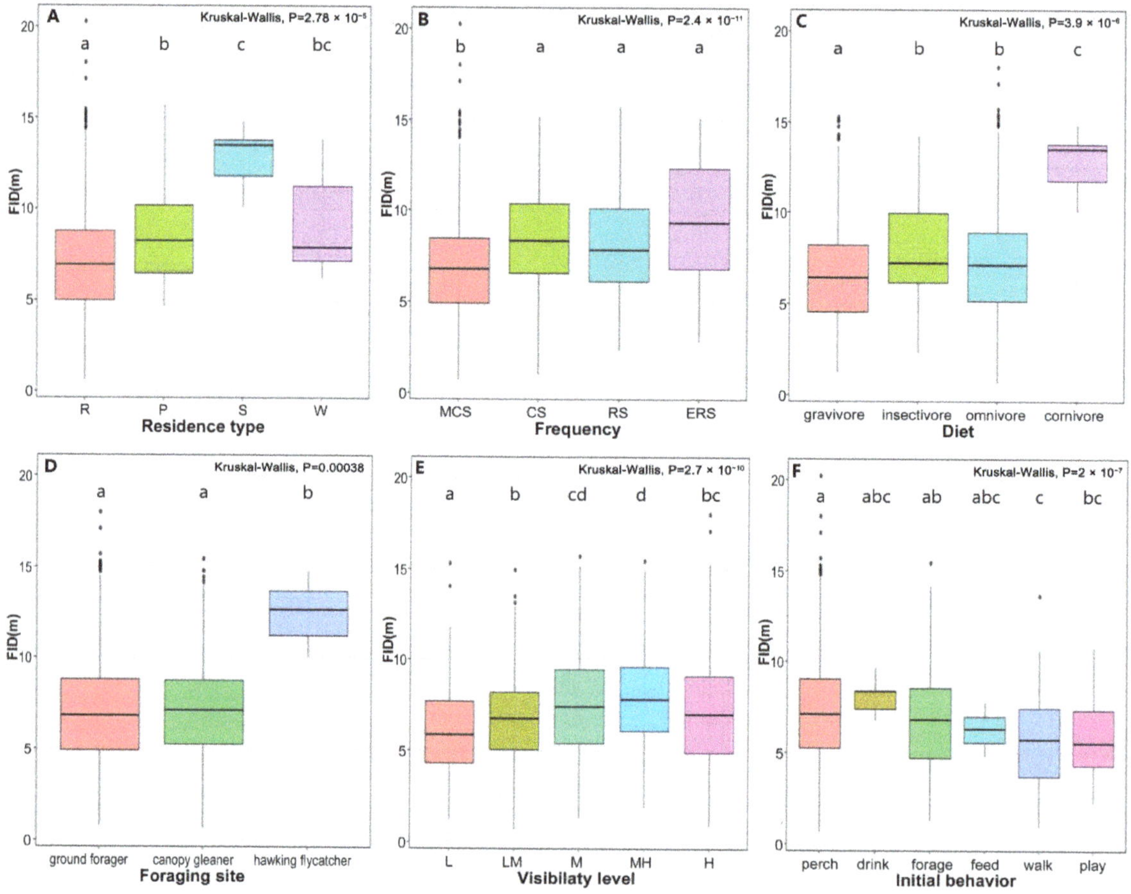

Figure 4. Comparison of FIDs among (**A**) status, (**B**) dominance level, (**C**) diet, (**D**) foraging site, (**E**) visibility level and (**F**) initial behaviour. The *y*-axis represents the value of FID (in meters). Box plots show the median (bar in the middle of rectangles), upper and lower quartiles, maximum and minimum values (vertical lines) and outliers (black dots). The letters above boxes indicates the significant difference among different groups ($p < 0.05$). Residence type: R = Resident, P = Passenger, S = Summer breeders, W = Winter visitors. Frequency: MCS = the most common species, CS = common species, RS = rare species, ERS = Extreme rare species. Visibility level: L = low, LM = low-medium, M = medium, MH = medium-high, H = high.

3.3. Birds FID in Different Initial Conditions

For individual initial conditions, FID also showed a significant difference, which gradually increased significantly with the visibility level of the individual initial location (Figure 4E). Different initial behaviours also affected the FID. Birds with quiet or immersive behaviours (perched, drinking, foraging) had significantly longer FIDs than individuals with dynamic behaviours such as walking and playing (Figure 4F). Birds showed different awareness and vigilance in different conditions.

3.4. Factors Influencing Birds FID

The best models were selected for total species, the most common species, and the other species (Table S4). For the three models, frequency had a significant negative effect on FID ($\beta_{total} = -1.22$, $p < 2 \times 10^{-16}$, Figure S1), while significance for the latter two models based on frequency was significantly reduced (Figures 5 and 6A–C). The most common species have an overwhelming impact on the total species. We mainly reported the most common species and the other species.

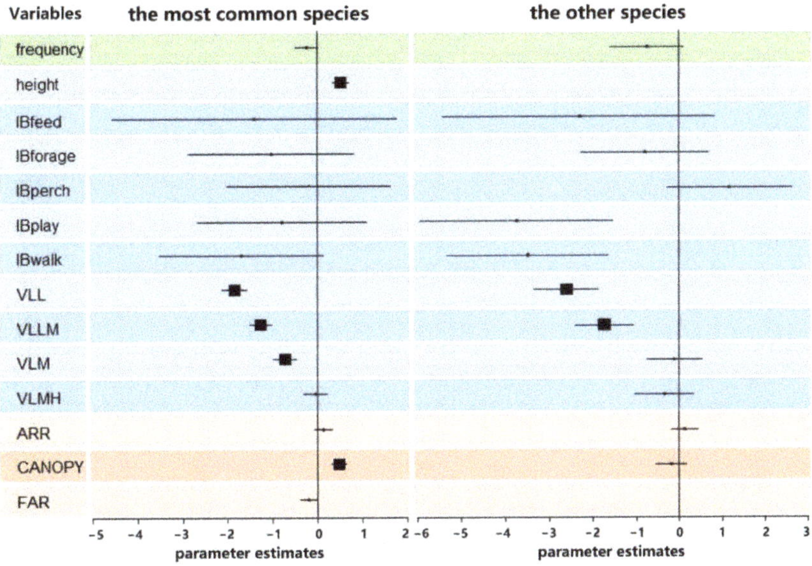

Figure 5. Effect sizes of variables are predicted based on the best models' conditional coefficient for birds FID. Estimates in the plot are shown using the mean values (black squares or points) and associated 95% CIs (black horizontal lines). The shape of mean values varies according to the p-value: $p < 0.05$ '■' and $p > 0.05$ '•'. The bigger black squares represent variables that have a significant effect on the averaged model. Green represents frequency variables, blue represents individual initial condition variables, and orange represents environment characteristic variables.

Individual initial conditions had significant effects on FID for the two models. The degree of negative effects decreased significantly with decreased visibility (Figure 5). Initial height had a significant positive effect on the most common species ($\beta = 0.53$, $p = 0.003$; Figures 5 and 6E), while there was no relationship for the other species without convergence (Figure 6F).

For environmental variables, ARR was not included in three best models (Table S4). FID of the most common species increased significantly with tree canopy coverage ($\beta = 0.48$, $p = 0.004$), while the other species decreased marginally (Figures 5 and 6G–I). Although not statistically significant ($p > 0.1$), the FID of the most common species still decreased with FAR (Figures 5 and 6K).

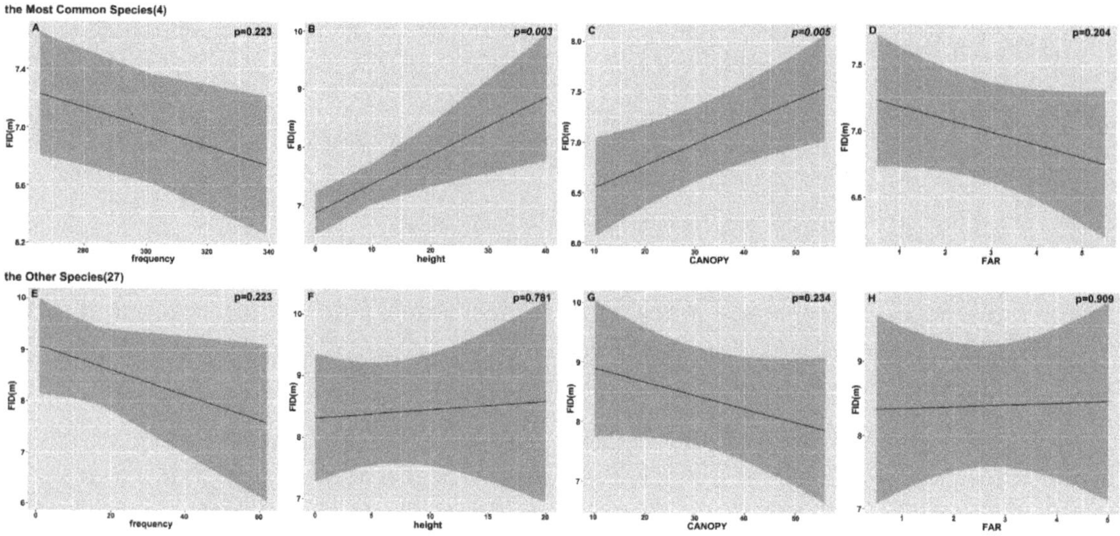

Figure 6. Relationships between variables and FID for (**A–D**) the most common species, (**E–H**) the other species. Black lines indicate estimates with 95% confidence intervals (grey area). The p values ($p < 0.05$) in italics indicates significant effect.

Height, tree canopy coverage, and FAR were all more significant for the most common bird species than the other species, which indicated that the variables were more applicable to common species than to rare species. The significance of the total species model was reduced by the addition of other species.

4. Discussion

4.1. Birds' FID in Residential Areas

Birds' FID in residential areas in this study was lower than that in other habitat types, such as forests, water bodies, gardens, parks [26,35,47,48]. Reducing FID is a strategy to fit urban characteristics, increase tolerance to human presence, and adjust the cost-benefit relationship between the risks of starvation and predation [49]. In the process of urbanization, birds in residential areas mostly play the role of exploiters and adaptors coping with frequent interactions with humans, which ultimately leads to them becoming more tolerant of humans and adapting to coexist [12]. In highly urbanized areas, due to landscape heterogeneity, and environmental complexity (lack of rich vegetation and full of a large number of artificial facilities or human interference), only a few bold birds can adapt to the complex environment and inhabit here [30]. Human commensals are likely to habituate human interference, whereas others are sensitive [50]. Living in urban environments for a long time makes these birds show lower vigilance to humans and shorter FIDs via learning, cognition, and rapid adjustment [32,36]. Food is an important factor in the reduction of FID in urban animals [14,22]. In residential areas, people feed grain or other food to birds, so that they have a friendlier relationship with humans and no longer fear them.

Our results indicated that the frequency of birds' presence in residential areas showed a negative correlation with FID, and the frequency of birds' presence was correlated with their residence types. Studies have shown that common birds have shorter FID than threatened birds [51], and FID is negatively correlated with the duration for which the birds have lived in the urban area. Migratory birds have longer FIDs than resident birds [21]. Resident birds (e.g., Spotted Dove, Light-vented Bulbul, and Common Magpie) that have been living in urban areas for a long time, especially in residential areas, have adapted to the environment full of human beings and various complex interferences (people, traffic, noise, etc.). For migratory birds that stay for only several months in summer or winter, the time to adapt is shortest and FID is the longest, while passengers pass through the same place in spring and autumn, so the FID lies in the middle. The FIDs of summer breeders are longer than those of winter visitors, even if they both stay for several months. This may be because summer breeders visit in the breeding season, and are busy hatching, and feeding their young. Additionally, parent birds will maintain higher vigilance to protect their offspring and improve their survival rate [10]. At the same time, when birds are able to meet their daily energy requirements easily in summer, the balance in the trade-off between avoidance of starvation and predation shifts toward greater FIDs [52]. Winter visitors may spend more time and energy on foraging and less on escape flight to reduce energy expenditure in winter when the weather is cold and food is scarce. The FID of birds was longer in autumn and winter [32,52,53]. In general, summer breeders come from lower latitudes, while winter visitors come from higher latitudes. Previous studies have shown that birds at higher latitudes have shorter FIDs than birds at lower latitudes [5,7], and our results also confirm this from another perspective.

In our study, bird species with different food sources and foraging sites showed different FIDs. Granivore species are the most intimate with humans. Some people sprinkle corn, millet, and other grains on their windowsills or on the ground in residential areas. Granivore species mainly forage on the ground, and they are the boldest because of their long adaptation to close contact with human beings [14]. In urban habitats, omnivorous and insectivorous species have shorter FIDs [54]. Insectivores and omnivores mostly live in trees (on trunks like woodpeckers or in the canopy) and eat flower buds, leaf buds, fruits, or insects, and are, therefore, less close to humans than the ground foragers. Aerial foraging species that are less vulnerable to predation have shorter FIDs than other species [39]. In our study, only the Brown Shrike (*Lanius cristatus*) was recorded among carnivores, which tends to perch on tree tops, pounce on prey, and return to its original position. They are hawking flycatcher-like hunters and exhibit greater alertness. Similar to this foraging pattern and predation behaviour, the Dark-sided Flycatcher (*Muscicapa sibirica*) was another insectivore we recorded among hawking flycatchers, which are agile and shy. These hawking flycatchers have longer FIDs, which may be due to their hunter-like foraging patterns and high vigilance to both prey and external interference.

4.2. Individual Initial Conditions That Affect Birds' FID

Our results demonstrated that the initial height of the bird significantly affected the FID. Some studies have shown that the higher the aboveground location of a bird, the shorter is its horizontal FID [16,55]. A higher aboveground might mean increased safety and a reduction in the perceived risk due to vertical relief from the threat [1,41]. Even if the threat does come, they have a longer time to escape because they are farther off the ground. However, in our study, FID increased with height, possibly because birds at higher positions have a wider field of vision and can detect threats in advance. Ground birds that lose the protection of height are exposed to open habitats, making them more likely to be disturbed. It should be noted that ground birds are usually closer to humans (e.g., Spotted Dove and Common Magpie) because foraging for small seeds at the soil surface in open sites makes them adapt to people via habituation [30,56]. Even when a person walks very close to the bird, the bird is completely immersed in the foraging process and is immune to interference.

Unlike previous studies [3], the FID of target birds varied with their initial behaviour. Birds' tolerance to urban environments is directly linked to habitat choice, resource requirements, reproductive strategy, and survival rate, and keeping vigilant constantly and taking flight are a huge drain and cost at the expense of other activities such as foraging or feeding due to limited energy and attention [18,19,22,49]. Birds may compromise the relationship between foraging or feeding behaviour and perceived risk. Birds exhibiting immersive behaviours (drinking, foraging, or feeding) have shorter FIDs than resting birds (perched) because their attention is used to supplement their energy rather than to keep alert or take flight. In contrast, resting birds have sufficient energy to perceive risk and have longer FIDs. Birds exhibiting dynamic behaviours (walking or playing) are bolder and have shorter FIDs. They are completely lost in their own world, unless people are in very close proximity.

The visibility level is a two-fold "see and been seen" concept, reflecting whether birds can detect people's approach and whether people can see the hidden bird easily. Birds perching in trees keep away from interference, and the canopy provides refuge for the birds to make them feel safer from being seen [24]. Dense canopies also block the sight of birds, making them less perceptive to dangers around them [40]. Low visibility means that birds in the canopy are not easy to see, and these birds have shorter FIDs, probably because they are not easy to detect. While high visibility means birds are exposed to the open habitat or tree canopy surface even if the tree canopy is without leaves, lack of refuge makes birds detect risks earlier and causes longer FID.

4.3. The Predictors of Birds' FID among Environmental Characteristics

The adjacent road ratio (ARR) of environmental factors was not included in the models. This may be because bird behaviour is more affected by the internal environmental factors of the residential areas, but is less related to circumjacent environmental factors. These predictors were more applicable to the most common birds than they were to rare birds. Common birds are likely to habituate, whereas rare birds are sensitized to human activity [50]. The most common birds appear more frequently in residential areas and show different adaptability and habituation to different levels of human activity intensity and habitat quality, which are represented by floor area ratio (FAR) and tree canopy coverage in residential areas. Most rare birds are urban avoiders, usually just passing through residential areas during migration. They show a strong fear of artificial environment or human interference and adapt to the environment poorly; therefore, their FIDs are weakly correlated with environmental characteristics. For common birds, the recorded data were sufficient, and the model was more reliable. For uncommon birds, the recorded data amount was smaller and the predictive power of variables decreased significantly, and a conclusion similar to that for common birds could not be drawn. Environmental characteristics can be used as predictors of FID for most common birds.

FAR and tree canopy coverage seem to be partly opposite environmental characteristics, but we cannot simply assume that the two variables have a negative correlation because FAR is affected by the building stories. Among the environmental factors, FAR reflects population density and human interference intensity in residential areas. Although not statistically significant, FID tended to decrease with an increase in FAR. Similar to previous studies [4,12,17,36,48,57], birds in habitats with high population densities may adapt to human interference and have shorter FIDs

Tree canopy coverage reflects the habitat quality of the residential areas. For most bird species, higher tree canopy coverage means more natural and suitable habitats like natural forests and fewer artificial facilities (buildings, roads, parking lots, etc.), making these birds have longer FIDs. At the same time, the tree canopy also provides refuge for birds when faced with risk [40]. Higher tree canopy coverage gives birds more refuge that can be chosen proactively in advance, resulting in a longer FIDs [24]. When tree canopy coverage is low, birds may be passively exposed to open habitats for a long time, and thus have to change their behaviour to adapt to human interference and gradually gain longer FIDs via behavioural flexibility and phenotypic plasticity [30,58,59].

It is worth noting that our predictors have certain limitations because birds' behaviour is affected by inherent genetic characteristics such as temperament and physiological properties [60]. Species could be a relatively important factor in explaining variations in FID [50]. Given that Eurasian Tree Sparrows (*Passer montanus*) are found in all residential areas and occupy an extremely dominant position, they are highly adapted to the human environment without distinction. Living closely with human beings for a long time, they have become accustomed to human interference. When people walk by them, or even approach them, such birds continue doing their tasks and are not perturbed. FIDs of sparrows are sometimes even less than one meter, and thus, are very different from those of other bird species, severely affecting the results of this study. Meanwhile, it is difficult to accurately measure the FIDs of sparrows because they like cluster activities and are naturally active and bold [61]. Therefore, the Eurasian Tree Sparrows were ignored in our study.

4.4. Cultivate Big Trees in Planning and Management

Residential green space is an important component of urban green space [62], which is an important activity space and habitat for many urban birds [63]. The urban pattern of plant diversity was significantly affected by land use, and larger tree canopy patches support more diverse species [64]. Urban tree canopy can provide a cooler downtown environment, privacy on larger lots, or reduced noise and pollution from a major road way [65]. Simultaneously, urban tree canopy can provide breeding or foraging habitat for some bird species [66]. Our results show that birds in higher canopy cover have longer FIDs. Studies have shown that conservation of small forest fragments and urban tree cover can benefit migrants including residential areas [67].

Residential area planning should plan green space rationally, cultivate big trees with larger tree canopies and leave enough space and time for small trees to grow. Local managers should ensure trees are healthier and not overdo the pruning. Branches more often being removed for thinning crowns results in narrow crown width [68], which is not benefit to conservation of urban bird diversity. Planners and managers should work together to make cities friendlier to bird.

5. Conclusions

To our knowledge, this is the first study of bird FIDs in residential areas that quantitatively measures environmental factors using FAR and tree canopy coverage. Our results indicate that bird FIDs in residential areas are different among different groups (residence type, frequency, diet, and foraging site). Birds that appear more frequently in residential areas or are ground foragers, insectivores, or omnivores are better adapted to human interference and have shorter FIDs. Further analyses revealed that individual initial conditions affect bird FID, and environmental characteristics (FAR, tree canopy coverage) can be used as predictors for the most common birds. Tree canopy coverage was found to positively affect FID, whereas FAR negatively affected FID.

Our results demonstrated that birds in residential areas have been adapting to the human environment, especially in places with high population density. Urban tree canopies can provide refuge for birds to avoid human interference, and a lower FAR is more habitable for both, birds and people. Our study focused on the response of bird FIDs to human interference and urban trees under high urbanization, which has substantial practical implications for urban residential planners and managers to improve habitat quality and help birds coexist with human beings. Furthermore, other potential factors and response variables should be thoroughly investigated to further our understanding of bird escape behaviour in residential areas.

Supplementary Materials: The following supporting information can be downloaded at: https://www.mdpi.com/article/10.3390/su15064994/s1, Table S1: Summary of sampling sites; Table S2: Summary of the birds' FIDs collected in residential areas; Table S3: Summary of the final best models within a ΔAICc < 2 for the FIDs of total species, resident species and migratory species; Table S4: Summary of the final best models within a ΔAICc < 2 for the FIDs of total species, the most common species and other species FIDs; Figure S1. Relationships between variables and FID for the total species.

Author Contributions: Conceptualization, C.W. and L.Y.; methodology, L.Y.; software, L.Y.; validation, C.W. and L.Y.; formal analysis, L.Y. and C.W.; investigation, L.Y. and W.H.; resources, C.W.; data curation, L.Y.; writing—original draft preparation, L.Y.; writing—review and editing, C.W. and C.Z.; visualization, L.Y.; Supervision, C.W.; project administration, C.W.; funding acquisition, C.W. and L.Y. All authors have read and agreed to the published version of the manuscript.

Funding: This research was funded by National Non-Profit Research Institutions of the Chinese Academy of F (CAFYBB2022SY006, CAFYBB2020ZB008).

Institutional Review Board Statement: Not applicable.

Informed Consent Statement: Not applicable.

Data Availability Statement: Not applicable.

Acknowledgments: We thank Dan Han for writing advice. We thank anonymous reviewers for comments and suggestions that greatly improved the manuscript.

Conflicts of Interest: The authors declare no conflict of interest.

References

1. Petrelli, A.R.; Levenhagen, M.J.; Wardle, R.; Barber, J.R.; Francis, C.D. First to Flush: The Effects of Ambient Noise on Songbird Flight Initiation Distances and Implications for Human Experiences with Nature. *Front. Ecol. Evol.* **2017**, *5*, 67. [CrossRef]
2. Morelli, F.; Mikula, P.; Benedetti, Y.; Bussière, R.; Jerzak, L.; Tryjanowski, P. Escape behaviour of birds in urban parks and cemeteries across Europe: Evidence of behavioural adaptation to human activity. *Sci. Total Environ.* **2018**, *631–632*, 803–810. [CrossRef] [PubMed]
3. Hall, M.J.; Burns, A.L.; Martin, J.M.; Hochuli, D.F. Flight initiation distance changes across landscapes and habitats in a successful urban coloniser. *Urban Ecosyst.* **2020**, *23*, 785–791. [CrossRef]
4. Gendall, J.; Lill, A.; Beckman, J. Tolerance of disturbance by humans in long-time resident and recent colonist urban doves. *Avian Res.* **2015**, *6*, 7. [CrossRef]
5. Samia, D.S.M.; Blumstein, D.T.; Díaz, M.; Grim, T.; Ibáñez-Álamo, J.D.; Jokimäki, J.; Tätte, K.; Markó, G.; Tryjanowski, P.; Møller, A.P. Rural-Urban Differences in Escape Behavior of European Birds across a Latitudinal Gradient. *Front. Ecol. Evol.* **2017**, *5*, 66. [CrossRef]
6. van Donselaar, J.L.; Atma, J.L.; Kruyf, Z.A.; LaCroix, H.N.; Proppe, D.S. Urbanization alters fear behavior in black-capped chickadees. *Urban Ecosyst.* **2018**, *21*, 1043–1051. [CrossRef]
7. Andrade, M.; Blumstein, D.T. Anti-predator behavior along elevational and latitudinal gradients in dark-eyed juncos. *Curr. Zool.* **2020**, *66*, 239–245. [CrossRef]
8. Charuvi, A.; Lees, D.; Glover, H.K.; Rendall, A.R.; Dann, P.; Weston, M.A. A physiological cost to behavioural tolerance. *Behav. Process.* **2020**, *181*, 104250. [CrossRef]
9. Coetzer, C.; Bouwman, H. Waterbird flight initiation distances at Barberspan Bird Sanctuary, South Africa. *Koedoe* **2017**, *59*, a1419. [CrossRef]
10. Dowling, L.; Bonier, F. Should I stay, or should I go: Modeling optimal flight initiation distance in nesting birds. *PLoS ONE* **2018**, *13*, e0208210. [CrossRef]
11. McFadden, T.N.; Herrera, A.G.; Navedo, J.G. Waterbird responses to regular passage of a birdwatching tour boat: Implications for wetland management. *J. Nat. Conserv.* **2017**, *40*, 42–48. [CrossRef]
12. Prestes, T.V.; Manica, L.T.; de Guaraldo, A.C. Behavioral responses of urban birds to human disturbance in urban parks at Curitiba, Paraná (Brazil). *Rev. Bras. Ornitol.* **2018**, *26*, 77–81. [CrossRef]
13. Weston, M.A.; McLeod, E.M.; Blumstein, D.T.; Guay, P.-J. A review of flight-initiation distances and their application to managing disturbance to Australian birds. *Emu–Austral Ornithol.* **2012**, *112*, 269–286. [CrossRef]
14. Feng, C.; Liang, W. Behavioral responses of black-headed gulls (Chroicocephalus ridibundus) to artificial provisioning in China. *Glob. Ecol. Conserv.* **2020**, *21*, e00873. [CrossRef]
15. Bateman, P.W.; Fleming, P.A. Who are you looking at? Hadeda ibises use direction of gaze, head orientation and approach speed in their risk assessment of a potential predator. *J. Zool.* **2011**, *285*, 316–323. [CrossRef]
16. Bötsch, Y.; Gugelmann, S.; Tablado, Z.; Jenni, L. Effect of human recreation on bird anti-predatory response. *PeerJ* **2018**, *6*, e5093. [CrossRef]
17. Díaz, M.; Møller, A.P.; Flensted-Jensen, E.; Grim, T.; Ibáñez-Álamo, J.D.; Jokimäki, J.; Markó, G.; Tryjanowski, P. The geography of fear: A latitudinal gradient in anti-predator escape distances of birds across Europe. *PLoS ONE* **2013**, *8*, e64634. [CrossRef]
18. Møller, A.P. The value of a mouthful: Flight initiation distance as an opportunity cost. *Eur. J. Ecol.* **2015**, *1*, 43–51. [CrossRef]
19. Lin, T.; Coppack, T.; Lin, Q.-X.; Kulemeyer, C.; Schmidt, A.; Behm, H.; Luo, T. Does avian flight initiation distance indicate tolerance towards urban disturbance? *Ecol. Indic.* **2012**, *15*, 30–35. [CrossRef]
20. Chamaillé-Jammes, S.; Blumstein, D.T. A case for quantile regression in behavioral ecology: Getting more out of flight initiation distance data. *Behav. Ecol. Sociobiol.* **2012**, *66*, 985–992. [CrossRef]

21. Mikula, P.; Díaz, M.; Møller, A.P.; Albrecht, T.; Tryjanowski, P.; Hromada, M. Migratory and resident waders differ in risk taking on the wintering grounds. *Behav. Process.* **2018**, *157*, 309–314. [CrossRef]
22. Fleming, P.A.; Bateman, P.W. Scavenging Opportunities Modulate Escape Responses over a Small Geographic Scale. *Ethology* **2017**, *123*, 205–212. [CrossRef]
23. Stansell, H.M.; Blumstein, D.T.; Yeh, P.J.; Nonacs, P. Individual variation in tolerance of human activity by urban dark-eyed juncos (*Junco hyemalis*). *Wilson J. Ornithol.* **2020**, *134*, 43–51. [CrossRef] [PubMed]
24. Tätte, K.; Møller, A.P.; Mänd, R. Towards an integrated view of escape decisions in birds: Relation between flight initiation distance and distance fled. *Anim. Behav.* **2018**, *136*, 75–86. [CrossRef]
25. Zhou, B.; Liang, W. Avian escape responses to observers wearing clothing of different colors: A comparison of urban and rural populations. *Glob. Ecol. Conserv.* **2020**, *22*, e00921. [CrossRef]
26. Bernard, G.E.; van Dongen, W.F.D.; Guay, P.-J.; Symonds, M.R.E.; Robinson, R.W.; Weston, M.A. Bicycles evoke longer flight-initiation distances and higher intensity escape behaviour of some birds in parks compared with pedestrians. *Landsc. Urban Plan.* **2018**, *178*, 276–280. [CrossRef]
27. Glover, H.K.; Guay, P.-J.; Weston, M.A. Up the creek with a paddle; avian flight distances from canoes versus walkers. *Wetl. Ecol. Manag.* **2015**, *23*, 775–778. [CrossRef]
28. McLeod, E.M.; Guay, P.-J.; Taysom, A.J.; Robinson, R.W.; Weston, M.A. Buses, cars, bicycles and walkers: The influence of the type of human transport on the flight responses of waterbirds. *PLoS ONE* **2013**, *8*, e82008. [CrossRef]
29. Vines, A.; Lill, A. Boldness and urban dwelling in little ravens. *Wildl. Res.* **2015**, *42*, 590. [CrossRef]
30. Battle, K.E.; Foltz, S.L.; Moore, I.T. Predictors of flight behavior in rural and urban songbirds. *Wilson J. Ornithol.* **2016**, *128*, 510–519. [CrossRef]
31. Guay, P.-J.; Lorenz, R.D.A.; Robinson, R.W.; Symonds, M.R.E.; Weston, M.A. Distance from Water, Sex and Approach Direction Influence Flight Distances Among Habituated Black Swans. *Ethology* **2013**, *119*, 552–558. [CrossRef]
32. Legagneux, P.; Ducatez, S. European birds adjust their flight initiation distance to road speed limits. *Biol. Lett.* **2013**, *9*, 20130417. [CrossRef] [PubMed]
33. Slater, C.; Cam, G.; Qi, Y.; Liu, Y.; Guay, P.-J.; Weston, M.A. Camera shy? Motivations, attitudes and beliefs of bird photographers and species-specific avian responses to their activities. *Biol. Conserv.* **2019**, *237*, 327–337. [CrossRef]
34. Jiang, X.; Liu, J.; Zhang, C.; Liang, W. Face masks matter: Eurasian tree sparrows show reduced fear responses to people wearing face masks during the COVID-19 pandemic. *Glob. Ecol. Conserv.* **2020**, *24*, e01277. [CrossRef]
35. Carrete, M.; Tella, J.L. Inter-individual variability in fear of humans and relative brain size of the species are related to contemporary urban invasion in birds. *PLoS ONE* **2011**, *6*, e18859. [CrossRef]
36. Davey, S.; Massaro, M.; Freire, R. Differences in flight initiation distance (FID) between rural and urban populations of two species of Australian birds. *Behaviour* **2019**, *156*, 1151–1164. [CrossRef]
37. Bjørvik, L.M.; Dale, S.; Hermansen, G.H.; Munishi, P.K.T.; Moe, S.R. Bird flight initiation distances in relation to distance from human settlements in a Tanzanian floodplain habitat. *J. Ornithol.* **2015**, *156*, 239–246. [CrossRef]
38. Møller, A.P. Basal metabolic rate and risk-taking behaviour in birds. *J. Evol. Biol.* **2009**, *22*, 2420–2429. [CrossRef]
39. Møller, A.P.; Garamszegi, L.Z. Between individual variation in risk-taking behavior and its life history consequences. *Behav. Ecol.* **2012**, *23*, 843–853. [CrossRef]
40. Blumstein, D.T.; Fernandez-Juricic, E.; LeDee, O.; Larsen, E.; Rodriguez-Prieto, I.; Zugmeyer, C. Avian Risk Assessment: Effects of Perching Height and Detectability. *Ethology* **2004**, *110*, 273–285. [CrossRef]
41. Blumstein, D.T. Developing an evolutionary ecology of fear: How life history and natural history traits affect disturbance tolerance in birds. *Anim. Behav.* **2006**, *71*, 389–399. [CrossRef]
42. Huang, Y.; Zhao, Y.; Li, S.; von Gadow, K. The Effects of habitat area, vegetation structure and insect richness on breeding bird populations in Beijing urban parks. *Urban For. Urban Green.* **2015**, *14*, 1027–1039. [CrossRef]
43. Bates, D.; Mächler, M.; Bolker, B.; Walker, S. Fitting Linear Mixed-Effects Models Using Lme4. *arXiv* **2015**, arXiv:1406.5823.
44. Bartoń, K. MuMIn: Multi-Model Inference. 2020. Available online: https://CRAN.R-project.org/package=MuMIn (accessed on 11 November 2022).
45. Mazerolle, M.J. AICcmodavg: Model Selection and Multimodel Inference Based on (Q)AIC(c). 2020. Available online: https://cran.r-project.org/package=AICcmodavg (accessed on 11 November 2022).
46. Gordon, M.; Lumley, T. Forestplot: Advanced Forest Plot Using 'Grid' Graphics. 2020. Available online: https://CRAN.R-project.org/package=forestplot (accessed on 11 November 2022).
47. Mayer, M.; Natusch, D.; Frank, S. Water body type and group size affect the flight initiation distance of European waterbirds. *PLoS ONE* **2019**, *14*, e0219845. [CrossRef] [PubMed]
48. Mikula, P. Pedestrian Density Influences Flight Distances of Urban Birds. *Ardea* **2014**, *102*, 53–60. [CrossRef]
49. Piratelli, A.J.; Favoretto, G.R.; de Almeida Maximiano, M.F. Factors affecting escape distance in birds. *Zoologia* **2015**, *32*, 438–444. [CrossRef]
50. Blumstein, D.T.; Anthony, L.L.; Harcourt, R.; Ross, G. Testing a key assumption of wildlife buffer zones: Is flight initiation distance a species-specific trait? *Biol. Conserv.* **2003**, *110*, 97–100. [CrossRef]
51. Jiang, Y.; Møller, A.P. Antipredator escape distances of common and threatened birds. *Behav. Ecol.* **2017**, *28*, 1498–1503. [CrossRef]

52. Collop, C.; Stillman, R.A.; Garbutt, A.; Yates, M.G.; Rispin, E.; Yates, T. Variability in the area, energy and time costs of wintering waders responding to disturbance. *Ibis* **2016**, *158*, 711–725. [CrossRef]
53. Poddubnaya, N.; Korotkova, T.; Vanicheva, P. Increasing corvid tolerance to humans in urban ecosystems with increasing latitude. *Biol. Commun.* **2020**, 64. [CrossRef]
54. Morelli, F.; Benedetti, Y.; Díaz, M.; Grim, T.; Ibáñez-Álamo, J.D.; Jokimäki, J.; Kaisanlahti-Jokimäki, M.; Tätte, K.; Markó, G.; Jiang, Y.; et al. Contagious fear: Escape behavior increases with flock size in European gregarious birds. *Ecol. Evol.* **2019**, *9*, 6096–6104. [CrossRef] [PubMed]
55. Chen, X.-M.; Xie, W.-T.; Shuai, L.-Y. Flush early and avoid the rush? It may depend on where you stand. *Ethology* **2020**, *126*, 987–992. [CrossRef]
56. Osorio-Beristain, M.; Rodríguez, A.; Martínez-Garza, C.; Alcalá, R.E. Relating flight initiation distance in birds to tropical dry forest restoration. *Zoologia* **2018**, *35*, 1–6. [CrossRef]
57. Gravolin, I.; Key, M.; Lill, A. Boldness of Urban Australian Magpies and Local Traffic Volume. *Avian Biol. Res.* **2014**, *7*, 244–250. [CrossRef]
58. Jablonszky, M.; Szász, E.; Markó, G.; Török, J.; Herczeg, G.; Garamszegi, L.Z. Escape ability and risk-taking behaviour in a Hungarian population of the collared flycatcher (*Ficedula albicollis*). *Behav. Ecol. Sociobiol.* **2017**, *71*, 54. [CrossRef]
59. Vincze, E.; Papp, S.; Preiszner, B.; Seress, G.; Bókony, V.; Liker, A. Habituation to human disturbance is faster in urban than rural house sparrows. *Behav. Ecol.* **2016**, *27*, 1304–1313. [CrossRef]
60. Jiang, Y.; Møller, A.P. Escape from predators and genetic variance in birds. *J. Evol. Biol.* **2017**, *30*, 2059–2067. [CrossRef]
61. García-Arroyo, M.; MacGregor-Fors, I. Tolerant to humans? Assessment of alert and flight initiation distances of two bird species in relation to sex, flock size, and environmental characteristics. *Ethol. Ecol. Evol.* **2020**, *32*, 445–456. [CrossRef]
62. Jiao, M.; Xue, H.; Yan, J.; Zheng, Z.; Wang, J.; Zhao, C.; Zhang, L.; Zhou, W. Tree abundance, diversity and their driving and indicative factors in Beijing's residential areas. Tree abundance, diversity and their driving and indicative factors in Beijing's residential areas. *Ecol. Indic.* **2021**, *125*, 107462. [CrossRef]
63. Yin, L.; Wang, C.; Han, W. Food source characteristics and diversity of birds based on feeding behavior in residential areas of Beijing. *Biodivers. Sci.* **2022**. [CrossRef]
64. Li, X.; Jia, B.; Zhang, W.; Ma, J.; Liu, X. Woody plant diversity spatial patterns and the effects of urbanization in Beijing, China. *Urban For. Urban Green.* **2020**, *56*, 126873. [CrossRef]
65. Sachs, A.L.; Boag, A.E.; Troy, A. Valuing urban trees: A hedonic investigation into tree canopy influence on property values across environmental and social contexts in Baltimore, Maryland. *Urban For. Urban Green.* **2023**, *80*, 127829. [CrossRef]
66. Chin, M.; Ruppert, W.; Shrestha, N.; Fortin, M. Urban Avian Conservation Planning Using Species Functional Traits and Habitat Suitability Mapping. *Land* **2022**, *11*, 1831. [CrossRef]
67. Amaya-Espinel, J.D.; Hostetler, M.E. The value of small forest fragments and urban tree canopy for Neotropical migrant birds during winter and migration seasons in Latin American countries: A systematic review. *Landsc. Urban Plan.* **2019**, *190*, 103592. [CrossRef]
68. Speak, A.F.; Salbitano, F. The impact of pruning and mortality on urban tree canopy volume. *Urban For. Urban Green.* **2023**, *79*, 127810. [CrossRef]

Disclaimer/Publisher's Note: The statements, opinions and data contained in all publications are solely those of the individual author(s) and contributor(s) and not of MDPI and/or the editor(s). MDPI and/or the editor(s) disclaim responsibility for any injury to people or property resulting from any ideas, methods, instructions or products referred to in the content.

sustainability

MDPI

Article

Prioritisation of Charismatic Animals in Major Conservation Journals Measured by the Altmetric Attention Score

Pavol Prokop [1,2,*], Rudolf Masarovič [1], Sandra Hajdúchová [1], Zuzana Ježová [1], Martina Zvaríková [1] and Peter Fedor [1]

[1] Department of Environmental Ecology and Landscape Management, Faculty of Natural Sciences, Comenius University, Ilkovičova 6, 842 15 Bratislava, Slovakia
[2] Institute of Zoology, Slovak Academy of Sciences, 845 06 Bratislava, Slovakia
* Correspondence: pavol.prokop@savba.sk

Abstract: Large, charismatic animals trigger human emotional responses, which consequently result in taxonomic biases that have been proven in various fields. In our research, we analysed the representation of animals and plants in scientific papers published in three major conservation journals (*Conservation Biology*, *Journal of Applied Ecology* and *Conservation Letters*) between 2011 and 2020. Furthermore, we examined the Altmetric Attention Score (AAS) and each paper's total number of citations focused exclusively on a single taxon (59% of all papers). Mammals were represented on journal cover pages significantly more frequently than other taxa, while reptiles, amphibians and fish were underrepresented. The total number of published papers and the AAS favoured mammals significantly, while reptiles, plants and amphibians received the lowest AAS. The AAS of mammals was positively influenced by the body mass and appeal score. Scientific citations showed a slight correlation with the AAS. Papers about mammals, invertebrates and amphibians received the most citations, followed by plants, fish, birds and reptiles. These results showed that there are taxonomic biases that favour large mammals over other taxa, both among scientists as well as the public. Therefore, publication policy should be changed in order to support the shift of scientists and, subsequently, public interest itself toward neglected taxa.

Keywords: body size; charisma; conservation biology; mammals; plant blindness

Citation: Prokop, P.; Masarovič, R.; Hajdúchová, S.; Ježová, Z.; Zvaríková, M.; Fedor, P. Prioritisation of Charismatic Animals in Major Conservation Journals Measured by the Altmetric Attention Score. *Sustainability* **2022**, *14*, 17029. https://doi.org/10.3390/su142417029

Academic Editor: Vasilios Liordos

Received: 24 November 2022
Accepted: 15 December 2022
Published: 19 December 2022

Publisher's Note: MDPI stays neutral with regard to jurisdictional claims in published maps and institutional affiliations.

Copyright: © 2022 by the authors. Licensee MDPI, Basel, Switzerland. This article is an open access article distributed under the terms and conditions of the Creative Commons Attribution (CC BY) license (https://creativecommons.org/licenses/by/4.0/).

1. Introduction

Scientists worldwide are traditionally evaluated by conventional metrics, including the journal impact factor, the total number of citations covered by WoS or Scopus or the Hirsch index. However, internet use rapidly increases and reaches near-universal access, allowing for open communication between scientists and laypeople [1]. Traditional scientometrics is supplied by the Altmetric Attention Score (AAS), which investigates the impact of various research activities on social media [2,3] and realistically reflects what the public wants to know [4]. The inconsistency between the public and scientists [5] is beautifully illustrated by moderate or non-significant relationships between conventional citations registered in scientific databases and the AAS [4,6–9]. Therefore, the AAS reflects the public's curiosity, ideas and knowledge about a particular research topic [10].

Environmental degradation and biodiversity loss capture public attention [11,12]. This is not surprising given that animal populations have experienced a decline of 69% since 1970 [13], which simultaneously affects the dispersal of plants' seeds by animals and reduces the capacity of plants to track climate change by 60% globally [14]. It is therefore crucial to better understand human attitudes toward living organisms to improve the effectiveness of nature protection campaigns using the acceptability and likeability of living things [15–17].

Human emotions toward animals strongly influence their preferences and attitudes toward them (reviewed by [16,18]). The most preferred species are phylogenetically closer to humans [19–22], colourful [23–25] and large [24,26–30]. Preferred animals also receive the greatest willingness to support their conservation by people (e.g., [25,31–35]).

Human preferences have non-trivial behavioural consequences; for instance, US conservation and nature magazines predominantly depict large-bodied mammals and birds on their covers, while invertebrates, fish, amphibians, reptiles or plants remain underrepresented (Clucas et al., 2008) [36]. In addition, large-bodied species are more frequently reared in zoos [24,37–39] and receive more donations than small, non-charismatic species [40–42]. Finally, plants are special because they receive lower conservation support than animals [43,44]. Although researchers are well informed about taxonomic biases, it is unclear whether these trends are pervasive or whether something has been changed.

In this study, we investigated the representation of living organisms in scientific articles published by major conservation journals and their preferences by laypeople, as estimated by the AAS. First, we hypothesised that mammals are depicted on the cover pages of the three journals more than other taxa. Second, we hypothesised that scientific papers about vertebrates are more prevalent and receive a higher AAS than those about invertebrates or plants. Third, mammals receive the greatest attention in terms of the total number of published articles and the AAS. Fourth, we hypothesised that the AAS of mammals is positively correlated with their body mass as an index of charisma (cf. [27]) and overall appeal score [45]. We do not predict that the total number of scientific citations correlates with the specific taxon, the body mass or the appeal score of mammals.

2. Materials and Methods

2.1. *Journal Selection*

We analysed all articles published between 2011 and 2020 from three major conservation journals: *Conservation Biology*, *Journal of Applied Ecology* and *Conservation Letters*. These journals were chosen because all of them are among the top quartile in Biodiversity Conservation according to the Journal Citation Reports (JCR) category (Thomson Reuters) in 2021. Furthermore, all these journals are published by the same publisher (John Wiley & Sons), preventing possible differences in AAS calculations among publishers.

2.2. *Cover Pages*

We analysed the taxa depicted on the cover pages in each issue. Taxa were categorised as mammals, birds, reptiles, amphibians, fish, invertebrates, plants and "others". In this case, we created a specific category for "humans" because we did not intend to mix humans with other mammals, since our focus was solely on non-humans.

2.2.1. AAS and the Total Number of Citations

Both the AAS and the total number of citations were obtained from journal web pages. We omitted calculations of articles that were not exclusively focused on one of the investigated taxa (mammals, birds, reptiles, amphibians, fish, invertebrates, plants).

2.2.2. Measuring of Species' Body Mass

The body mass of each species was calculated using data from Jones et al. [46]. The data were logarithmically transformed, following the recommendations of Smith et al. [47]. We found a match of $n = 356$ species for the investigated mammals.

2.2.3. Measuring of Species Appeal

The appeal scores for the mammal species were taken from the available data for 4320 species of mammals [45]. These scores reflect participants' preferences for each species in the context of conservation. We found a match of $n = 238$ for the investigated mammals. Higher scores mean a greater appeal (range = 0.77 to 5.01).

2.3. Statistical Analyses

Differences in the frequency of the appearance of cover pages of each taxon between the three journals were calculated with the Generalised Linear Mixed Model (GLMM) with a Poisson distribution, where the total number of occurrences was defined as a dependent variable, and journal identity was defined as a random effect. Finally, the AAS and citations were analysed with the GLMM with a Poisson distribution, where the journal identity and the matter of whether the paper was open-access or not were defined as random effects. Open-access papers have more citations than non-open-access papers in ecology [48]; thus, it was necessary to control our analysis with this variable. Note that all articles published in Conservation Letters are open-access; therefore, we did not compare the AAS or the total number of citations between open-access and non-open-access papers. The taxon was always defined as a fixed factor. Post hoc tests were performed with contrast analysis. All statistical analyses were performed in SPSS ver. 26.

3. Results

3.1. Journals' Cover Pages

There were significant differences in the total number of cover pages representing various taxa (GLMM, $F(8,18) = 9.58$, $p < 0.001$). Mammals received significantly greater attention compared with all other taxa (contrast analyses, all $p < 0.001$), followed by birds (comparison with all other taxa, all $p < 0.001$) (Figure 1). Reptiles, amphibians, plants and fish were significantly the least frequently depicted on journal cover pages.

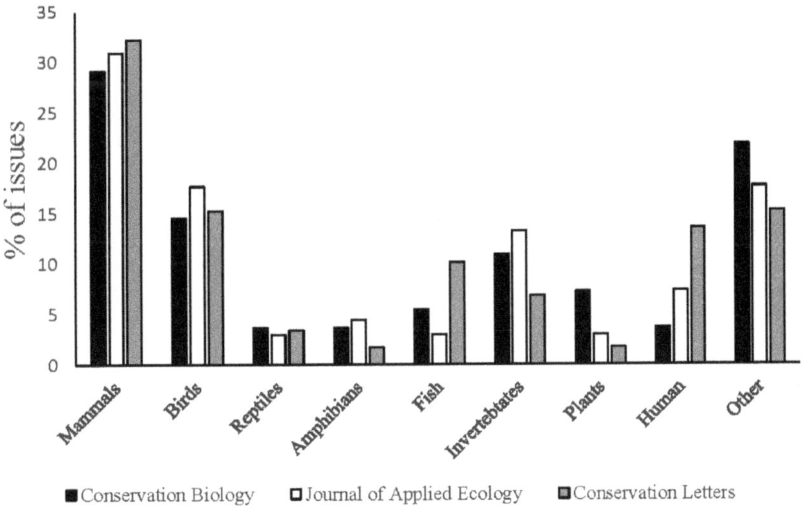

Figure 1. Occurrence of the investigated taxa on the cover pages in the three journals between 2011 and 2020.

3.2. The Influence of the Major Organisms Group on Altmetrics and Citations

Among all 3647 papers that were analysed, 2152 (59%) focused on one taxon (e.g., vertebrates). There was a small but significant correlation between the altmetrics and the total number of citations (Spearman $r = 0.21$, $p < 0.001$).

There were apparent differences in the distribution of papers among the three groups of organisms (plants, invertebrates and vertebrates). The articles on vertebrates ($n = 1351$, 62.8%) were the most frequent, followed by the articles on plants ($n = 410$, 19.1%) and invertebrates ($n = 390$, 18.1%). Additionally, there were significant differences in the AAS between the three groups of organisms (GLMM, $F(2,2145) = 2038.14$, $p < 0.001$, Figure 2).

Pair-wise analysis of contrasts showed that vertebrates received the highest AAS, followed by invertebrates and plants.

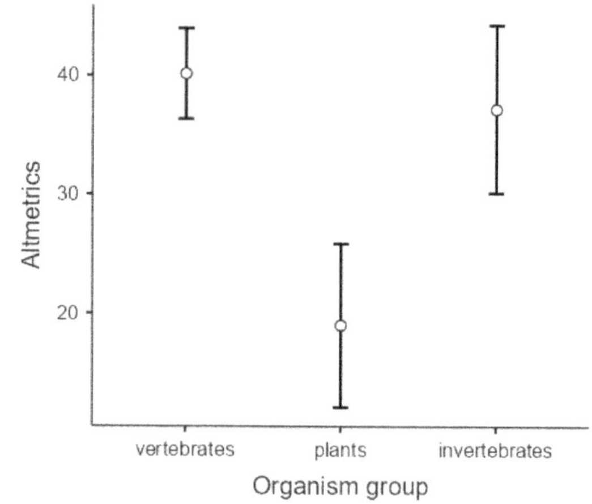

Figure 2. Differences in altmetrics (AAS) among the major taxa (means ± 95% CI).

Concerning the total number of citations, invertebrates received significantly more citations than vertebrates or plants, while there was no difference between the latter two (GLMM, $F(2,2146) = 40.2$, $p < 0.001$, Figure 3).

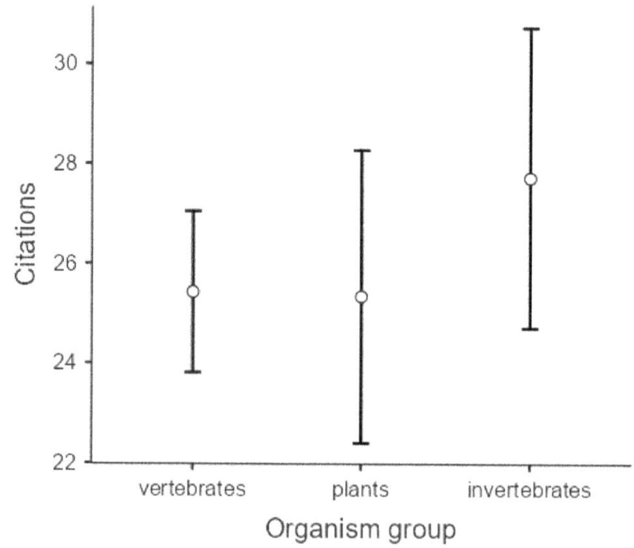

Figure 3. Differences in the total number of citations among the major taxa (means ± 95% CI).

3.3. Differences in the Class Level

Subsequent analyses focus on differences between vertebrate classes and invertebrates and plants. Most papers were published about mammals, followed by plants, birds and invertebrates. Fish, amphibians and reptiles received the lowest attention (Table 1).

Table 1. The frequency of papers published on each taxon.

	n	%
mammals	527	25.5
plants	408	19.7
birds	402	19.4
invertebrates	389	18.8
fish	218	10.5
amphibians	76	3.7
reptiles	49	2.4

There were significant differences in the AAS between the three groups of organisms (GLMM, F(6,2059) = 1196.75, $p < 0.001$). Pairwise contrast analysis showed that vertebrates received the highest AAS, followed by invertebrates and plants. Mammals received the highest AAS compared to other groups of organisms (all $p < 0.001$, Table 2). Mammals were followed by invertebrates, fish and birds, while reptiles, plants and amphibians received the lowest AAS.

Table 2. Descriptive statistics for the AAS for all taxa (means and ±95% CI). Differences in the AAS between all taxonomic groups are significant at $p < 0.01$ and less.

	Mammals	**Invertebrates**	**Fish**	**Birds**	**Reptiles**	**Plants**	**Amphibians**
Mean	52	42.4	37.4	38.8	31.3	23.9	22.9
Lower CI	39.9	30.3	25.3	26.7	19.2	11.9	10.8
Upper CI	64	54.5	49.3	50.9	43.4	36	35

Regarding the total number of citations, the differences between the groups were significant (GLMM, F(6,2060) = 94.02, $p < 0.001$). Mammals, invertebrates and amphibians received the most citations, followed by plants, fish, birds and reptiles (Table 3).

Table 3. Descriptive statistics for the total number of citations for all taxa (means and ±95% CI). Different letters denote significant differences based on contrast analysis. All differences are significant at $p < 0.001$.

	Mammals ^{a}	**Invertebrates** ^{a}	**Fish** ^{b}	**Birds** ^{c}	**Reptiles** ^{d}	**Plants** ^{e}	**Amphibians** ^{a}
Mean	28.6	27.9	23.9	22.2	19.7	25.7	27.5
Lower CI	27.4	26.7	22.7	21.6	18.1	24.5	25.9
Upper CI	29.7	29.1	25.2	23.3	21.3	26.8	29

3.4. *Body Mass and Appeal Score as Predictors of the AAS?*

Because mammals received the greatest attention in terms of the total number of papers and altmetrics scores, we proceeded with calculations of the relationships between the body mass of mammals and their altmetrics and citation scores. Mammalian body mass was a significant positive predictor of the AAS (GLMM, F(1,354) = 446.82, coefficient = 4.6, $p < 0.001$). Because appeal scores were found for a reduced number of mammals ($n = 238$), we repeated the GLMM with the inclusion of an appeal score together with mammalian body mass in the AAS. Both the appeal score and body mass significantly and positively influenced the AAS (GLMM, F(1,125) = 124.95 and 36.6, coefficient = 4.22 and 2.5, both $p < 0.001$). The appeal scores were correlated with mammalian body mass (Spearman r = 0.7, $p < 0.001$, $n = 238$). To address possible problems with multicollinearity, we repeated the GLMM without mammalian body mass. The influence of the appeal score on the AAS remained significant (F(1,236) = 461.05, coefficient = 5.58, $p < 0.001$).

Mammalian body mass negatively and significantly influenced the total number of citations (GLMM, F(1,354) = 14.7, coefficient = −0.71, $p < 0.001$). The repeated GLMM with the inclusion of the appeal score presented the appeal score positively and the mammalian body mass negatively while influencing the total number of citations at the same time ($F(1,235) = 175.8$ and 125.6, coefficient = 3.65 and −4.04, both $p < 0.001$). Citation scores did not correlate with mammalian body mass (Spearman $r = 0.04$, $p = 0.54$, $n = 238$). To address possible problems with multicollinearity, we repeated the GLMM without mammalian body mass. The influence of the appeal score on the total number of citations remained significant ($F(1,236) = 52.02$, coefficient = 1.42, $p < 0.001$).

4. Discussion

This study investigated publication biases toward charismatic species in three major conservation journals on cover pages and published scientific papers during the past ten years. To the best of our knowledge, this is the first study that has attempted to associate the charisma of the animal, which was estimated by the body mass and appeal score, with the AAS. Our results support the idea that taxonomic bias in conservation journals toward large, charismatic animals (predominantly mammals) exists. Furthermore, charismatic animals received greater attention from the public, at least according to the calculations obtained from the AAS.

Our first hypothesis dealt with the presentation of mammals on the covers of three conservation journals. This hypothesis was supported because mammals significantly exceed all other taxa in these journals. By investigating the covers of ten representative US conservation and nature magazines, Clucas et al. (2008) [36] showed that mammals were used more frequently than other taxa, followed by birds. In their study, reptiles, amphibians, fish and plants were underrepresented to a similar extent as in our current research.

Many reptiles are traditionally considered disgusting and dangerous [49,50], while amphibians are viewed as slimy (and therefore disgusting) [51,52], and fish are of little interest to people [53]. People are generally less interested in plants than in animals [43,44,54,55], and photographs of plants in textbooks are less numerous and less diverse than images of animals [55,56]. In general, these traits seem to be responsible for the biases toward more charismatic organisms. Unfortunately, many plant species, particularly Cactaceae, Asparagaceae, Crassulaceae, Orchidaceae and Bromeliaceae, are endangered due to land use changes and illegal trading [57]. However, considering journal covers, editorial boards primarily contribute to taxonomic biases, because the chosen covers are selected by the editors despite readers' desire. The editors' specialisation may further influence the selection of covers. Moreover, it is expected that mammals are more represented on the covers since they are also more represented in the papers published in each journal.

Our second hypothesis suggested that scientific papers on vertebrates are more prevalent and receive a higher AAS than invertebrates or plants. This hypothesis received partial support because vertebrates received a higher AAS than plants, although there were no differences in the mean AAS scores between vertebrates and invertebrates. As discussed above, these results are not surprising in the case of plants; however, in the case of invertebrates, phylogenetically distant and less attractive animals receive less empathy and conservation support than vertebrates [21,22,25]. Interestingly, certain flagship species, such as colourful butterflies, dragonflies or insects that provide ecological services, are perceived positively by people [58,59]. Therefore, a deeper investigation of preferred invertebrates, as measured by the AAS, is a challenge for future research.

Our final hypotheses dealt with the superiority of mammals, as measured by the total number of published papers, with a relatively higher AAS compared to other taxa (Hypothesis 3) and with a positive association between mammalian body mass, appeal score and AAS. These hypotheses received statistical support. Since the AAS favours large mammals, it can be assumed that people are more likely to talk about larger animals on blogs and share information about them through social networks (such as Twitter and Facebook) rather than small-bodied species. These results correspond with research showing that people prefer larger-bodied animals, particularly exotic terrestrial mammals [27,41,42,60]. Alternatively, some exotic mammals are currently studied and published in conservation journals (e.g., *Sus scrofa*, *Callithrix jacchus*, domestic cats and dogs on islands); thus, our results could be influenced by a temporal bias rather than by charisma. This idea requires deeper attention.

It seems that taxonomic biases, which favour the keeping of large mammals in zoos [37–39], exist in conservation science as well as in the exchange of information through social networks by the public. Finally, our findings are derived from the additional influence of the appeal score on AAS and the positive correlation between mammalian body mass and appeal scores. In other words, large-bodied mammals are perceived as more appealing using people's interest, as measured with AAS.

We did not expect significant shifts in the number of scientific citations among the taxa examined. However, there was a considerable bias favouring mammals, invertebrates and amphibians over plants, fish, birds and reptiles. The interest in invertebrates would be acceptable, since invertebrates comprise 97% of all animals on our planet [61]. However, plants, which exceed 300,000 extant species [62], and fish, with more than 30,000 species (which is more than all other vertebrates combined) [63], are heavily underestimated by scientists. Interestingly, even though mammalian body mass showed a negative influence, appeal positively influenced the total number of citations. Citations accumulate more slowly than AAS [64], and AAS is slow for smaller mammals (this study). This is one possible explanation for why smaller mammals receive more citations relative to AAS; it also supports the low consistency between the public and scientists [5]. A small correlation between AAS and the total number of citations suggests that public opinion does not necessarily reflect scientists' opinions [4,6–9].

A positive influence of the appeal score on the total number of citations could reflect the intrinsic interest in particular animals among scientists, which is corroborated by the greater overall number of publications favouring mammals, as shown in the present study. However, more research is necessary in order to investigate whether publications and citations of papers about each species are influenced by conservation needs or scientists' attraction to charismatic species. Additional predictors that need to be considered are life history strategies. Common, abundant species with fast reproduction (e.g., fish, insects, plants) are less vulnerable to extinction; thus, research about them has lower changes in terms of being published in conservation journals. On the other hand, slowly reproducing animals (certain mammals, birds, etc.) have little capacity to recover [65], and their research requires attention by conservation journals.

Limitations

The main limitation of our research is that we were able to investigate taxonomic biases for only 59% of all published papers. The remaining papers focused on other topics and/or various unexplored taxa. Still, our analyses are based on more than 2000 papers published in three major, influential conservation journals which have a non-trivial impact on creating public opinion. Second, we did not control the geographical distribution of the studied species, the author(s) affiliations and the richness of the country where the research was carried out. Tahamtan et al. (2016) [66], for example, showed that US papers received more citations than papers from other countries, which may reflect the reputation of the country or the research team. Although we acknowledge this shortcoming of our methodological approach, we do not believe that taxonomic bias on journal cover pages or the influences of mammalian body mass and appeal scores on AAS could be confounded by these factors. Finally, our analysis is stratified on relatively recent papers published in the selected conservation journals, which could influence the representation of certain taxa. For instance, during the 1990s–2000s, an invertebrate golden mussel (*Limnoperna fortunei*) was frequently represented in conservation journals as a "current" threat to biodiversity. To address these shortcomings, further research should involve more conservation journals (e.g., *Biological Conservation*, *Biodiversity and Conservation*, *Environmental Conservation*), taking articles published earlier into account.

5. Conclusions

In conclusion, our research showed that publication biases, measured in terms of the total number of papers and citations and the public interest calculated with AAS, favour certain taxonomic groups of animals (particularly large-bodied mammals) over others. Surprisingly, these trends, which had been previously investigated by different methods, persist, even in professional conservation journals. Publication policies that favour neglected taxa are therefore necessary to improve current trends shaped by animal charisma. For instance, the editors should support special issues regarding the conservation of plants, fish or reptiles, which would contribute to a more balanced situation without any apparent focus on just a few taxonomic groups. Furthermore, improving the publication bias may influence communication between scientists and laypeople [1] and public opinion about living organisms.

Author Contributions: Conceptualisation, P.P.; Methodology, P.P., R.M. and P.F.; Software, P.P.; Validation, P.P., R.M., P.F. and M.Z.; Formal analysis, R.M.; Investigation, P.P., R.M., S.H., Z.J. and M.Z.; Resources, P.F.; Data curation, P.P. and R.M.; Writing—original draft preparation, P.P.; Writing—review & editing, R.M., S.H., Z.J., M.Z. and P.F.; Visualisation, P.P.; Supervision, P.P. All authors have read and agreed to the published version of the manuscript.

Funding: This work was supported by the Scientific Grant Agency of the Ministry of Education of the Slovak Republic, Grant No. VEGA 1/0286/20 and 1/0007/21. P.P. was also supported by the Operation Program of Integrated Infrastructure for the project UpScale of Comenius University Capacities and Competence in Research, Development and Innovation, ITMS2014+: 313021BUZ3, co-financed by the European Regional Development Fund.

Institutional Review Board Statement: Not applicable.

Informed Consent Statement: Not applicable.

Data Availability Statement: Raw data and visual material are available upon request.

Acknowledgments: We thank the three anonymous referees for their constructive comments.

Conflicts of Interest: The authors declare that they have no conflict of interest.

References

1. Cho, J. Altmetrics analysis of highly cited academic papers in the field of library and in formation science. *Scientometrics* **2021**, *126*, 7623–7635. [CrossRef]
2. Costas, R.; Zahedi, Z.; Wouters, P. Do "altmetrics" correlate with citations? Extensive comparison of altmetric indicators with citations from a multidisciplinary perspective. *J. Assoc. Inf. Sci. Technol.* **2015**, *66*, 2003–2019. [CrossRef]
3. Erdt, M.; Nagarajan, A.; Sin, S.-C.J.; Then, Y.-L. Altmetrics: An Analysis of the State-Of-The-Art in Measuring Research Impact on Social Media. *Scientometrics* **2016**, *109*, 1117–1166. [CrossRef]
4. Kim, Y.; Kim, J.E.; Kim, Y.H.; Yoon, D.Y.; Kim, Y.J.; Bae, J.S. Social attention and scientific articles on stroke: Altmetric analysis of top-50 articles. *Clin. Neurol. Neurosurg.* **2019**, *183*, 105386. [CrossRef] [PubMed]
5. Araujo, R.; Sorensen, A.A.; Konkiel, S.; Bloem, B.R. Top Altmetric scores in the Parkinson's disease literature. *J. Park. Dis.* **2017**, *7*, 81–87. [CrossRef]
6. Kolahi, J.; Khazaei, S. Altmetric: Top 50 dental articles in 2014. *Br. Dent. J.* **2016**, *220*, 569–574. [CrossRef]
7. Ezema, I.J.; Ugwu, C.I. Correlating research impact of library and information science journals using citation counts and altmetrics attention. *Inf. Discov. Deliv.* **2019**, *47*, 143–153. [CrossRef]
8. Baek, S.; Yoon, D.Y.; Lim, K.J.; Hong, J.H.; Moon, J.Y.; Seo, Y.L.; Yun, E.J. Top-cited articles versus top Altmetric articles in nuclear medicine: A comparative bibliometric analysis. *Acta Radiol.* **2020**, *61*, 1343–1349. [CrossRef]
9. Nabavi, M. An analysis of journalism articles achieving high Altmetric attention scores. *Learn. Publ.* **2022**, *35*, 617–624. [CrossRef]
10. Chen, M.; Wang, L. An Altmetrics and citation analysis of selected predatory journals in library and information science field. *J. Acad. Libr.* **2022**, *48*, 102618. [CrossRef]
11. Grodzińska-Jurczak, M.; Cent, J. Can public participation increase nature conservation effectiveness? *Innovation* **2011**, *24*, 371–378. [CrossRef]
12. Donkersley, P.; Ashton, L.; Lamarre, G.P.; Segar, S. Global insect decline is the result of wilful political failure: A battle plan for entomology. *Ecol. Evol.* **2022**, *12*, e9417. [CrossRef]
13. WWF. *Living Planet Report 2022—Building a Nature-Positive Society*; Almond, R.E.A., Grooten, M., Juffe Bignoli, D., Petersen, T., Eds.; WWF: Gland, Switzerland, 2022; pp. 1–118.

14. Fricke, E.C.; Ordonez, A.; Rogers, H.S.; Svenning, J.C. The effects of defaunation on plants' capacity to track climate change. *Science* **2022**, *375*, 210–214. [CrossRef]
15. Martín-López, B.; Montes, C.; Benayas, J. The non-economic motives behind the willingness to pay for biodiversity conservation. *Biol. Conserv.* **2007**, *139*, 67–82. [CrossRef]
16. Prokop, P.; Randler, C. Biological predispositions and individual differences in human attitudes toward animals. In *Ethnozoology: Animals in Our Lives*; Alves, R.R.N., Albuquerque, U.P., Eds.; Academic Press: London, UK, 2018; pp. 447–466.
17. Lévesque, A.; Gagné, L.; Dupras, J. Expressing citizen preferences on endangered wildlife for building socially appealing species recovery policies: A stated preference experiment in Quebec, Canada. *J. Nat. Conserv.* **2022**, *69*, 126255. [CrossRef]
18. Castillo-Huitrón, N.M.; Naranjo, E.J.; Santos-Fita, D.; Estrada-Lugo, E. The importance of human emotions for wildlife conservation. *Front. Psychol.* **2020**, *11*, 1277. [CrossRef]
19. Plous, S. The role of animals in human society. *J. Soc. Issue* **1993**, *49*, 11–52. [CrossRef]
20. Serpell, J.A. Factors influencing human attitudes to animals and their welfare. *Anim. Welf.* **2004**, *13*, S145–S152.
21. Miralles, A.; Raymond, M.; Lecointre, G. Empathy and compassion toward other species decrease with evolutionary divergence time. *Sci. Rep.* **2019**, *9*, 1955. [CrossRef]
22. Prokop, P.; Zvaríková, M.; Zvarík, M.; Pazda, A.; Fedor, P. The effect of animal bipedal posture on perceived cuteness, fear, and willingness to protect them. *Front. Ecol. Evol.* **2021**, *9*, 681241. [CrossRef]
23. Marešová, J.; Landová, E.; Frynta, D. What makes some species of milk snakes more attractive to humans than others? *Theor. Biosci.* **2009**, *128*, 227–235. [CrossRef] [PubMed]
24. Frynta, D.; Lišková, S.; Bültmann, S.; Burda, H. Being attractive brings advantages: The case of parrot species in captivity. *PLoS ONE* **2010**, *5*, e12568. [CrossRef] [PubMed]
25. Prokop, P.; Fančovičová, J. Does colour matter? The influence of animal warning colouration in human emotions and willingness to protect them. *Anim. Conserv.* **2013**, *16*, 458–466. [CrossRef]
26. Landová, E.; Poláková, P.; Rádlová, S.; Janovcová, M.; Bobek, M.; Frynta, D. Beauty ranking of mammalian species kept in the Prague Zoo: Does beauty of animals increase the respondents' willingness to protect them? *Sci. Nat.* **2018**, *105*, 69. [CrossRef] [PubMed]
27. Berti, E.; Monsarrat, S.; Munk, M.; Jarvie, S.; Svenning, J.C. Body size is a good proxy for vertebrate charisma. *Biol. Conserv.* **2020**, *251*, 108790. [CrossRef]
28. Frynta, D.; Marešová, E.; Landová, E.; Lišková, S.; Šimková, O.; Tichá, I.; Zelenková, M.; Fuchs, R. Are animals in zoos rather conspicuous than endangered? In *Endangered Species—New Research*; Columbus, A.M., Kuznetsov, L., Eds.; Nova Science Publishers, Inc.: New York, NY, USA, 2009; pp. 299–341.
29. Lišková, S.; Frynta, D. What determines bird beauty in human eyes? *Anthrozoös* **2013**, *26*, 27–41. [CrossRef]
30. Ward, P.I.; Mosberger, N.; Kistler, C.; Fischer, O. The relationship between popularity and body size in zoo animals. *Conserv. Biol.* **1998**, *12*, 1408–1411. [CrossRef]
31. Samples, K.C.; Dixon, J.A.; Gowen, M.M. Information disclosure and endangered species valuation. *Land Econ.* **1986**, *62*, 306–312. [CrossRef]
32. Small, E. The new Noah's Ark: Beautiful and useful species only. Part 1. Biodiversity conservation issues and priorities. *Biodiversity* **2011**, *12*, 232–247. [CrossRef]
33. Small, E. The new Noah's Ark: Beautiful and useful species only. Part 2. The chosen species. *Biodiversity* **2012**, *13*, 37–53. [CrossRef]
34. Prokop, P.; Fančovičová, J. Animals in dangerous postures enhance learning, but decrease willingness to protect animals. *Eurasia J. Math. Sci. Technol. Educ.* **2017**, *13*, 6069–6077. [CrossRef]
35. Knight, A.J. "Bats, snakes and spiders, Oh my!" How aesthetic and negativistic attitudes, and other concepts predict support for species protection. *J. Environ. Psychol.* **2008**, *28*, 94–103. [CrossRef]
36. Clucas, B.; McHugh, K.; Caro, T. Flagship species on covers of US conservation and nature magazines. *Biodivers. Conserv.* **2008**, *17*, 1517–1528. [CrossRef]
37. Fa, J.E.; Gusset, M.; Flesness, N.; Conde, D.A. Zoos have yet to unveil their full conservation potential. *Anim. Conserv.* **2014**, *17*, 97–100. [CrossRef]
38. Balmford, A.; Leader-Williams, N.; Green, M.J.B. Parks or arks: Where to conserve threatened mammals? *Biodivers. Conserv.* **1995**, *4*, 595–607. [CrossRef]
39. Frynta, D.; Šimková, O.; Lišková, S.; Landová, E. Mammalian collection on Noah's ark: The effects of beauty, brain and body size. *PLoS ONE* **2013**, *8*, e63110. [CrossRef]
40. Walpole, M.J.; Leader-Williams, N. Tourism and flagship species in conservation. *Biodivers. Conserv.* **2002**, *11*, 543–547. [CrossRef]
41. Colléony, A.; Clayton, S.; Couvet, D.; Saint Jalme, M.; Prévot, A.C. Human preferences for species conservation: Animal charisma trumps endangered status. *Biol. Conserv.* **2017**, *206*, 263–269. [CrossRef]
42. Fančovičová, J.; Prokop, P.; Repáková, R.; Medina-Jerez, W. Factors Influencing the Sponsoring of Animals in Slovak Zoos. *Animals* **2021**, *12*, 21. [CrossRef]
43. Balding, M.; Williams, K.J. Plant blindness and the implications for plant conservation. *Cons. Biol.* **2016**, *30*, 1192–1199. [CrossRef]
44. Fančovičová, J.; Prokop, P.; Kubíčková, M. The Effect of Aposematic Signals of Plants on Students' Perception and Willingness to Protect Them. *Sustainability* **2022**, *14*, 9121. [CrossRef]

45. Macdonald, E.A.; Hinks, A.; Weiss, D.J.; Dickman, A.; Burnham, D.; Sandom, C.J.; Malhi, Y.; Macdonald, D.W.; Macdonald, D.W. Identifying ambassador species for conservation marketing. *Glob. Ecol. Conserv.* **2017**, *12*, 204–214. [CrossRef]
46. Jones, K.E.; Bielby, J.; Cardillo, M.; Fritz, S.A.; O'Dell, J.; Orme, C.D.L.; Safi, K.; Sechrest, W.; Boakes, E.H.; Carbone, C.; et al. PanTHERIA: A species-level database of life history, ecology, and geography of extant and recently extinct mammals: Ecological Archives E090-184. *Ecology* **2009**, *90*, 2648. [CrossRef]
47. Smith, R.J.; Veríssimo, D.; Isaac, N.J.; Jones, K.E. Identifying Cinderella species: Uncovering mammals with conservation flagship appeal. *Conserv. Lett.* **2012**, *5*, 205–212. [CrossRef]
48. Tang, M.; Bever, J.D.; Yu, F.H. Open access increases citations of papers in ecology. *Ecosphere* **2017**, *8*, e01887. [CrossRef]
49. Prokop, P.; Özel, M.; Uşak, M. Cross-cultural comparison of student attitudes toward snakes. *Soc. Anim.* **2009**, *17*, 224–240.
50. Rádlová, S.; Janovcová, M.; Sedláčková, K.; Polák, J.; Nácar, D.; Peléšková, Š.; Frynta, D.; Landová, E. Snakes represent emotionally salient stimuli that may evoke both fear and disgust. *Front. Psychol.* **2019**, *10*, 1085. [CrossRef]
51. Prokop, P.; Fančovičová, J. Tolerance of amphibians in Slovakian people: A comparison of pond owners and non-owners. *Anthrozoos* **2012**, *25*, 277–288. [CrossRef]
52. Frynta, D.; Peléšková, Š.; Rádlová, S.; Janovcová, M.; Landová, E. Human evaluation of amphibian species: A comparison of disgust and beauty. *Sci. Nat.* **2019**, *106*, 41. [CrossRef]
53. Carr, N. An analysis of zoo visitors' favourite and least favourite animals. *Tour. Manag. Perspect.* **2016**, *20*, 70–76. [CrossRef]
54. Kinchin, I.M. Investigating secondary-school girls' preferences for animals or plants: A simple 'head-to-head' comparison using two unfamiliar organisms. *J. Biol. Educ.* **1999**, *33*, 95–99.
55. Balas, B.; Momsen, J.L. Attention "blinks" differently for plants and animals. *CBE Life Sci. Educ.* **2014**, *13*, 437–443. [CrossRef]
56. Link-Perez, M.A.; Dollo, V.H.; Weber, K.M.; Schussler, E.E. What's in a Name: Differential Labelling of Plant and Animal Photo-graphs in Two Nationally Syndicated Elementary Science Textbook Series. *Int. J. Sci. Educ.* **2010**, *32*, 1227–1242. [CrossRef]
57. Flores-Palacios, A.; Valencia-Díaz, S. Local illegal trade reveals unknown diversity and involves a high species richness of wild vascular epiphytes. *Biol. Cons.* **2007**, *136*, 372–387. [CrossRef]
58. Oberhauser, K.; Guiney, M. Insects as flagship conservation species. *Terr. Arthropod. Rev.* **2009**, *1*, 111–123. [CrossRef]
59. Schlegel, J.; Rupf, R. Attitudes towards potential animal flagship species in nature conservation: A survey among students of different educational institutions. *J. Nat. Conserv.* **2010**, *18*, 278–290. [CrossRef]
60. Albert, C.; Luque, G.M.; Courchamp, F. The twenty most charismatic species. *PLoS ONE* **2018**, *13*, e0199149. [CrossRef]
61. May, R.M. How many species are there on earth? *Science* **1988**, *241*, 1441–1449. [CrossRef]
62. Christenhusz, M.J.; Byng, J.W. The number of known plants species in the world and its annual increase. *Phytotaxa* **2016**, *261*, 201–217. [CrossRef]
63. Froese, R.; Pauly, D. Fish Base—World Wide Electronic Publication (ver. 6/2022). Available online: https://www.fishbase.se/search.php (accessed on 18 November 2022).
64. Abramo, G.; Cicero, T.; D'Angelo, C.A. A sensitivity analysis of researchers' productivity rankings to the time of citation observation. *J. Informetr.* **2012**, *6*, 192–201. [CrossRef]
65. González-Suárez, M.; Revilla, E. Variability in life-history and ecological traits is a buffer against extinction in mammals. *Ecol. Lett.* **2013**, *16*, 242–251. [CrossRef] [PubMed]
66. Tahantan, I.; Afshar, A.S.; Ahamdzadeh, K. Factors affecting number of citations: A comprehensive review of the literature. *Scientometrics* **2016**, *107*, 1195–1225. [CrossRef]

sustainability

MDPI

Article

Wildlife Knowledge and Attitudes toward Hunting: A Comparative Hunter–Non-Hunter Analysis

Georgia Raftogianni, Vasileios J. Kontsiotis and Vasilios Liordos *

Department of Forest and Natural Environment Sciences, International Hellenic University, P.O. Box 172, 66100 Drama, Greece
* Correspondence: liordos@for.ihu.gr

Abstract: Assessing the public's attitudes toward hunting and knowledge about wildlife is critical for successfully managing and conserving resources. This need is further emphasized by the increase in urbanization, resulting in decreasing participation in outdoor activities, such as hunting. This study aimed at investigating the attitudes toward hunting and the wildlife knowledge of Greek residents and at understanding the variation among hunters and non-hunters. Respondents to on-site, face-to-face surveys (n = 461; hunters, 146; non-hunters, 315) were asked to rate their acceptance of the motives for hunting and of hunting as a management tool and their knowledge about the ecology, biology, and behavior of wildlife. The hunters were highly motivated for hunting and supported it as a management tool. The non-hunters' attitudes were, however, neutral to negative. The hunters had greater knowledge about wildlife species, both game and non-game, than the non-hunters. The more experienced hunters with greater knowledge about wildlife were generally more positive toward hunting. Older, male, non-hunters who have a greater knowledge about wildlife and who consume game meat and have hunters in the family or among their friends were generally more positive toward hunting. The findings suggested that hunting is a controversial social issue. Policies aimed at informing public groups about good hunting practices and at increasing the public's engagement in outdoor activities would reduce such controversies, improve human health and well-being, and reinforce nature and wildlife stewardship and support for biodiversity conservation.

Keywords: hunting motives; wildlife management; experiential knowledge; sociodemographics; northeast Mediterranean

Citation: Raftogianni, G.; Kontsiotis, V.J.; Liordos, V. Wildlife Knowledge and Attitudes toward Hunting: A Comparative Hunter–Non-Hunter Analysis. *Sustainability* **2022**, *14*, 14541. https://doi.org/10.3390/su142114541

Academic Editors: Iain J. Gordon and Alberto Meriggi

Received: 25 July 2022
Accepted: 2 November 2022
Published: 5 November 2022

Publisher's Note: MDPI stays neutral with regard to jurisdictional claims in published maps and institutional affiliations.

Copyright: © 2022 by the authors. Licensee MDPI, Basel, Switzerland. This article is an open access article distributed under the terms and conditions of the Creative Commons Attribution (CC BY) license (https://creativecommons.org/licenses/by/4.0/).

1. Introduction

In modern times, and especially in western societies, hunting is exercised as a recreational activity [1,2]. Recreational hunting, hereafter just hunting, is defined as a pastime without a commercial or subsistence component, carried out voluntarily and involving the active pursuit and killing of wild vertebrate animals other than fish [3,4]. Hunting is also an important socioeconomic activity [5] and management tool, having been used as a means of controlling overabundant wildlife populations; in addition, the proceeds from hunting license fees provide funding for the conservation of wildlife species [6,7]. However, participation in hunting has declined, especially in developed countries, e.g., [8–11]. This decline has been attributed to the increase in the endorsement of more protectionist, mutualism values (seeing wildlife as part of one's social community and deserving of rights like humans) and the weakening of utilitarian, domination values (treating wildlife as a resource to be used for human benefit) due to increased urbanization, educational level, and income [12–14]. Hunting is controversial among the public, with varying degrees of acceptance among the public, both as an activity [15–17] and as a management tool [18,19]. Moreover, hunting is an outdoor activity and those participating in such activities have high experiential knowledge about wildlife and are supportive of wildlife conservation and management issues [20]. In addition, the ever-increasing urbanization has promoted the gradual alienation from nature and wildlife, termed as the "extinction of experience" [21], leading to the deterioration of human health and well-being, collectively described as the 'nature-deficit disorder' [22]. Studying people's attitudes toward hunting and their level of knowledge about wildlife would provide critical information for the assessment of the acceptability of the practice of hunting and its use as a management tool and of people's degree of contact with nature. This would allow for the better management of resources for the conservation of biodiversity and the reversal of the extinction of experience [23,24].

Attitudes may vary from positive to negative and represent "an association, in memory, of an evaluation with an object or activity" (p. 341 in Fazio et al. [25]). Attitudes toward hunting may address the way that people perceive the reasons why someone would hunt, namely the motives for hunting, and also the evaluation of the usefulness of hunting as a management tool. Hunters regard hunting as an activity offering opportunities for excitement, exercise, enjoyment of nature, and learning about wildlife, and one which allows for the reinforcing of relationships with friends and family and the reduction in everyday stress [15–17]. In these studies, the non-hunters had more negative attitudes toward hunting than the hunters. The hunters are supporters of wildlife management, especially when it positively affects their favorite game [18,19]. They also have a long tradition of helping to conserve animal species, especially game, and their habitats in many countries [26,27]. The hunters accepted hunting more than the non-hunters as a useful wildlife management technique in different situations [15].

Knowledge refers to the collection of facts, information, and experience that people acquire, retain, and use through complex cognitive processes, such as belief, perception, communication, association, and reasoning [28]. Hunters have a greater knowledge about biology, ecology, populations, and the conservation status of wildlife species than the general public; their knowledge is similar to that of birdwatchers and members of nature protection organizations [20]. Such knowledge has been found to positively affect people's attitudes toward wildlife conservation and management [29–31].

Among the demographic characteristics, age, gender, level of education, pet ownership, meat consumption, and having hunters in the family or among friends have been proposed as important factors influencing attitudes toward hunting. In general, men, older people, less educated people, and pet owners are more favorable toward hunting when compared with women, younger people, more educated people, and non-pet owners [15,32,33]. The consumption of game meat is positively associated with hunting [34]. Friends and family members who hunt also positively affect attitudes toward hunting [35,36].

The hunting population in Greece follows the international trend, having decreased from 344,000 in 1985 to 230,000 in 2010 [37] and to around 170,000 in 2019 (1.6% of the total population; [38]). Greece is an already highly urbanized country, with its level of urbanization expected to rise from 79% in 2018 to 88% by 2050 [39], a trend that could further negatively affect hunting participation. Along with urbanization, the steep decline in hunting participation after 2010 has also been attributed to the onset of the Greek debt crisis and the consequent decrease in the national gross domestic product (GDP) per capita [40]. The assessment of public attitudes toward hunting and knowledge about wildlife is important if we are to understand whether hunting is controversial and also understand the public's degree of involvement with wildlife and contact with nature in general. In this study, we aimed at: (1) identifying differences in the attitudes toward the motives for hunting and of hunting as a management tool among Greek hunters and non-hunters; (2) assessing the level of factual knowledge about wildlife among Greek hunters and non-hunters; (3) examining how sociodemographic characteristics affect the attitudes of the Greek public toward hunting and the factual knowledge about wildlife; and (4) discussing the implications of our findings for hunting, wildlife management and the extinction of experience, in the light of the decreasing hunting participation and the increasing urbanization.

2. Materials and Methods

2.1. Sampling Protocol

The study was carried out in north Greece, in the districts of Central Macedonia and Eastern Macedonia and Thrace (Figure 1), an area with a population of roughly 2,490,000 people [41]. Data were collected by on-site, face-to-face surveys of residents of north Greece between June and September 2018. A pretest of the survey (n = 30 random people) was conducted to test question clarity and completion time. Cities, towns, and villages were visited in all the districts during open market hours (9.00–15.00 and 17.00–21.00, from Monday to Saturday). Every fifth person passing in front of the researcher was asked to participate by completing a questionnaire [42]. In the cases in which more than five persons had passed while a questionnaire was being completed, the first person encountered upon completion was selected. Hunting clubs within the study area were also visited to ensure the representation of hunters in the sample. It took respondents 40 min on average to orally complete the questionnaire with the assistance of the interviewer.

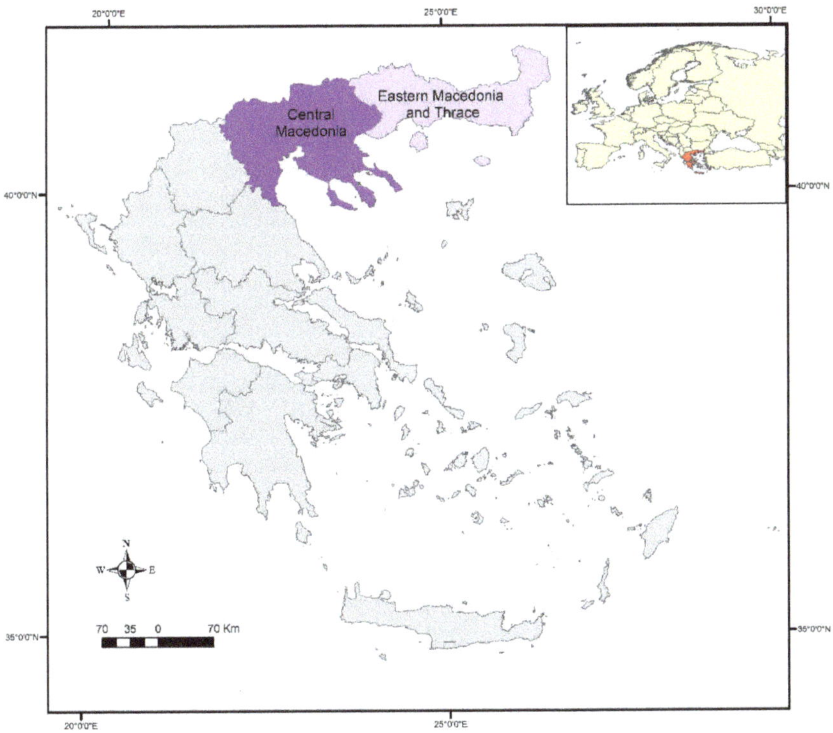

Figure 1. Map showing the region of Greece in which the study was carried out.

2.2. Research Design

The survey participants, classified as hunters, male non-hunters, and female non-hunters, were asked a series of questions about their sociodemographic characteristics, their knowledge about wildlife, and their attitudes toward hunting. The sociodemographic characteristics included gender (female or male), age, educational level (recorded as higher and lower [including elementary and secondary education]), pet ownership (yes or no), consumption of game meat (yes or no), having hunters in the family or among friends (yes or no), and hunting experience (in years; for hunters).

The participants' acceptance of the motives for hunting were assessed with 9 statements, while their attitudes toward hunting as a management tool were assessed with another 9 statements. We ranked the mean responses of each participant for the motive and attitude scales as: (a) negative (mean 1–2.49), (b) neutral (mean 2.50–3.49), and (c) positive (mean 3.50–5). The factual knowledge about wildlife was assessed with 20 statements relating to the ecology, biology, and behavior of wildlife species, 10 of which concerned game species, while 10 concerned non-game species. The survey participants were asked to rate each motive, attitude, and knowledge statement on a 5-point scale (5 = strongly agree, 4 = agree, 3 = neither disagree nor agree, 2 = disagree, 1 = strongly disagree).

2.3. Data Analysis

The hunters were all male, and the genders usually differ in their perceptions and attitudes toward hunting [32]. Therefore, we made comparisons among the hunters, male non-hunters, and female non-hunters. One-way analysis of variance (ANOVA) with pairwise Tamhane post hoc tests to account for heteroscedasticity, applying Bonferroni correction to adjust for multiple testing, was used for comparing the mean responses of the hunters and non-hunters regarding the acceptability of the motives for hunting, the attitudes toward hunting as a management tool, and the factual knowledge about wildlife statements.

Next, we were interested in assessing the effects of sociodemographic characteristics on the acceptability of the motives for hunting, attitudes toward hunting as a management tool, and factual wildlife knowledge. First, we investigated whether motive, attitude, and knowledge statements could adequately describe constructs (factors) [42], using principal component exploratory factor analysis. The following criteria were used for the factor and variable selection: (a) factor eigenvalue ≥ 1, (b) communality of a variable ≥ 0.5, (c) factor loading of a variable ≥ 0.4, and (d) exclusion of a variable when factor loadings exceeded 0.400 in two or more factors [43]. Cronbach's alpha was used to determine whether the statements included in a factor reliably measured the respondent's acceptance, attitude, or knowledge, with a value greater than 0.7 considered acceptable [44]. The relationships between the determined motive, attitude, and knowledge factors and the sociodemographic characteristics were then assessed with multiple linear regression models. Multicollinearity among the independent variables was tested with the variance inflation factor (VIF).

All analyses were performed in SPSS Statistics (version 21.0, IBM Corp., Armonk, NY, USA, 2012). The significance level was set at $\alpha = 0.05$.

3. Results

3.1. Sociodemographics

A total of 461 questionnaires were completed, with 55 refusals, yielding a response rate of 89% (the required size for a confidence level of 95% and a margin of error of 4.56%). Among the survey participants, 146 were hunters and 315 were non-hunters. The study area's population has a 51.3% female/48.7% male gender ratio; the age ratio, after excluding those under 18, is 39.7%/36.3%/34.0% in the age classes of 18–34, 35–54, and 55+ years old, respectively, and the lower/higher educational ratio is 77.5%/22.5% [41]. The non-hunters' gender, (50.5% female/49.5% male), age (36.3%/33.5%/30.2% in the age classes of 18–34, 35–54, and 55+ years old, respectively), and the educational level (72.6%/27.4%) structure was not different to that of the population (gender: $\chi^2 = 0.059$, df = 1, $p = 0.765$, age: $\chi^2 = 4.583$, df = 2, $p = 0.101$; educational level: $\chi^2 = 2.734$, df = 2, $p = 0.083$).

The hunters were all male; their age ratio was 19.2%/53.8%/26.9% in the age classes of 18–34, 35–54, and 55+ years old, respectively, and their lower/higher educational ratio was 80.2%/19.8%.

3.2. Acceptability of Motives for Hunting

The hunters did not accept hunting for trophies or for the game meat (Table 1). However, they were mostly involved in hunting for contacting with nature, for the excitement, for socializing, and as a stress-reducing recreational activity. Both the male and the female non-hunters perceived hunting as a source of pride, reducing stress, offering excitement, and promoting contact with nature and as providing opportunities to socialize and identify with hunters, while they did not accept hunting for trophies, recreation, and game meat. The differences between the hunters and non-hunters were significant with regard to all the statements on the acceptability of the motives for hunting, except for hunting as identity between hunters and both male and female non-hunters, hunting for trophies between hunters and male non-hunters, and hunting for meat between hunters and female non-hunters. The acceptance of hunting for meat was significantly lower in females than males.

Table 1. Comparisons regarding the acceptability of motives for hunting among hunters and non-hunters.

Motive Statements [a] Hunting Is Acceptable Because...	Hunters (n = 146)	Non-Hunters (n = 315) Male (n = 156)	Non-Hunters (n = 315) Female (n = 159)	$F_{2,458}$	Factor Loadings [b] Hunters	Factor Loadings [b] Non-Hunters
It promotes contact with nature.	4.88 ± 0.38 A	3.77 ± 1.13 B	3.45 ± 1.25 B	84.803 ***	0.91	0.72
It is exciting.	4.69 ± 0.67 A	3.75 ± 0.96 B	3.65 ± 0.93 B	68.953 ***	0.93	0.62
It provides identity.	3.13 ± 1.41 A	3.31 ± 1.03 A	3.29 ± 1.10 A	2.314	0.69	0.65
It is an important means of socializing.	4.21 ± 1.09 A	3.45 ± 1.13 B	3.09 ± 1.14 B	32.472 ***	0.88	0.64
It is a source of pride.	3.42 ± 1.38 A	4.35 ± 0.73 B	4.06 ± 0.99 B	49.556 ***	0.80	—
It offers peace and quiet and helps in reducing stress.	4.81 ± 0.63 A	3.92 ± 0.93 B	3.66 ± 0.74 B	89.677 ***	0.93	0.66
It is a recreational activity.	4.77 ± 0.65 A	2.38 ± 1.61 B	2.09 ± 1.45 B	182.185 ***	0.75	0.83
It is done for collecting trophies.	1.83 ± 1.12 A	1.52 ± 0.98 AB	1.34 ± 0.78 B	8.465 ***	0.66	0.72
It provides meat.	2.00 ± 1.05 A	2.78 ± 1.37 B	2.26 ± 1.27 A	19.140 ***	0.69	—

[a] Mean ± SD; 5-point scale: 1 (strongly disagree)–5 (strongly agree). *** $p < 0.001$. [b] Factor loadings were determined by principal component factor analysis for the hunter and non-hunter (males and females combined) groups. Note: one-way ANOVAs were used for comparisons. Means not sharing a common letter (A or B) are significantly different ($p < 0.05$; Tamhane post hoc tests with Bonferroni correction).

Overall, 70.5% of the hunters were positive, and 29.5% were neutral toward the motives for hunting (mean 3.75 ± 0.46 SD), while 27.4%, 66.3%, and 6.3 % of the male non-hunters were positive, neutral, and negative, respectively (3.25 ± 0.60), and 16.9%, 66.7%, and 15.4% of the female non-hunters were positive, neutral, and negative, respectively (2.99 ± 0.59). The differences were significant among all the groups (ANOVA $F_{2,458} = 60.678$, $p < 0.001$; $p < 0.05$, pairwise Tamhane post hoc tests with Bonferroni correction; Figure 2).

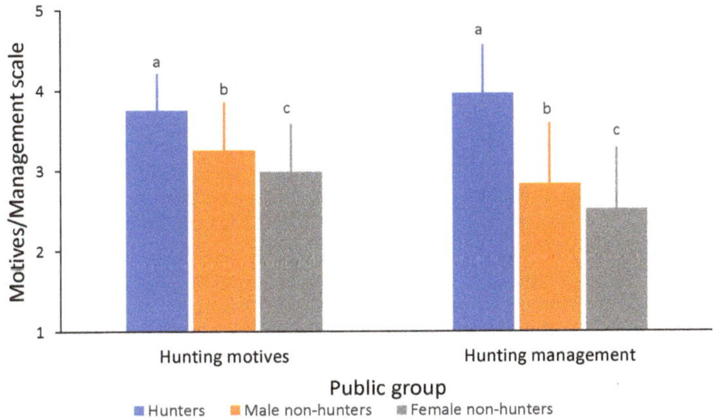

Figure 2. Comparisons of the acceptability of motives for hunting and attitudes toward hunting as a management tool (mean + SD) among hunters (n = 146), male (n = 156), and female (n = 159) non-hunters. In each public group, mean responses not sharing a lowercase letter (a, b, c) are significantly different ($p < 0.05$; pairwise Tamhane post hoc tests with Bonferroni correction).

The factor analysis determined one factor of the acceptability of the motives for hunting for: (a) hunters, including nine statements, with an eigenvalue of 5.9 and accounting for 65.6% of the common variance, and (b) non-hunters, including seven statements, with an eigenvalue of 3.4 and accounting for 48.1% of the common variance (Table 1). The Cronbach's α was 0.924 and 0.785 for the hunters and non-hunters, respectively. These factors were used in the subsequent analyses.

3.3. *Attitudes toward Hunting as a Management Tool*

Differences were significant in all the attitudes toward hunting as a management tool statements, except for the acceptance of hunting abundant game populations between hunters and male non-hunters (Table 2). Male and female hunters did not significantly differ in their attitudes toward hunting as a management tool, except for the acceptance of hunting abundant game populations and hunting as a wildlife habitat management tool. Overall, 81.5% of hunters were positive, and 18.5% were neutral toward hunting as a management tool (3.96 ± 0.60), while 24.2%, 33.7%, and 42.1 % of the male non-hunters were positive, neutral, and negative, respectively (2.83 ± 0.60), and 15.4%, 21.5%, and 63.1% of the female non-hunters were positive, neutral, and negative, respectively (2.52 ± 0.76). The differences were significant among all groups (ANOVA $F_{2,458} = 117.095$, $p < 0.001$; $p < 0.05$, pairwise Tamhane post hoc tests with Bonferroni correction; Figure 2).

Table 2. Comparisons of attitudes toward hunting as a management tool among hunters and non-hunters.

Attitude Statements [a]	Hunters (n = 146)	Non-Hunters (n = 315)		$F_{2,458}$	Factor Loadings [c]	
		Male (n = 156)	Female (n = 159)		Hunters	Non-Hunters
It is acceptable to hunt animals when their populations are abundant.	3.88 ± 1.11 [A]	3.55 ± 1.14 [A]	2.83 ± 1.23 [B]	35.767 ***	0.84	0.64
It is acceptable to hunt animals that were reared and released by people.	3.52 ± 1.50 [A]	2.84 ± 1.01 [B]	2.49 ± 1.09 [B]	30.497 ***	0.80	0.84
Hunting helps keep nature in balance.	4.21 ± 1.00 [A]	3.23 ± 1.21 [B]	2.86 ± 1.18 [B]	55.662 ***	0.92	0.65
Hunting helps reduce agricultural damage by reducing animal populations.	4.08 ± 1.17 [A]	3.00 ± 1.41 [B]	2.63 ± 1.27 [B]	60.763 ***	0.84	0.72
Hunting helps control predators such as foxes and martens.	4.13 ± 1.07 [A]	2.47 ± 1.49 [B]	2.51 ± 1.51 [B]	82.335 ***	0.82	0.79
Hunting commonly results in a species becoming threatened or endangered. [b]	3.77 ± 1.20 [A]	1.75 ± 0.91 [B]	1.80 ± 1.13 [B]	146.762 ***	0.62	0.65
Hunting helps control wildlife diseases by reducing animal populations.	4.02 ± 1.18 [A]	3.11 ± 0.79 [B]	2.89 ± 0.66 [B]	60.078 ***	0.78	—
Hunting provides funds used to manage other wildlife species that are not hunted.	3.94 ± 1.29 [A]	2.63 ± 1.44 [B]	2.26 ± 1.33 [B]	65.546 ***	0.85	0.83
The demand for hunting maintains wildlife habitats.	4.10 ± 0.93 [A]	2.92 ± 1.02 [B]	2.39 ± 1.00 [C]	116.385 ***	0.86	—

[a] Mean ± SD; 5-point scale: 1 (strongly disagree)–5 (strongly agree). [b] Reverse-coded. *** $p < 0.001$. [c] Factor loadings were determined by principal component factor analysis for the hunter and non-hunters (males and females combined) group. Note: one-way ANOVAs were used for comparisons. Means not sharing a common letter (A, B or C) are significantly different ($p < 0.05$; Tamhane post hoc tests with Bonferroni correction).

The factor analysis determined one factor of the attitudes toward hunting as a management tool for: (a) hunters, including nine statements, with an eigenvalue of 6.0 and accounting for 66.9% of the common variance, and (b) non-hunters, including seven statements, with an eigenvalue of 3.8 and accounting for 54.1% of the common variance (Table 2).

Cronbach's α was 0.938, and 0.752 for hunters and non-hunters respectively. These factors were used in the subsequent analyses.

3.4. Knowledge about Wildlife

The knowledge of hunters about the ecology and biology of wildlife species, both non-game (the first 10 statements in Table 3) and game (the last 10 statements in Table 3), was generally high for both the males and the females. On the other hand, wildlife knowledge was generally medium among non-hunters. The knowledge about wildlife was significantly higher for hunters than for non-hunters in 16 of the 20 statements, for both the males and the females, while gender differences were not observed between non-hunters in any of the knowledge statements (Table 3).

Table 3. Comparisons of knowledge about wildlife among hunters and non-hunters.

Knowledge Statements [a]	Hunters (n = 146)	Non-Hunters (n = 315)		$F_{2,458}$	Factor Loadings [c]	
		Male (n = 156)	Female (n = 159)		Hunters	Non-Hunters
Brown bears mostly eat meat. [b]	3.63 ± 1.30 [A]	3.54 ± 1.37 [A]	3.92 ± 1.25 [A]	1.077	—	—
Black storks nest in trees.	3.71 ± 0.65 [A]	2.75 ± 0.85 [B]	2.95 ± 1.08 [B]	31.766 ***	0.84	—
Common European adders are male nose-horned vipers. [b]	3.52 ± 1.02 [A]	2.93 ± 0.93 [B]	2.85 ± 0.85 [B]	22.375 ***	0.58	0.54
Eurasian otters are rodents. [b]	2.50 ± 1.21 [A]	2.27 ± 1.25 [A]	2.43 ± 1.36 [A]	1.002	0.61	0.59
Eurasian otters mostly eat cultivated seeds and fruits. [b]	4.35 ± 0.99 [A]	3.03 ± 1.19 [B]	2.66 ± 1.09 [B]	103.861 ***	—	0.52
Northern, white-breasted hedgehogs mostly eat leaves and grasses. [b]	3.25 ± 1.48 [A]	2.16 ± 0.95 [B]	2.28 ± 1.00 [B]	39.597 ***	—	0.50
Red foxes might carry rabies	4.54 ± 0.94 [A]	4.36 ± 0.74 [A]	4.32 ± 0.81 [A]	1.381	0.73	—
Red and roe deer shed their antlers each year.	4.38 ± 0.97 [A]	3.23 ± 1.12 [B]	3.09 ± 1.43 [B]	49.912 ***	0.65	0.63
Roe deer are monogamous.	2.52 ± 1.16 [A]	2.73 ± 1.18 [A]	2.92 ± 1.15 [A]	1.945	—	0.53
Turtles are a common sight in winter. [b]	3.90 ± 1.40 [A]	2.24 ± 1.09 [B]	2.23 ± 0.96 [B]	89.655 ***	0.77	0.67
Brown hares nest in burrows. [b]	3.77 ± 1.64 [A]	2.33 ± 1.29 [B]	2.51 ± 1.45 [B]	43.714 ***	—	0.53
Female brown hares give birth to one young each year. [b]	4.56 ± 1.02 [A]	3.88 ± 1.26 [B]	3.80 ± 1.38 [B]	22.587 ***	—	0.56
Ducks feed during the day and sleep during the night. [b]	4.12 ± 1.23 [A]	2.34 ± 1.20 [B]	2.54 ± 1.25 [B]	98.231 ***	0.67	0.59
Female ducks have colorful plumage. [b]	4.48 ± 1.00 [A]	3.40 ± 1.27 [B]	3.65 ± 1.32 [B]	35.901 ***	0.57	—
Eurasian woodcocks prefer wet, densely vegetated habitats.	4.79 ± 0.54 [A]	3.72 ± 1.03 [B]	3.75 ± 0.97 [B]	76.066 ***	0.68	—
Rock partridges are galliforms.	4.40 ± 0.87 [A]	3.65 ± 0.88 [B]	3.58 ± 0.97 [B]	37.197 ***	0.81	—
Rock partridges form pairs at the end of winter.	4.10 ± 0.85 [A]	2.91 ± 0.64 [B]	2.89 ± 0.72 [B]	134.868 **	—	0.59
Turtle doves are migratory birds.	4.71 ± 0.89 [A]	3.43 ± 1.15 [B]	3.68 ± 1.11	73.934 ***	—	—
Wild boars can mate with domestic pigs.	4.56 ± 0.83 [A]	3.35 ± 1.12 [B]	3.26 ± 0.96 [B]	78.025 ***	0.77	0.65
Wild boars take mud baths to cool themselves. [b]	2.85 ± 1.67 [A]	1.90 ± 0.91 [B]	2.08 ± 1.04 [B]	22.342 ***	0.83	—

[a] Mean ± SD; 5-point scale: 1 (strongly disagree)–5 (strongly agree). [b] Reverse-coded. ** $p < 0.01$. *** $p < 0.001$. [c] Factor loadings were determined by principal component factor analysis for the hunter and non-hunter (males and females combined) groups. Note: one-way ANOVAs were used for comparisons. Means not sharing a common letter (A or B) are significantly different ($p < 0.05$; Tamhane post hoc tests with Bonferroni correction).

The overall knowledge about wildlife was high for hunters with regard to game species (4.23 ± 0.55), non-game species (3.63 ± 0.50), and for all species (3.93 ± 0.46) (Figure 3). Conversely, wildlife knowledge was generally medium for non-hunters (male: 3.09 ± 0.36, 2.92 ± 0.33, 3.01 ± 0.28; female: 3.17 ± 0.42, 2.97 ± 0.38, 3.07 ± 0.32 for game, non-game, and all species, respectively). The differences in knowledge about wildlife were significant between hunters and both male and female non-hunters for game, non-game, and all species (all $F_{2,458} > 127.071$, $p < 0.001$; $p < 0.05$, pairwise Tamhane post hoc tests with Bonferroni correction). In contrast, the differences in knowledge about wildlife between male and female non-hunters were not significant for game, non-game, and all species ($p > 0.05$, pairwise Tamhane post hoc tests with Bonferroni correction).

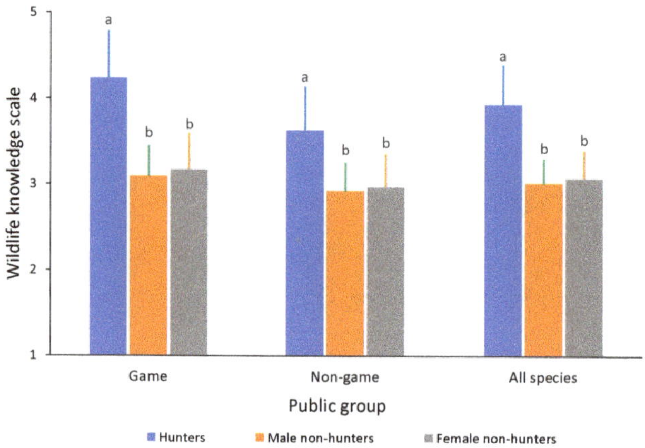

Figure 3. Comparisons of knowledge about wildlife (mean + SD) among hunters (n = 146) and male (n = 156) and female (n = 159) non-hunters, by game species (10 statements), non-game species (10 statements), and all species (20 statements). In each species group, mean responses not sharing a lowercase letter (a, b, c) are significantly different ($p < 0.05$; pairwise Tamhane post hoc tests with Bonferroni correction).

Within the groups, knowledge was significantly higher for game than for non-game species among the hunters (paired $t_{145} = -14.285$, $p < 0.001$) and male (paired $t_{155} = -5.543$, $p < 0.001$) and female (paired $t_{158} = -5.139$, $p < 0.001$) non-hunters.

The factor analysis determined one factor of knowledge about wildlife for: (a) hunters, including 12 statements, with an eigenvalue of 6.1 and accounting for 50.6% of the common variance, and (b) non-hunters, including 12 statements, with an eigenvalue of 4.0 and accounting for 33.3% of the common variance (Table 3). The Cronbach's α was 0.854 and 0.712 for hunters and non-hunters, respectively. These factors were used in the subsequent analyses.

3.5. *Effects of Sociodemographic Factors*

Multicollinearity among the independent variables was not detected in all the multiple regression models, with VIF values of 2.579 or lower. The hunters that were more highly motivated, more knowledgeable about wildlife, and had more years of hunting experience also held more positive attitudes toward hunting as a management tool than those with a lower motivation for hunting, less wildlife knowledge, and less hunting experience (Table 4). Hunters with more hunting experience had higher knowledge about wildlife than those with less hunting experience.

The non-hunters that were older, had higher acceptance of hunting motives, had more knowledge about wildlife, ate game, and had hunters in their family or as friends were more positive toward hunting as a management tool than younger females, who did not eat game meat and did not have hunters in their family or as friends. Older, male non-hunters, with more wildlife knowledge, and with hunters in their family or as friends had more acceptance of the motives for hunting than those that were younger, female, had less knowledge about wildlife, and did not have hunters in their family or as friends. Older, more educated non-hunters with hunters in their family or as friends had more wildlife knowledge than those who were older, less educated, and did not have hunters in their family or as friends.

Table 4. Relationships between the attitudes toward hunting as a management tool, the acceptability of motives for hunting, wildlife knowledge, and sociodemographic factors, as assessed by hunter and non-hunter groups.

	Hunters (n = 146)			Non-Hunters (n = 315)		
	Hunting and Management	**Hunting Motives**	**Wildlife Knowledge**	**Hunting and Management**	**Hunting Motives**	**Wildlife Knowledge**
Hunting motives	0.388 ***	-	-	0.489 ***	-	-
Wildlife knowledge	0.208 ***	0.056	-	0.208 ***	0.134 *	-
Age	−0.078	−0.028	−0.016	0.244 ***	0.558 ***	0.521 ***
Gender (female)	-	-	-	−0.123 *	−0.147 *	0.076
Education (higher)	0.061	0.007	0.057	−0.05	0.033	0.234 ***
Pet ownership	0.373 ***	0.023	0.074	0.001	−0.009	0.031
Eat game	-	-	-	0.118 *	0.074	−0.027
Hunters' kin/friends	0.078	0.205 ***	0.075	0.109 *	0.151 *	0.172 **
Hunting experience	0.279 ***	0.054	0.329 ***	-	-	-
Constant	1.569	3.933 ***	4.184 ***	0.873 **	1.351 ***	1.991 ***
adj. R^2	0.337	0.194	0.219	0.715	0.527	0.290

* $p < 0.05$; ** $p < 0.01$; *** $p < 0.001$. Note: multiple linear regression models were used. Dummy variables: gender (female = 1), education (higher = 1), pet ownership (yes = 1), eat game (yes = 1), hunters' family/friends (yes = 1). Standardized regression coefficients and adjusted R^2 are given.

4. Discussion

4.1. Attitudes toward Hunting

Greek hunters were generally highly motivated toward hunting, considering their pastime as a valuable wildlife management tool. In contrast, most non-hunters, both male and female, were neutral in the acceptance of the motives for hunting and displayed negative attitudes toward hunting as a management tool. Their opinions were significantly more negative than those of the hunters for both hunting motives and management. The results from other similar studies indicated that the attitudes toward hunting are related to its purpose. Ljung et al. [34], in a study in Sweden, reported that 80% of non-hunters had a favorable attitude toward hunting. This rate of approval was strongly associated with game meat consumption. We also asked about hunting in general, and meat consumption was also positively associated with attitudes toward hunting in our study. Furthermore, Ljung et al. [34] did not study male and female non-hunters separately; however, their sample's gender composition was similar to ours (49% male); hence, our datasets could be compared. In our sample, 21% of non-hunters, including males and females, had positive attitudes toward hunting, with most being negative (51%), and 23% accepted the motives for hunting, with most being neutral (67%). In a Danish study, the majority of non-hunters had a positive attitude toward recreational hunting (43%), while 25% displayed negative attitudes [35]. In Denmark again, only 25% of non-hunters expressed a positive attitude toward recreational hunting [17]. Grandy et al. [45] reported that in the U.S. the approval for hunting for meat was higher than 80%; the approval for hunting for recreation and meat was a little higher than 60%, while approval for hunting only for recreation was below 40%. Ljung et al. [34] also reported that 63% of non-hunters considered hunting for "sport and recreation" cruel. In responses to specific statements, 66%, 90%, and 61% of non-hunters rejected recreation, sport, and game meat procurement as motives for hunting. In contrast, the vast majority of Greek hunters declared that they hunted for recreation (98%) and not for sport (12%) or the game meat (10%). As hunting is a recreational activity in Greece and at the same time the main way for procuring game meat, we consider the results from similar European countries to be comparable to ours. These results suggested that the attitudes toward hunting are more negative in Greece compared to other, mainly European, countries. Hunting is a complex social phenomenon and several factors, such as differences in hunting tradition, culture and management, and game meat use (non-commercial in Greece, commercial in other countries) might explain the observed differences, but not before further research.

Both the hunters and the non-hunters ranked similar hunting motives (i.e., contact with nature, excitement, and socializing) and management functions (i.e., keeps nature in balance and helps control wildlife diseases) as the most important. Other studies that compared hunting attitudes between hunters and non-hunters also reported differences in the perception of hunting motives and management functions and similarities in the rankings of the hunting motives and management functions that were similar to our findings [15,16]. Hunters of other European countries and of North America also see themselves as stewards of nature [46,47]. Hunters participated in other outdoor activities, both consumptive (e.g., berry picking and fishing) and non-consumptive (e.g., birdwatching and hiking), more often than non-hunters [48]. Cooper et al. [49] found that public groups with an interest in wildlife, such as hunters, were 4–5 times more likely than those with no such interest to participate in pro-environmental behaviors.

Hunting has helped in mitigating agricultural damage by regulating overabundant wildlife populations [5,7,50,51]. In other cases, hunting was responsible for the considerable decline of wildlife populations and for biodiversity loss [52]. Hunters often engage in wildlife conservation and management, especially when game species are involved [26,27]. Furthermore, they are more positive toward wildlife management when the game species are not negatively affected. Kontsiotis et al. [18], in a study of the public opinion in the region of eastern Macedonia and Thrace, north Greece, found that hunters were less willing to accept management strategies involving the reduction in the game wild boar (*Sus scrofa*) than of the non-game European badger (*Meles meles*) populations, when both damage crops. Similarly, hunters in central Italy accepted non-lethal and rejected lethal strategies to prevent wild boar damage to crops [53].

4.2. Knowledge about Wildlife

Hunters had greater knowledge about the ecology, biology, and behavior of wildlife species, both game and non-game, than non-hunters. Kellert [20] measured the public's knowledge about predatory animal species. In the survey, he included, among others, questions about the taxonomy, biology, superstitions, and folk knowledge concerning wildlife. His results revealed that hunters had great knowledge of predators, similar to that of birdwatchers and environmental protection organization members. In contrast, anti-hunters and zoo enthusiasts, although they expressed strong affection and support for protecting predators, had relatively low knowledge about them. As hunters participate in outdoor activities, both consumptive and non-consumptive, more than non-hunters, they have the opportunity for a hands-on experience of nature and wildlife and gain direct knowledge about several aspects of the life history of wildlife species, both game and non-game [48]. Zoo enthusiasts might focus their interest on the exotic species that zoos most often host.

4.3. The Effect of Sociodemographics

The acceptability of the hunting motives and knowledge about wildlife were positively associated toward hunting for managing wildlife, for both hunters and non-hunters. Hunters are consumptive users of wildlife that gain knowledge through experience. Moreover, the more experienced among hunters are more strongly attached to their favorite pastime [54]. Non-hunters hold variable utilitarian and animal rights convictions [55]. It seems that increasing knowledge about wildlife and the associated management issues shifts the balance toward human benefits instead of animal welfare, resulting in higher support for hunting. Having friends or family members who hunt positively affected the acceptability of the hunting motives for both hunters and non-hunters, as well as attitudes toward hunting as a management tool for non-hunters. Previous research also found similar trends, suggesting that social interactions with hunters positively affect attitudes toward hunting [34–36]. These studies also concluded that, because socializing is an important motive for hunting, those with hunters in their social network are more likely to become hunters and to be more attached to their activity as they have the opportunity to hunt in close-knit groups.

Hunting experience positively affected the hunters' attitudes toward hunting as a management tool. In [56], it was found that more experienced hunters were more efficient in bagging willow grouse (*Lagopus lagopus*) than less experienced hunters. Harvest success was directly related to satisfaction with white-tailed deer (*Odocoileus virginianus*) management in Pennsylvania [57]. The more experienced hunters also had more knowledge about wildlife species. The more time someone spends in the outdoors, the more experiential knowledge about the ecology, biology, and behavior of wildlife species they gain [15,32,33].

Older non-hunters had more positive attitudes toward hunting motives and hunting as a management tool than younger people. Females were more negative toward hunting as a management tool and hunting motives than males. Previous research also found that older age is positively associated and female gender is negatively associated with hunting [15,32,33]. Non-hunters who eat game meat were more positive toward hunting as a management tool than those who do not. Game meat consumption was an important reason for the positive attitudes toward hunting expressed by non-hunters in Sweden, where it is legally sold [34]. However, the sale of game meat is illegal in Greece and can only be accessed through friends and family. Older, more educated non-hunters who had friends or family members who hunt were more knowledgeable about wildlife. Experience comes with age, while education has been related to an increased interest in wildlife and support for wildlife conservation and management [58,59]. Socializing with hunters allows non-hunters to learn about wildlife species, both game and non-game, through the narration of outdoor ventures [34–36].

4.4. Management Implications

Attitudes toward hunting were controversial, being more positive for Greek hunters than for non-hunters. For example, the hunters stated that they hunt mostly for recreation, while non-hunters largely rejected hunting as a recreational activity but justified it for meat procurement. Moreover, participation in hunting has declined in Greece [37,38] and is predicted to continue to decline because of the continually increasing rates of urbanization and the consequently greater proportion of the public subscribing to mutualism value orientations toward wildlife [13,14,39]. Wildlife managers must adapt to this change. The necessary funds for wildlife conservation and management should be collected through the promotion of other, non-consumptive outdoor activities, such as wildlife watching and photography. Outreach programs should aim at informing public groups about good hunting practices and the utility of hunting for addressing specific wildlife issues. Our findings suggested that priority groups for outreach programs should be anti-hunters and young females without connections with hunters and with little knowledge about wildlife and wildlife-related issues. Such information should reduce the controversy among the stakeholders and allow for the successful management of the conflicts concerning the good practice of hunting and its use as a management tool.

Our findings also revealed that hunters had a greater knowledge than non-hunters about wildlife, both game and non-game. As the participation in hunting is currently declining and expected to further decline [13,14] and also because people are disconnected from nature due to urbanization, an ever-decreasing proportion of the population will acquire experiential knowledge about wildlife and understand nature and wildlife-related problems [48]. This may result in weaker attitudes toward environmental issues and therefore in fewer people acting as advocates of nature and wildlife. The alienation from nature and wildlife is also responsible for the deterioration of human health and well-being [21,22]. Therefore, there is an urgent need for wildlife managers to act to stop and reverse these trends and thus secure support for wildlife conservation and management and improve human health and well-being. Educational programs should aim at increasing the public's knowledge about the ecology, biology, and behavior of wildlife species [60]. Research has shown that outdoor educational programs involving experiential activities increased wildlife knowledge and improved attitudes toward wildlife species [61–64]. Therefore, environmental educators should aim at introducing urban people to outdoor activities that would allow for the increase in wildlife knowledge, improving health and well-being and reinforcing nature and wildlife stewardship through direct experience thereof (e.g., wildlife watching, photography, and animal tracking through the seasons). In this context, hunters could be used as educators to teach people how to experience nature and wildlife. This contact of hunters with non-hunters could also reduce the antipathy toward hunters and hunting.

5. Conclusions

We used several statements to reveal the differences in the acceptance of the motives for hunting and of hunting as a management tool and in the knowledge about wildlife among hunters and non-hunters. Hunters had high motivations for hunting, which they also considered as a valuable management tool. On the other hand, both male and female non-hunters were significantly more negative than hunters in their acceptance of the motives for hunting and in their attitudes toward hunting as a management tool. In addition, hunters had greater knowledge about wildlife species than non-hunters. Our findings suggested that hunting is highly controversial between hunters and non-hunters in the Greek society and as such they would be valuable for wildlife managers and policy makers in their efforts to manage this conflict. Further research on the perceptions and knowledge of specific public groups, such as vegetarians and outdoor enthusiasts, would also help reaching better decisions.

Author Contributions: Conceptualization, G.R. and V.L.; investigation, G.R.; methodology, G.R., V.L. and V.J.K.; software, G.R. and V.L.; validation, V.L. and V.J.K.; formal analysis, G.R., V.L. and V.J.K.; resources, G.R., V.L. and V.J.K.; data curation, V.L. and V.J.K.; writing—original draft preparation, G.R. and V.L.; writing—review and editing, G.R., V.L. and V.J.K.; visualization, V.L.; supervision, G.R. and V.L.; project administration, V.L. All authors have read and agreed to the published version of the manuscript.

Funding: This research received no external funding.

Institutional Review Board Statement: The study was conducted according to the guidelines of the Declaration of Helsinki and adhered to the ethical standards laid out by the Research and Academic Committee of the International Hellenic University.

Informed Consent Statement: We sought informed consent from all the participants and maintained anonymity at all stages of the research.

Data Availability Statement: The data presented in this study are available on reasonable request from the corresponding author.

Acknowledgments: We thank the survey participants for sharing their time and opinions with us.

Conflicts of Interest: The authors declare no conflict of interest.

References

1. Heberlein, T.A.; Ericsson, G. Ties to the countryside: Accounting for urbanites attitudes toward hunting, wolves, and wildlife. *Hum. Dimens. Wildl.* **2005**, *10*, 213–227. [CrossRef]
2. Øian, H.; Skogen, K. Property and possession: Hunting tourism and the morality of landownership in rural Norway. *Soc. Nat. Resour.* **2016**, *29*, 104–118. [CrossRef]
3. Plummer, R. *Outdoor Recreation*; Routledge: New York, NY, USA; London, UK, 2009.
4. Wightman, A.; Higgins, P.; Jarvie, G.; Nicol, R. The cultural politics of hunting: Sporting estates and recreational land use in the highlands and Islands of Scotland. *Cult. Sport Soc.* **2002**, *5*, 53–70. [CrossRef]
5. Fischer, A.; Sandström, C.; Delibes-Mateos, M.; Arroyo, B.; Tadie, D.; Randall, D.; Hailu, F.; Lowassa, A.; Msuha, M.; Kereži, V.; et al. On the multifunctionality of hunting—An institution analysis of eight cases from Europe and Africa. *J. Environ. Plan. Manag.* **2013**, *56*, 531–552. [CrossRef]
6. Heffelfinger, J.R.; Geist, V.; Wishart, W. The role of hunting in North American wildlife conservation. *Int. J. Environ. Stud.* **2013**, *70*, 399–413. [CrossRef]

7. Jenkins, L.H.; Jenkins, M.A.; Webster, C.R.W.; Zollner, P.A.; Shields, J.M. Herbaceous layer response to 17 years of controlled deer hunting in forested natural areas. *Biol. Conserv.* **2015**, *175*, 119–128. [CrossRef]
8. Brown, T.L.; Decker, D.J.; Siemer, W.E.; Enck, J.W. Trends in hunting participation and implications for management of game species. In *Trends in Outdoor Recreation, Leisure and Tourism*; Gartner, W.C., Lime, D.W., Eds.; CAB International: Oxon, UK; New York, NY, USA, 2000; pp. 145–154.
9. Leonard, J. *Fishing and Hunting Recruitment and Retention in the US from 1990 to 2005. Addendum to the 2001 National Survey of Fishing, Hunting, and wildlife-Associated Recreation*; Report 2001-11; US Fish & Wildlife Service: Arlington, VA, USA, 2007.
10. FACE. Annual Report 2009–2010. Available online: http://www.face.eu/sites/deafult/files/attachments/data_hunters-region_sept_2010.pdf (accessed on 15 June 2022).
11. Pinet, J.-M. The Hunters in Europe. *Federation of Associations for Hunting and Conservation of the EU. FACE*. 1995. Available online: https://www.kora.ch/malme/05_library/5_1_publications/P_and_Q/Pinet_1995_The_hunters_in_Europe.pdf (accessed on 15 June 2022).
12. Cerri, J.; Ferretti, M.; Coli, L. Where the wild things are: Urbanization and income affect hunting participation in Tuscany, at the landscape scale. *Eur. J. Wildl. Res.* **2018**, *64*, 23. [CrossRef]
13. Manfredo, M.J.; Sullivan, L.; Don Carlos, A.W.; Dietsch, A.M.; Teel, T.L.; Bright, A.D.; Bruskotter, J.T. *America's Wildlife Values: The Social Context of Wildlife Management in the US*; Colorado State University, Department of Human Dimensions of Natural Resources: Fort Collins, CO, USA, 2018.
14. Manfredo, M.J.; Teel, T.L.; Berl, R.E.W.; Bruskotter, J.T.; Kitayama, S. Social value shift in favour of biodiversity conservation in the United States. *Nat. Sustain.* **2021**, *4*, 323–330. [CrossRef]
15. Byrd, E.; Lee, J.G.; Widmar, N.J.O. Perceptions of hunting and hunters by US respondents. *Animals* **2017**, *7*, 83. [CrossRef]
16. Daigle, J.J.; Hrubes, D.; Ajzen, I. A comparative study of beliefs, attitudes, and values among hunters, wildlife viewers, and other outdoor recreationists. *Hum. Dimens. Wildl.* **2002**, *7*, 1–19. [CrossRef]
17. Gamborg, C.; Jensen, F.S.; Sandøe, P. Attitudes to the shooting of rear and release birds among landowners, hunters and the general public in Denmark. *Land Use Policy* **2016**, *57*, 296–304. [CrossRef]
18. Kontsiotis, V.J.; Vadikolios, G.; Liordos, V. Acceptability and consensus for the management of game and nongame crop raiders. *Wildl. Res.* **2020**, *47*, 296–308. [CrossRef]
19. Liordos, V.; Kontsiotis, V.J.; Emmanouilidou, F. Understanding stakeholder preferences for managing red foxes in different situations. *Ecol. Process.* **2020**, *9*, 20. [CrossRef]
20. Kellert, S.R. Public perceptions of predators, particularly the wolf and coyote. *Biol. Conserv.* **1985**, *31*, 167–189. [CrossRef]
21. Soga, M.; Gaston, K.J. Extinction of experience: The loss of human–nature interactions. *Front. Ecol. Environ.* **2016**, *14*, 94–101. [CrossRef]
22. Louv, R. *The Nature Principle: Human Restoration and the end of Nature-deficit Disorder*; Algonquin Books: Chapel Hill, NC, USA, 2011.
23. Echeverri, A.; Karp, D.S.; Naidoo, R.; Tobias, J.A.; Zhao, J.; Chan, K.M.A. Can avian functional traits predict cultural ecosystem services? *People Nat.* **2020**, *2*, 138–151. [CrossRef]
24. McCleery, R.A.; Moorman, C.E.; Peterson, M.N. (Eds.) *Urban Wildlife Conservation: Theory and Practice*; Springer: New York, NY, USA, 2014.
25. Fazio, R.H.; Chen, J.; McDonel, E.C.; Sherman, S.J. Attitude accessibility, attitude behavior consistency, and the strength of the object-evaluation association. *J. Pers. Soc. Psychol.* **1982**, *47*, 277–286. [CrossRef]
26. Holsman, R.H. Goodwill hunting? Exploring the role of hunters as ecosystem stewards. *Wildl. Soc. Bull.* **2000**, *28*, 808–816.
27. Loveridge, A.J.; Reynolds, J.C.; Milner-Gulland, E.J. Does sport hunting benefit conservation? In *Key Topics in Conservation Biology*; Macdonald, D.W., Service, K., Eds.; Blackwell Publishing: Oxford, UK, 2007; pp. 224–241.
28. Bouthillier, F.; Shearer, K. Understanding knowledge management and information management: The need for an empirical perspective. *Inf. Res.* **2002**, *8*, 141. Available online: http://InformationR.net/ir/8-1/paper141.html (accessed on 25 May 2022).
29. García-López, R.; Villegas, A.; Pacheco-Coronel, N.; Gómez-Álvarez, G. Traditional use and perception of snakes by the Nahuas from Cuetzalan del Progreso, Puebla, Mexico. *J. Ethnobiol. Ethnomed.* **2017**, *13*, 6. [CrossRef]
30. Liordos, V.; Kontsiotis, V.J.; Nevolianis, C.; Nikolopoulou, C.E. Stakeholder preferences and consensus associated with managing an endangered aquatic predator: The Eurasian otter (Lutra lutra). *Hum. Dimens. Wildl.* **2019**, *24*, 446–462. [CrossRef]
31. Pradhan, A.; Yonle, R. Socio-ecological assessment of squamate reptiles in a human-modified ecosystem of Darjeeling, Eastern Himalaya. *Hum. Dimens. Wildl.* **2022**, *27*, 134–150. [CrossRef]
32. Duda, M.D.; Jones, M.F.; Criscione, A. *The Sportsman's Voice: Hunting and Fishing in America*; Venture Publishing: State College, PA, USA, 2010.
33. Teel, T.L.; Krannich, R.S.; Schmidt, R.H. Utah stakeholders' attitudes toward selected cougar and black bear management practices. *Wildl. Soc. Bull.* **2002**, *30*, 2–15. Available online: https://www.jstor.org/stable/3784630 (accessed on 25 May 2022).
34. Ljung, P.E.; Riley, S.J.; Heberlein, T.A.; Ericsson, G. Eat prey and love: Game-meat consumption and attitudes toward hunting. *Wildl. Soc. Bull.* **2012**, *36*, 669–675. [CrossRef]
35. Gamborg, C.; Søndergaard Jensen, F. Attitudes toward recreational hunting: A quantitative survey of the general public in Denmark. *J. Outdoor Recreat. Tour.* **2017**, *17*, 20–28. [CrossRef]
36. Stedman, R.C.; Decker, D.J. Illuminating an overlooked hunting stakeholder group: Nonhunters and their interest in hunting. *Hum. Dimens. Wildl.* **1996**, *1*, 29–41. [CrossRef]

37. MEECC. *Forest Service Activities for the Year 2010 (in Greek)*; Ministry of Environment, Energy and Climate Change Report: Athens, Greece, 2012.
38. Sokos, C.K. The Disappearance of Greek Hunters: Predicting Hunting Licences. Available online: https://www.ihunt.gr/%CE%B7-%CE%B5%CE%BE%CE%B1%CF%86%CE%AC%CE%BD%CE%B9%CF%83%CE%B7-%CF%84%CF%89%CE%BD-%CE%B5%CE%BB%CE%BB%CE%AE%CE%BD%CF%89%CE%BD-%CE%BA%CF%85%CE%BD%CE%B7%CE%B3%CF%8E%CE%BD-%CF%80%CF%81%CF%8C%CE%B2%CE%BB/ (accessed on 15 June 2022). (In Greek)
39. United Nations, Department of Economic and Social Affairs, Population Division. World Urbanization Prospects: The 2018 Revision (ST/ESA/SER.A/420). *United Nations*. Available online: https://population.un.org/wup/Publications/Files/WUP2018-Report.pdf (accessed on 15 June 2022).
40. Liordos, V. Membership trends and attitudes of a Greek hunting community. *Eur. J. Wildl. Res.* **2014**, *60*, 821–826. [CrossRef]
41. Hellenic Statistical Authority. Population Census 2011. Available online: http://www.statistics.gr/portal/page/portal/ESYE/PAGE-census2011 (accessed on 15 June 2022). (In Greek)
42. Vaske, J.J. *Survey Research and Analysis: Applications in Parks, Recreation and Human Dimensions*; Venture Publishing Inc.: State College, PA, USA, 2008.
43. Hair, J.F.; Babin, B.J.; Anderson, R.E.; Black, W.C. *Multivariate Data Analysis*, 8th ed.; Cengage Learning EMEA: London, UK, 2018.
44. Nunnally, J.C. *Psychometric Theory*, 2nd ed.; McGraw-Hill: New York, NY, USA, 1978.
45. Grandy, J.W.; Stallman, E.; Macdonald, D.W. The science and sociology of hunting: Shifting practices and perceptions in the United States and Great Britain. In *The State of the Animals II*; Salem, D.J., Rowan, A.N., Eds.; Humane Society Press: Washington, DC, USA, 2003; pp. 107–130.
46. MacMillan, D.C.; Leitch, K. Conservation with a gun: Understanding landowner attitudes to deer hunting in the Scottish Highlands. *Hum. Ecol.* **2008**, *36*, 473–484. [CrossRef]
47. Treves, A.; Martin, K.A. Hunters as stewards of wolves in Wisconsin and the northern Rocky Mountains, USA. *Soc. Nat. Resour.* **2011**, *24*, 984–994. [CrossRef]
48. Ericsson, G.; Heberlein, T.A. "Jägare talar naturens språk" (Hunters speak nature's language): A comparison of outdoor activities and attitudes toward wildlife among Swedish hunters and the general public. *Z. Jagdwiss.* **2002**, *48*, 301–308. [CrossRef]
49. Cooper, C.; Larson, L.; Dayer, A.; Stedman, R.; Decker, D. Are wildlife recreationists conservationists? *Linking hunting, birdwatching, and pro-environmental behavior. J. Wildl. Manag.* **2015**, *79*, 446–457. [CrossRef]
50. Estrada, A.; Delibes-Mateos, M.; Caro, J.; Viñuela, J.; Díaz-Fernández, S.; Casas, F.; Arroyo, B. Does small-game management benefit steppe-birds of conservation concern? A field study in central Spain. *Anim. Conserv.* **2015**, *18*, 567–575. [CrossRef]
51. Giménez-Anaya, A.; Herrero, J.; García-Serrano, A.; García-González, R.; Prada, C. Wild boar battues reduce crop damages in a protected area. *Folia Zool.* **2016**, *65*, 214–220. [CrossRef]
52. Benítez-López, A.; Alkemade, R.; Schipper, A.M.; Ingram, D.J.; Verweij, P.A.; Eikelboom, J.A.J.; Huijbregts, M.A.J. The impact of hunting on tropical mammal and bird populations. *Science* **2017**, *356*, 180–183. [CrossRef] [PubMed]
53. Frank, B.; Monaco, A.; Bath, A.J. Beyond standard wildlife management: A pathway to encompass human dimension findings in wild boar management. *Eur. J. Wildl. Res.* **2015**, *61*, 723–730. [CrossRef]
54. Liordos, V. Sociodemographic analysis of hunters' preferences: A Greek hunting club perspective. *Zool. Ecol.* **2014**, *24*, 389–396. [CrossRef]
55. Lund, T.B.; McKeegan, D.E.F.; Cribbin, C.; Sandøe, P. Animal ethics profiling of vegetarians, vegans and meat-Eaters. *Anthrozoös* **2016**, *29*, 89–106. [CrossRef]
56. Asmyhr, L.; Willebrand, T.; Hörnell-Willebrand, M. General experience rather than of local knowledge is important for grouse hunters bag size. *Hum. Dimens. Wildl.* **2012**, *17*, 437–445. [CrossRef]
57. Miller, C.A.; Graefe, A.R. Effect of harvest success on hunter attitudes toward white-tailed deer management in Pennsylvania. *Hum. Dimens. Wildl.* **2001**, *6*, 189–203. [CrossRef]
58. Knight, A.J. 'Bats, snakes and spiders, oh my!' How aesthetic and negativistic attitudes, and other concepts predict support for species protection. *J. Environ. Psychol.* **2008**, *28*, 94–103. [CrossRef]
59. Liordos, V.; Kontsiotis, V.J.; Anastasiadou, M.; Karavasias, E. Effects of attitudes and demography on public support for endangered species conservation. *Sci. Total Environ.* **2017**, *595*, 25–34. [CrossRef] [PubMed]
60. Liordos, V.; Kontsiotis, V.J.; Kokoris, S.; Pimenidou, M. The two faces of Janus, or the dual mode of public attitudes towards snakes. *Sci. Total Environ.* **2018**, *621*, 670–678. [CrossRef] [PubMed]
61. Morgan, J.N.; Gramann, J.H. Predicting effectiveness of wildlife education programs: A study of students' attitudes and knowledge toward snakes. *Wildl. Soc. Bull.* **1989**, *17*, 501–509. Available online: https://www.jstor.org/stable/3782720 (accessed on 25 May 2022).
62. Ballouard, J.M.; Provost, G.; Barre, D.; Bonnet, X. Influence of a field trip on the attitude of schoolchildren toward unpopular organisms: An experience with snakes. *J. Herpetol.* **2012**, *46*, 423–428. [CrossRef]
63. Espinosa, S.; Jacobson, S.K. Human-Wildlife Conflict and Environmental Education: Evaluating a Community Program to Protect the Andean Bear in Ecuador. *J. Environ. Educ.* **2012**, *43*, 55–65. [CrossRef]
64. Greving, H.; Bruckermann, T.; Schumann, A.; Straka, T.M.; Lewanzik, D.; Voigt-Heucke, S.L.; Marggraf, L.; Lorenz, J.; Brandt, M.; Voigt, C.C.; et al. Improving attitudes and knowledge in a citizen science project about urban bat ecology. *Ecol. Soc.* **2022**, *27*, 24. [CrossRef]

sustainability

Review

Conceptualising Drivers of Illegal Hunting by Local Hunters Living in or Adjacent to African Protected Areas: A Scoping Review

Paul Zyambo [1,*], Felix K. Kalaba [2], Vincent R. Nyirenda [3] and Jacob Mwitwa [4]

1 School of Postgraduate, University of Lusaka, Lusaka P.O. Box 36711, Zambia
2 Department of Plant and Environmental Sciences, School of Natural Resources, The Copperbelt University, Kitwe P.O. Box 21692, Zambia
3 Department of Zoology and Aquatic Sciences, School of Natural Resources, The Copperbelt University, Kitwe P.O. Box 21692, Zambia
4 School of Applied Sciences, Kapasa Makasa University, Chinsali P.O. Box 480195, Zambia
* Correspondence: paulzya@yahoo.com; Tel.: +260-978-290-175

Abstract: Illegal hunting of wildlife by community members abutting African protected areas contributes to unsustainable use of wildlife, resulting in significant declines in wildlife populations. Contemporary intervention measures have largely been ineffective, leading to pervasive and persistent illegal hunting. Such illegal hunting of wildlife is partly exacerbated by poor understanding of what motivates people to hunt illegally. Applying a scoping review approach, this study aims at developing concepts for drivers of illegal hunting and how they influence illegal hunting behaviour by local hunters living in or adjacent to African protected areas. A total of 30 publications were included for review analysis from 1014 publications retrieved using data base searches on Google Scholar and ScienceDirect. The study identified 12 proximate and five underlying drivers, which were categorised into 10 thematic drivers of illegal hunting by local hunters. The need for survival and sustaining livelihoods was conceptualised as the key thematic driver of illegal hunting by local hunters. The study represents a novel work of conceptualising drivers of illegal hunting by local hunters with implications on the persistence of illegal hunting in Africa.

Keywords: Africa; drivers of illegal hunting; illegal hunters' behaviour; local hunters; survival and sustaining livelihoods; wildlife

Citation: Zyambo, P.; Kalaba, F.K.; Nyirenda, V.R.; Mwitwa, J. Conceptualising Drivers of Illegal Hunting by Local Hunters Living in or Adjacent to African Protected Areas: A Scoping Review. *Sustainability* **2022**, *14*, 11204. https://doi.org/10.3390/su141811204

Academic Editor: Vasilios Liordos

Received: 25 July 2022
Accepted: 16 August 2022
Published: 7 September 2022

Publisher's Note: MDPI stays neutral with regard to jurisdictional claims in published maps and institutional affiliations.

Copyright: © 2022 by the authors. Licensee MDPI, Basel, Switzerland. This article is an open access article distributed under the terms and conditions of the Creative Commons Attribution (CC BY) license (https://creativecommons.org/licenses/by/4.0/).

1. Introduction

Illegal hunting of wildlife is prevalent in Africa and has reached crisis levels, as wildlife populations are decimated in 52% of forests, 62% of wilderness areas and 20% of protected areas, thereby threatening sustainability in biodiversity conservation and community livelihoods [1,2]. Illegal hunting refers to any capturing, shooting, killing or extraction of wildlife that is not explicitly sanctioned by the state or private owner of wildlife [3–5], and has possibly persisted in Africa because intervention measures or responses to illegal hunting have been less effective [6]. The sustained illegal hunting is attributed to poor understanding of illegal hunting and what motivates people to hunt illegally [6] and emanates mainly from inadequate empirical information on illegal hunting, a narrow view that it is only a conservation matter and the assertion that it is mainly driven by poverty [5]. However, some evidence does not support these assertions. Duffy et al. [5] indicated that the perspectives on illegal hunting were framed by certain understandings of poverty and that motivations for illegal hunting, such as those arising from complex historical context in regard to the outlaw of community, have not been adequately understood. Travers et al. [7] also found that a lack of alternative employment choices might be a more significant driver for hunters in Uganda than material poverty, which is contrary to the narrative that people hunt illegally because of poverty. Thus, the narratives on illegal hunting and its drivers may have been inadequate and simplistic. It is for this reason that Duffy et al. [5] and Travers et al. [7] advocated for a much broader understanding of complex illegal hunting and its drivers in order to design effective interventions. This justifies the need for conceptual views that provide a broader understanding on illegal hunting and what motivates people to hunt illegally.

Previous studies by Milner-Gulland and Leader-Williams [8], Hofer et al. [9], Damania et al. [10] and Keane et al. [11] developed models of relationships between illegal hunting and costs, benefits, sanctions, rewards and incentives. These models are mostly depicted in monetary form for benefits and highlighted law enforcement efforts (regulation) as a cost to illegal hunting. However, as observed by von Essen et al. [4], there are other non-monetary factors such as socio-political and normative values and beliefs that can influence hunters' illegal hunting behaviour. The economic models may not be robust enough to effectively represent the reality of the illegal hunting phenomenon. Therefore, despite providing some knowledge on the dynamics of illegal hunting, these models have application inadequacies in identifying relevant research variables and designing effective intervention measures against illegal hunting. Recently, Carter et al. [12] developed a conceptual framework for understanding illegal killing of large carnivores, which includes socio-economic, ecological and psychological factors and illustrates the complexity of illegal hunting. However, the study is focused on large carnivores which have specific illegal hunting risks and influences on the motivation to hunt illegally that might be different for other taxa. Thus, the conceptual framework by Carter et al. [12] requires validation of its applicability to wider taxa.

The literature on the illegal hunting phenomenon has emphasised instrumental economic theories despite other available perspectives, such as psychological and social-political, which also influence illegal hunting behaviour [4]. However, the significance of the psychological perspective in influencing illegal hunting behaviour is depicted in the Theory of Planned Behaviour, which holds that beliefs (attitude, subjective norms and perceived control) are determinants of both intentions and behaviour, with behavioural intentions being the most proximal determinant of social behaviour [13,14]. In relation to illegal hunting, the implication of the Theory of Planned Behaviour is that beliefs, norms and values towards illegal hunting determine the intention to hunt illegally and the illegal hunting behaviour. As a theoretical framework, the Theory of Planned Behaviour is therefore important to this study in facilitating the building of concepts on drivers of illegal hunting and in conceptualising how drivers of illegal hunting influence local hunters into illegal hunting behaviour. Considering that local hunters and the natural system and environment are linked and interdependent, the Socio-Ecological System (SES), as proposed by Ostrom [15], is adapted as this study's conceptual framework. Based on this conceptual framework, the local hunters are actors who are influenced by factors such as drivers of illegal hunting and others, and manifest illegal hunting behaviour within the system. Within the SES, the Theory of Planned Behaviour provides a theoretical underpinning for how drivers of illegal hunting influence local hunters' illegal hunting behaviour.

Notwithstanding a few studies on illegal hunting in Africa [1], studies in eastern, central, southern and western Africa that have identified drivers of illegal hunting provide some valuable information that can enhance cohesive conceptual understanding on the persistence of poaching on the continent. African countries may easily relate to one another in regard to the illegal hunting phenomenon because they have shared historical, socio-economic and political contexts. Therefore, the study uses a scoping review approach to provide an overview of the available evidence of what drives local hunters to hunt illegally in Africa based on the lived experiences of local hunters and not on perceptions from non-hunters. This study aims at developing concepts for drivers of illegal hunting and how they influence local hunters' illegal hunting behaviour in or adjacent to African protected areas. The study investigates the research question: what conceptual views can describe how drivers of illegal hunting influence local hunters' behaviours and the persistence of illegal hunting in or adjacent to protected areas? The review-based conceptual framework of drivers of illegal hunting may provide a basis for contributing to a broader understanding of drivers of illegal hunting in Africa and help in identifying relevant elements for designing effective intervention measures that ensure sustainable wildlife conservation. Based on our knowledge, this is the first time drivers of illegal hunting by local hunters are conceptualised as underlying, proximate and thematic drivers of illegal hunting which are developed into a conceptual framework and used to explain the persistent illegal hunting in Africa.

2. Materials and Methods

2.1. Study Area

This scoping review is based on the studies that were conducted in five African regions (eastern, northern, western, central and southern). The islands in the Atlantic and Indian Oceans, such as Cape Verde, Comoros, Madagascar, Mauritius, Réunion and Seychelles, are part of the study area. The study focal sites are areas in or adjacent to forest or wildlife protected areas where hunting of wildlife without a permit is considered illegal.

2.2. Review Protocol

Prior to the review process, the review protocol was set to guide the process. Firstly, the review objectives and questions are: to identify available evidence on what drives local hunters to hunt illegally and to identify concepts on drivers of illegal hunting and how they influence illegal hunting using available evidence in Africa. The sources of available evidence are peer-reviewed articles, such as journal research papers, PhD and master's theses and book chapters. The review questions include: what concepts does the available evidence provide on drivers of illegal hunting by local hunters, and what conceptual framework can be developed from the available evidence to depict how drivers of illegal hunting influence illegal hunting behaviour? Secondly, to address these review objectives and questions, the study uses online searching of databases on drivers of illegal hunting in Africa with Google Scholar and ScienceDirect. The protocol has a pre-determined search strategy for identifying articles in the databases and criteria for inclusion and exclusion of identified articles (see below for details). Thirdly, the protocol addresses the extraction and presentation of data. Relevant data from included articles are identified during review and indicated in respective rows for each item in the table (see Supplementary Materials: Table S1). Simple frequency calculations are done for each identified item and the summarised data are presented in the results table. Fourthly, in fostering transparency, the search and identification of articles and summarising of data are done by the first author and the co-authors verify and approve the process and results.

2.3. Search Strategy

The Google Scholar and ScienceDirect database search engines were used to search for relevant studies on the drivers of illegal hunting by local hunters in Africa. The database was searched using phrases or words such as "bushmeat hunting", "drivers of illegal hunting", "hunters", "illegal hunting", "illegal killing", "motivation for illegal hunting", "motivation for poaching" and "poaching". The identified publications were initially screened for relevance to the objective of this study.

2.4. Inclusion and Exclusion Criteria

The relevance of the identified publications was further screened by checking if they met the inclusion and exclusion criteria indicated in Table 1. The publications were included or excluded depending on whether they met or failed to meet the criteria, respectively. Sampling local hunters in the publications was a critical criterion because hunters or resource users may have different experiences of what motivates them to hunt illegally from perceptions of non-hunters [16].

Table 1. Criteria for inclusion and exclusion of articles in the scoping review process.

Criteria	Included	Excluded	Justification for the Criteria Used
Date of publication	2000 to 2021	Before 2000 and beyond 2021.	For current perspectives on drivers of illegal hunting and to access increased publications during this period.
Language of publication	English	All languages that are not English.	Researchers' proficiency in English language and to ensure increased readability.
Location of study	Publication on African countries.	Publications on non-African countries.	To maintain specific relevance and scope of the review.
Article availability	Available full papers identified though Google Scholar and ScienceDirect.	Full papers not accessible.	To access full/entire research findings from papers.
Type of articles	Peer-reviewed research journal articles, book chapters and PhD/master's theses.	Articles that are not peer-reviewed.	To ensure quality and validity of findings.
Publication content	Papers with drivers/motivations/ reasons for engaging in illegal hunting.	Papers without drivers/motivations/ reasons for engaging in illegal hunting.	To be specific and focused on the scope of the review.
Sampling methodology	Sampling local hunters through direct observations, interviews, questionnaires and focus group discussions.	Sampling non-hunters only.	To identify drivers of illegal hunting from lived experiences and not from perceptions by non-hunters.

2.5. *Identification and Analysis of Drivers of Illegal Hunting*

The included publications were reviewed to identify drivers of illegal hunting by local hunters. The number and frequency of identification of each driver of illegal hunting were recorded to indicate levels of pervasiveness of respective drivers of illegal hunting in Africa. The included publications were also qualitatively analysed to identify behavioural intentions by local hunters to hunt illegally. Behavioural intentions to hunting illegally were expressed beliefs (behavioural, normative and perceived control) towards illegal hunting based on the Theory of Planned Behaviour [13,14]. According to the Theory of Planned Behaviour, beliefs (attitude, subjective norms and perceived control) are determinants of both intentions and behaviour, with behavioural intentions being the most proximal determinant of social behaviour [13,14]. Therefore, responses by local hunters in questionnaires, interviews and group discussions, as reported or quoted in the included publications, that depicted behavioural, normative or perceived control beliefs were used to determine behavioural intention towards illegal hunting. Beliefs expressed as 'hunting is our birth right or cultural right', 'hunting is the only way to support my family', 'we have no other option apart from hunting' or 'we have other ways to outwit anti-poaching measures' were indicative of the behavioural intention to hunt illegally by local hunters. In this study, behavioural intentions to hunt illegally were considered drivers of illegal hunting based on the Theory of Planned Behaviour [17].

2.6. *Proximate, Underlying and Thematic Drivers of Illegal Hunting*

The identified drivers of illegal hunting from reviewed publications were categorised into proximate and underlying drivers. This categorisation follows the descriptions of proximate and underlying drivers adapted from those on tropical deforestation and conversion of natural vegetation to agricultural land use in Africa by Geist and Lambin [18] and Jellason et al. [19], respectively. Based on these adaptations, we characterised proximate drivers of illegal hunting as any immediate desires, feelings, shortages or needs by humans at a local level that directly trigger them to hunt illegally. Similarly, we characterised underlying drivers as factors that underpin, enhance or enable proximate drivers and may also work at the local level or have an indirect influence from the national or global levels. Further, the identified drivers of illegal hunting were qualitatively analysed and categorised into thematic drivers. Thematic drivers were determined by considering similarities or related characteristics of both proximate and underlying drivers and assigning them appropriate respective thematic driver categories.

2.7. Conceptual Framework of Drivers of Illegal Hunting by Local Hunters

The conceptual framework of drivers of illegal hunting was developed to depict the process of how underlying and proximate drivers, working with other social and ecological influences and constraints, affect illegal hunting behaviour. The conceptual framework was informed by and adapted from the Socio-Ecological System (SES) framework by Ostrom [15]. The relevance of SES to the development of a conceptual framework of drivers of illegal hunting by local hunters is that it considers human and natural systems as being linked and interdependent. Therefore, illegal hunting behaviour manifests in a natural system with complex, linked and interdependent components. The development of the conceptual framework of how drivers of illegal hunting affect illegal hunting behaviour was based on the Theory of Planned Behaviour [17].

3. Results

3.1. Search Results

The scoping review process identified 30 publications that were included for review from the initial 1014 articles identified using Google Scholar (n = 995) and ScienceDirect (n = 19) (see Figure 1). A total of 948 articles were excluded from 997 retrieved duplicate free articles, owing to the studies not being conducted in Africa, inaccessible full papers, and not addressing drivers or motivations for illegal hunting. A total of 19 full-text articles were excluded from 49 full-text articles which were assessed for eligibility for not sampling local hunters. The 30 studies which were included for review were conducted in 17 countries, with 13 studies from eastern, 10 from southern, 4 from western and 3 from central regions of Africa (Supplementary Materials: Table S1).

3.2. Identified Drivers of Illegal Hunting

A total of 17 drivers of illegal hunting were identified in the publications that were included for review (see Table 2). The need to generate income/no income source was the most frequently identified driver of illegal hunting by 26 studies (86.7%), followed by the need/preference for bushmeat consumption, identified by 25 studies (83.3%). The third-ranking drivers of illegal hunting were cultural needs/rights and preventative/retaliatory killing, both identified by 11 studies (36.7%). Poverty and weak/inadequate law enforcement were identified by 6 (20.0%) and 4 (13.3%) studies and ranked seventh and eighth among drivers of illegal hunting, respectively. Notably, defiance/protest as a driver of illegal hunting was identified by studies conducted in southern African regions only (see Supplementary Materials: Table S1).

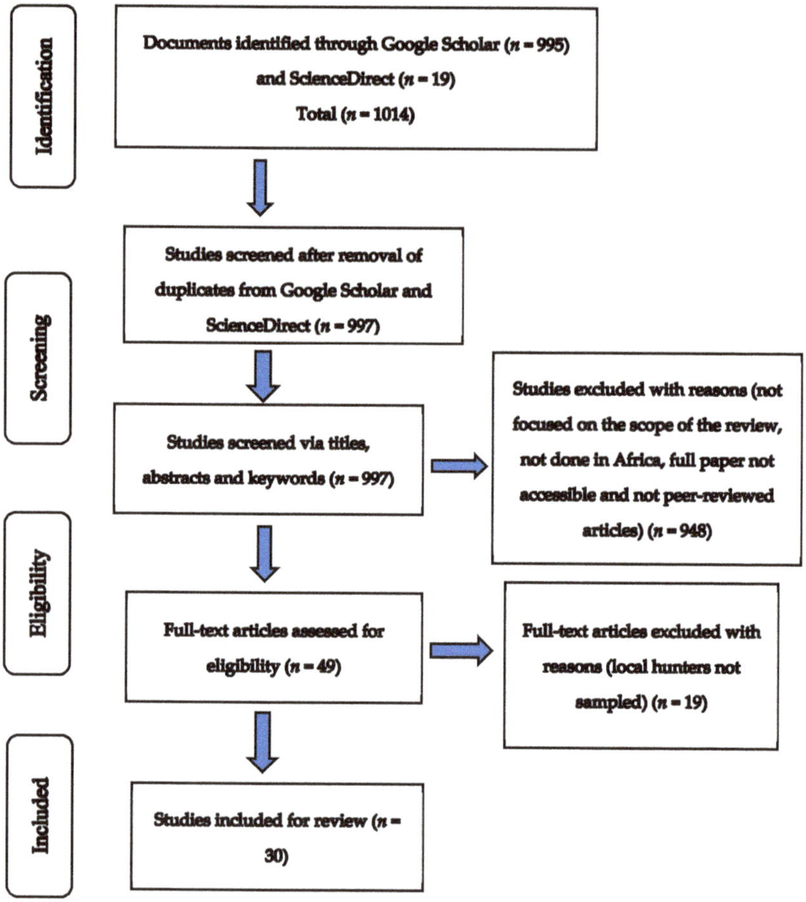

Figure 1. Flowchart indicating number of publications included and excluded during the scoping review process (based on Moher et al. [20]).

Table 2. Identified, proximate, underlying and thematic drivers of illegal hunting of wildlife derived from scoping review of publications (published from 2000 to 2021) that surveyed experiences of local hunters living in or adjacent to African protected areas.

Identified Drivers of Illegal Hunting	No. of Publications with Identified Drivers	Frequency of Identification of Drivers by Publications	Classification of Driver (Proximate or Underlying)	Thematic Drivers
Need to generate income/ no income source	26	86.7%	Proximate	Need for survival and sustaining livelihoods
Need/preference for bushmeat consumption	25	83.3%	Proximate	Need for survival and sustaining livelihoods
Cultural needs/rights	11	36.7%	Proximate	Cultural needs/significance

Table 2. *Cont.*

Identified Drivers of Illegal Hunting	No. of Publications with Identified Drivers	Frequency of Identification of Drivers by Publications	Classification of Driver (Proximate or Underlying)	Thematic Drivers
Preventative/retaliatory killing of wildlife	11	36.7%	Proximate	Human–wildlife conflict
Behavioural intention to hunt illegally	9	30.0%	Proximate	Behavioural intention to hunt illegally
Lack of employment/ livelihoods	8	26.7%	Proximate	Need for survival and sustaining livelihoods
Shortage/expensive/lack of protein source	7	23.3%	Proximate	Need for survival and sustaining livelihoods
Poverty	6	20.0%	Underlying	Need for survival and sustaining livelihoods
Weak/inadequate law enforcement	4	13.3%	Underlying	Inadequate legislation/ enforcement
Defiance/protest	3	10.0%	Proximate	Defiance/protest
Political instability/armed warfare	3	10.0%	Underlying	Political/armed conflicts
Demand for wildlife products	2	6.7%	Underlying	Market demand for wildlife products
Recreational/sports needs	2	6.7%	Proximate	Recreational needs
Population influx/increase	2	6.7%	Underlying	Demographic growth
Need for traditional medicine	2	6.7%	Proximate	Need for survival and sustaining livelihoods
Social status identity	1	3.3%	Proximate	Cultural needs/significance
Need for skins and bones	1	3.3%	Proximate	Cultural needs/significance

3.3. Proximate, Underlying and Thematic Drivers of Illegal Hunting

Among the identified drivers of illegal hunting, 12 (70.6%) were categorised as proximate drivers and 5 (29.4%) as underlying drivers. The first seven most frequently identified drivers of illegal hunting were proximate drivers and included the need for income generation (86.7%), the need for bushmeat consumption (83.3%), cultural needs/rights (36.7%), preventative/retaliatory killing of wildlife (36.7%), behavioural intention to hunt illegally (30.0%), lack of employment/livelihood (26.7%) and shortage/expensive/lack of protein source (23.3%). The underlying drivers were among the last 10 least frequently identified drivers of illegal hunting. Poverty (20.0%) and weak/inadequate law enforcement (13.3%) were the most frequently identified drivers among the underlying drivers.

The identified drivers of illegal hunting were further categorised into 10 thematic drivers (Table 2). Six identified drivers of illegal hunting were thematically categorised as need for survival and sustaining livelihoods, and five of these were ranked among the eight most frequently identified drivers by the included publications. The six identified drivers that contributed to the thematic driver of the need for survival and sustaining livelihoods were basically socio-economic drivers and included the need to generate income, need to consume bushmeat, lack of employment/livelihoods, shortage/expensive/lack of protein source, poverty and the need for traditional medicine. The second-ranking thematic driver of illegal hunting was cultural needs/significance and included cultural needs/rights, social status identity and the need for skins and bones.

The human–wildlife conflict was the third most identified thematic category of drivers of illegal hunting, with preventative/retaliatory killing of wildlife being a contributing driver. Next, after behavioural intention to hunt illegally, is the fifth-ranked thematic driver of illegal hunting, categorised as inadequate legislation/enforcement.

3.4. Conceptual Framework of Drivers of Illegal Hunting by Local Hunters

The conceptual framework was developed to show how underlying, proximate and most proximate drivers (behavioural intentions) influence illegal hunting behaviour (Figure 2). The drivers of illegal hunting behaviour are influenced and constrained by socio-ecological factors within the SES.

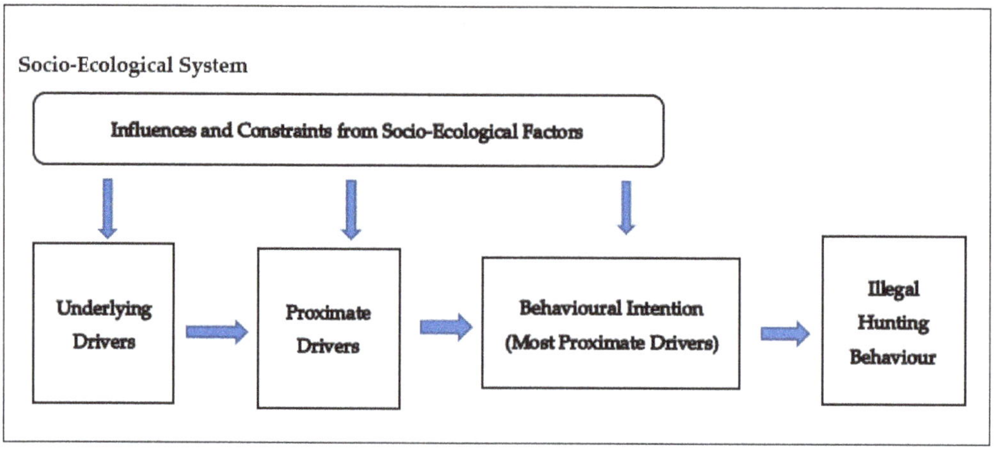

Figure 2. Conceptual framework of how underlying and proximate drivers and behavioural intentions influence illegal hunting behaviour of local hunters living in or adjacent to protected areas in Africa (based on Ostrom [15]).

4. Discussion

4.1. Identified Drivers of Illegal Hunting

This study shows that the needs to generate income (or lacking income sources) (86.7%) and consume bushmeat (83.3%) are the two most prevalent drivers of illegal hunting identified by studies among local hunters in Africa. Since illegal hunting is a wildlife crime that is considered an integral part of the illegal wildlife trade [21], it can be argued, therefore, that the illegal wildlife trade has thrived among local people in Africa primarily because local hunters are mostly driven to hunt illegally by the needs to generate income and consume bushmeat. Despite being less prevalently identified in publications, the other drivers of illegal hunting, such as cultural needs (36.7%), lack of employment (26.7%), lack of protein sources (23.3%), poverty (20.0%), demand for wildlife products (6.7%) and need for traditional medicine (6.7%), may play complementary roles to the two most prevalent drivers of the illegal wildlife trade among local hunters in Africa.

The study identified behavioural intention to hunt illegally (in 30% of included publications) as a driver of illegal hunting by analysing further the responses from hunters in the included publications. The survey and identification of behavioural intention to hunt illegally as one of the drivers of illegal hunting were not planned in the included publications for review. Other studies on drivers of illegal hunting have also not identified behavioural intention to hunt illegally as a driver of illegal hunting. Therefore, the identification and inclusion of behavioural intention as one of the drivers of illegal hunting is a novelty that may have implications on enhancing understanding on illegal hunting behaviour and potential intervention measures. The Theory of Planned Behaviour [14,17] provides support and a basis for adopting behavioural intention to hunt illegally as a driver of illegal hunting behaviour. The theory is particularly relevant, as it has been used in investigating potential predictors of illegal hunting and as a framework for assessing intervention measures against illegal hunting [22,23].

The defiance or protest against injustice and illegitimate authorities or rules as a driver of illegal hunting appear to be associated with the southern African region. This may be attributed to the historical background in southern Africa, where local people were racially discriminated against, dispossessed of land and disenfranchised from accessing land, wildlife and forest resources by colonial and militarised conservation authorities [24]. Local people hunt wildlife in defiance of rules and authorities and in protest against any perceived injustices that deny them to use resources that are considered a birthright [25].

4.2. Proximate, Underlying and Thematic Drivers of Illegal Hunting

The study here conceptualised the need for survival and sustaining livelihoods as the key driver of illegal hunting in Africa. Studies in Africa have provided empirical evidence that illegal hunting is a strategy employed by local hunters for survival and sustaining livelihoods [26–30]. Accordingly, this study has not conceptualised poverty as the key driver of illegal hunting in Africa, but as one of the underlying drivers which contributes to the major thematic driver of illegal hunting—the need for survival and sustaining livelihoods. Whereas some studies have indicated that poverty is a key driver of illegal hunting [31,32], others have shown that there is no evidence that poverty is the major driver of illegal hunting in some areas in Africa [7,33]. This implies that the role of poverty as a driver of illegal hunting varies across Africa. The contribution of poverty to the key thematic driver of poaching may be major or less significant, depending on the socio-economic situation of a study area. Therefore, conceptualising poverty as one of the contributing drivers to the key thematic driver of illegal hunting (need for survival and sustaining livelihoods) provides a unifying framework for both views.

In this study, cultural needs/significance was the second most identified thematic driver of illegal hunting among local hunters. A recent global systematic review by Lavadinović et al. [34] also identified socio-cultural influence as the third most prevalent identified driver of poaching. The corroborative findings of studies by Lavadinović et al. [34] and this study suggest that cultural factors (values, beliefs and norms towards wildlife) may have profound effects on hunter's behaviour towards wildlife species and their habitats and on societal responses to illegal hunting [35,36]. However, data on and understanding of socio-cultural values, beliefs and norms towards wildlife, bushmeat and the environment have been low [37]. It is only recently that recognition of the importance of considering how socio-cultural contexts influence illegal hunting and strategies for curbing the poaching problem is being made [36,38].

This study found that human–wildlife conflict was the third most identified thematic driver of illegal hunting in Africa, and thereby confirmed its importance in threatening conservation in Africa [39]. Human–wildlife conflict has direct major consequences for humans in rural people's development, income, health, food security and livelihoods [40], making it appropriate to be categorised under the thematic driver—the need for survival and sustaining livelihoods. This may additionally augment the need for survival and sustaining livelihoods as the key driver of illegal hunting in Africa. However, because human–wildlife conflict also has direct consequences on the conservation of individual animals, species and the broader ecosystem and biodiversity in African landscapes [39,40], it has been categorised under a separate thematic driver of illegal hunting for the need for survival and sustaining livelihoods. Considering human–wildlife conflict as a separate category, which is the third most identified thematic driver of illegal hunting, highlights it as a major African concern that has consequences on development, livelihoods and conservation.

A few studies (4, 13.3% frequency) identified weak/inadequate law enforcement as a driver of illegal hunting by local hunters, implying that it may not be a prevalent driver of illegal hunting to local hunters in Africa. Weak/inadequate law enforcement ranked as the eighth most identified driver of illegal hunting by local hunters. The low frequency of identification of weak/inadequate law enforcement as a driver of poaching by this study is surprising considering that other studies show that law enforcement is the most prioritised in terms of investments and the most studied among intervention measures against illegal hunting in Africa [1,41,42]. Ideally, the most prioritised intervention efforts in terms of investments should be targeted at addressing the most prevalent driver of illegal hunting by local hunters. However, the less prevalent driver of illegal hunting (weak/inadequate law enforcement) reported here is apparently targeted by highly prioritised efforts. In the identified studies, few local hunters indicated weak/inadequate law enforcement as a motivation for hunting illegally, probably because, as an underlying driver, it is indirect and less relevant to local hunters in influencing them to hunt illegally. It is similarly surprising that local hunters in the included publications did not indicate lack of/inadequate involvement of local communities in the management of wildlife as a driver of their poaching behaviour. However, it has been argued that involvement of local communities in wildlife management should be prioritised as a potentially effective intervention for addressing illegal hunting [3,43,44]. The probable explanation for this apparent conflicting situation again is that lack of/inadequate involvement in the management of wildlife is an underlying driver which may not directly influence local hunters to hunt illegally. This underscores the importance of understanding what motivates local hunters to hunt illegally, which might be different from perceptions of non-hunters located in or adjacent to African wildlife protected areas [16].

4.3. Conceptual Framework of Drivers of Illegal Hunting by Local Hunters

The proposed conceptual framework provides a relevant and enhanced understanding of illegal hunting behaviour of local hunters who live in or adjacent to African protected areas. Firstly, the novel inclusion of a behavioural intention to hunt illegally component as driver of illegal hunting behaviour may provide an expanded understanding of drivers of illegal hunting behaviour. Previous studies had not considered behavioural intention to hunt illegally as a driver of poaching, and thus it has had no specific intervention measures to address it. The Theory of Planned Behaviour [17] posits that behavioural intention is the most proximate determinant of an illegal behaviour, which therefore mediates proximate drivers and the illegal hunting behaviour in the conceptual framework. Behavioural intention is a function of beliefs which result from psychological aspects such as attitudes, subjective norms and perceived control. Based on the Theory of Planned Behaviour, the most proximate psychological drivers of illegal hunting in the proposed conceptual framework are probably the most critical drivers that energise other various drivers of illegal hunting to influence local hunters into illegal hunting behaviour. Secondly, the proposed conceptual framework may not be complicated, but represents complex linkages and interdependent components from underlying, proximate and most proximate drivers and socio-ecological factors. The socio-ecological factors and drivers of illegal hunting may be social, political, economic, psychological and ecological in nature. The linkages and interdependence of components in the framework underscores the appropriateness of basing the construction of the proposed framework on SES. The local hunters, as actors in the framework, are influenced by factors such as underlying and proximate drivers of illegal hunting, other socio-ecological factors and behavioural intentions to manifest illegal hunting behaviour. Therefore, this proposed conceptual framework may provide further understanding of how drivers of illegal hunting influence illegal hunting behaviour and what components should be targeted when tackling illegal hunting.

4.4. Implications on Interventions and Persistence of Illegal Hunting in Africa

The findings of this study have practical implications on the effectiveness of intervention measures and the persistence of illegal hunting in Africa. Firstly, the prioritised and increased efforts of law enforcement [1,41,42] may not address the main driver of illegal hunting by local hunters in Africa. Whereas the key driver of illegal hunting by local hunters is the need for survival and sustaining livelihoods, the main intervention measure is instead law enforcement. Local hunters who are motivated by the need to survive and sustain livelihoods may not be deterred by law enforcement but continue to hunt illegally by changing hunting tactics to those that are not easily detectable by law enforcement workers [45,46]. This is because law enforcement is used as a measure for dealing with illegal hunting activities and not the drivers. Further, Milner-Gulland and Leader-Williams [8] found that very high levels of enforcement deterred outsider-organised hunters and not local hunters and reported that local hunters responded positively to community schemes that provided employment for addressing poaching. The positive response to community schemes is because it addressed the main driver of illegal hunting among local hunters. Therefore, using law enforcement to address illegal hunting that is mainly driven by the need to survive and sustain livelihoods among local hunters is a mismatched intervention measure. Under these circumstances, illegal hunting by local hunters may persist decimating wildlife in African protected areas.

Secondly, behavioural intention to hunt illegally has not been considered one of the drivers of poaching, and therefore no specific intervention measures have been designed to address it. Behavioural intention is a function of beliefs (behavioural or attitude towards hunting, normative and perceived control) and are crucial in influencing illegal hunting behaviour. People living in the same location and influenced by the same underlying and proximate drivers may exhibit opposite behaviours (some hunting illegally and others not hunting) owing to differences in their behavioural intentions towards illegal hunting. Therefore, illegal hunting may have persisted in Africa due to lacking specific intervention measures for addressing behavioural intentions to hunt illegally.

Thirdly, each identified underlying, proximate and most proximate (behavioural intention) driver provides the contextual bases for identifying relevant elements and designing specific intervention measures against illegal hunting motivations. Each identified driver of illegal hunting should be addressed by specific intervention measures to ensure poaching is effectively tackled. Since prevalence levels of the drivers of illegal hunting are different, as observed in this study, levels of efforts for intervention measures should be distributed accordingly. However, the efforts and investments for addressing the most prevalent drivers of illegal hunting, such as the need for survival and sustaining livelihoods, have been left disproportionately less than those of law enforcement [41,42]. As a result, the main driver of illegal hunting among local hunters may not be effectively addressed and this may consequently facilitate the persistence of illegal hunting in Africa.

Fourthly, despite being the second most identified thematic driver of illegal hunting among local hunters in Africa, cultural needs/significance, and particularly the cultural constructions that shape values, attitudes, beliefs and norms towards wildlife in respective contexts, are inadequately known and understood [37]. Inadequate data on and understanding of socio-cultural factors are major constraints in designing specific intervention measures for addressing or mitigating the cultural needs/significance as a driver of poaching. As such, intervention measures that are based on inadequate understanding are likely to fail in addressing illegal hunting that is driven by the cultural needs/significance among local hunters and may thereby contribute to persistence of illegal hunting in Africa.

Fifthly, the third most identified thematic driver of illegal hunting in Africa, the human–wildlife conflict, is escalating globally because of competition for space and resources, such as water and food, by wildlife and growing human populations and expanded cultivation and livestock husbandry, which is influencing increased illegal preventative and retaliatory killings due to crop and livestock depredation by wildlife [39,40,47]. However, efforts to address human–wildlife conflict have been failing, as the level of the solutions does not match the level of the problem, and they are usually not applied to scale and holistically [40]. The escalating human–wildlife conflict is probably another reason contributing to persistence of illegal hunting in Africa.

Based on the foregoing, we postulate that the persistent (and prevalent) illegal hunting (and, implicitly, the illegal wildlife trade) by local hunters in or adjacent to African wildlife protected areas may be associated mainly with two factors: the first factor is the prevalence of drivers of illegal hunting that are related to the need for survival and sustaining livelihoods among local hunters. These drivers of illegal hunting include the need to generate income, need for bushmeat consumption, lack of employment, poverty and the need for traditional medicine. As discussed earlier, the human–wildlife conflict may also appropriately be included among drivers that relate to the need for survival and sustaining livelihoods among local hunters. Secondly, the persistent illegal hunting by local hunters in Africa is also probably associated with the prevalence of unaddressed or ineffectively addressed drivers of illegal hunting. The unaddressed or ineffectively addressed drivers of poaching in this regard relate mostly to the need for survival and sustaining livelihoods among local hunters.

4.5. Limitations of the Study

The study did not consider hunters who were not living in nor adjacent to protected areas and perceptions of local people or stakeholders who are non-hunters on what drives local hunters to hunt illegally in areas. Therefore, the study may not provide further understanding of whether perceptions of non-hunters on drivers of illegal hunting are different to drivers indicated by local hunters. Consequently, the study may also not determine whether surveying perceptions from non-hunters on drivers of illegal hunting is valid for use in designing intervention measures. Another limitation is the exclusion criterion for articles not published in the English language. This could have biased results because there are countries in Africa with official languages that are not English. This could have been mitigated by using translators for articles published in other languages. However, it was assumed that numbers of articles published on the subject matter in non-English languages were very small and it was observed in this study that there were articles published in English from regions with non-English official languages. Furthermore, another limitation of this study is that searches on Google Scholar databases retrieved more articles than those on ScienceDirect databases; in particular, Google Scholar has limitations for use as a single database search source [48]. Using both Google Scholar and ScienceDirect would suffice for a scoping review which is an overview study where assessments for quality and effectiveness assessment may not be required.

4.6. Future Directions

We recommend that comparable studies that consider hunters that live farther away from protected areas and non-hunters who live in or adjacent to protected areas be conducted to determine drivers and perceptions, respectively. The studies would compare findings on drivers from these populations and determine if perceptions by non-hunters are different from the experiences of local and distant hunters. Site-specific studies may be conducted for testing, comparison and validation of the conceptual views reported in this study. We also recommend for studies to enhance understanding of the identification of the behavioural intention to hunt illegally, as the most proximate driver of illegal hunting, and of how the adoption of intervention measures for addressing it affects illegal hunting behaviour. Furthermore, we recommend that when designing intervention measures, it is critical to ensure that measures are not designed to deal with superficial illegal hunting activities, but the causes which are drivers of illegal hunting behaviour. Therefore, all drivers of illegal hunting (including behavioural intentions to hunt illegally) should be identified first and then specific intervention measures for each driver should be designed.

5. Conclusions

The problem of inadequate evidence-based information has led to ineffective and restricted choices of intervention measures for tackling illegal hunting that is persistently decimating wildlife populations in African protected areas. However, we believe this study has contributed to enhancing understanding of illegal hunting and what motivates people to hunt illegally. Firstly, the study identified behavioural intention to hunt illegally as one of the 17 drivers of illegal hunting by local hunters in Africa, which hitherto had not been considered a driver of illegal hunting by previous studies. Secondly, the study conceptualised identified drivers of illegal hunting as proximate, underlying and thematic. Consequently, the need for survival and sustaining livelihoods was conceptualised as the main thematic driver of illegal hunting in Africa, a narrative that is different from the one that highlights poverty as a key driver. Thirdly, the conceptual framework suggested by this study represents how drivers of illegal hunting influence illegal hunting behaviour and provides a novel aspect that might enhance further understanding on illegal hunting by local hunters in Africa. The findings of this study may be helpful to researchers and conservationists in providing concepts, statistics and frameworks on drivers of illegal hunting for application in identifying relevant variables for designing research projects, intervention measures and strategies for tackling illegal hunting drivers and ensuring sustainability in wildlife conservation.

Supplementary Materials: The following supporting information can be downloaded at: https://www.mdpi.com/article/10.3390/su141811204/s1, Table S1: Summary of included publications with drivers of illegal hunting by local hunters in Africa identified using a scoping review of literature published from 2000 to 2021. References [7,24,25,27,28,30,32,33,43,49–69] are cited in the Supplementary Materials.

Author Contributions: P.Z.—Conceptualisation of the study; design of methodology; conducting literature search; data analysis; interpretation of results, writing, reviewing and editing the draft manuscript. F.K.K.—Providing advice on conceptualisation of the study, supervision of the study, interpretation of results and making corrections to the draft. J.M.—Providing advice on conceptualisation of the study, scholarly supervision of the study, interpretation of results and making corrections to the draft. V.R.N.—Providing advice on conceptualisation of the study, interpretation of results, making comments and corrections to the draft. All authors have read and agreed to the published version of the manuscript.

Funding: This research received no external funding.

Institutional Review Board Statement: This is not applicable, as this study is a review of already peer-reviewed publications, and the study does not involve contact with animal and human targets.

Informed Consent Statement: This is not applicable, as this study is a review of already peer-reviewed publications and the study does not involve contact with animal and human targets.

Data Availability Statement: Data are contained within the article and Supplementary Material: Table S1.

Acknowledgments: This study is part of a PhD study by the first author, and the academic staff in the School of Postgraduate Studies at the University of Lusaka provided valuable advice and guidance.

Conflicts of Interest: The authors declare no conflict of interest.

References

1. van Velden, J.; Wilson, K.; Biggs, D. The evidence for the bushmeat crisis in African savannas: A systematic quantitative literature review. *Biol. Conserv.* **2018**, *221*, 345–356. [CrossRef]
2. Benítez-López, A.; Santini, L.; Schipper, A.M.; Busana, M.; Huijbregts, M.A.J. Intact but empty forests? Patterns of hunting-induced mammal defaunation in the tropics. *PLOS Biol.* **2019**, *17*, e3000247. [CrossRef] [PubMed]
3. Kahler, S.J.; Roloff, G.; Gore, L.G. Poaching risks in a Community-Based Natural Resource System. *J. Conserv. Biol.* **2013**, *27*, 177–186. [CrossRef]
4. von Essen, E.; Hansen, H.P.; Källström, H.N.; Peterson, M.N.; Peterson, T.R. Deconstructing the poaching phenomenon. *Br. J. Criminol.* **2014**, *54*, 632–651. [CrossRef]
5. Duffy, R.; John, F.S.; Büscher, B.; Brockington, D. Toward a new understanding of the links between poverty and illegal wildlife hunting. *Conserv. Biol.* **2015**, *30*, 14–22. [CrossRef] [PubMed]
6. Challender, D.; Macmillan, D.C. Poaching is more than an Enforcement Problem. *Conserv. Lett.* **2014**, *7*, 484–494. [CrossRef]
7. Travers, H.; Archer, L.J.; Mwedde, G.; Roe, D.; Baker, J.; Plumptre, A.J.; Rwetsiba, A.; Milner-Gulland, E. Understanding complex drivers of wildlife crime to design effective conservation interventions. *Conserv. Biol.* **2019**, *33*, 1296–1306. [CrossRef]
8. Milner-Gulland, E.; Leader-Williams, N. A Model of Incentives for the Illegal Exploitation of Black Rhinos and Elephants: Poaching Pays in Luangwa Valley, Zambia. *J. Appl. Ecol.* **1992**, *29*, 388. [CrossRef]
9. Hofer, H.; Campbell, K.; East, M.; Huish, S. Modelling the spatial costs and benefits of illegal game meat hunting in the Serengeti. *Nat. Resour. Model.* **2000**, *13*, 151–177. [CrossRef]
10. Damania, R.; Milner-Gulland, E.; Crookes, D.J. A bioeconomic analysis of bushmeat hunting. *Proc. R. Soc. B Boil. Sci.* **2005**, *272*, 259–266. [CrossRef]

11. Keane, A.; Jones, J.P.G.; Milner-Gulland, E.J. Modelling the effect of individual strategic behaviour on community-level outcomes of conservation interventions. *Environ. Conserv.* **2012**, *39*, 305–315. [CrossRef]
12. Carter, N.H.; López-Bao, J.V.; Bruskotter, J.T.; Gore, M.; Chapron, G.; Johnson, A.; Epstein, Y.; Shrestha, M.; Frank, J.; Ohrens, O.; et al. A conceptual framework for understanding illegal killing of large carnivores. *Ambio* **2016**, *46*, 251–264. [CrossRef] [PubMed]
13. Ajzen, I. From intentions to actions: A theory of planned behaviour. In *Action-Control: From Cognition to Behaviour*; Juhi, J., Beckmann, J., Eds.; Springer: Berlin/Heidelberg, Germany, 1985; pp. 11–39.
14. Ajzen, T. The theory of planned behaviour. *Organ. Behav. Hum. Decis. Process* **1991**, *50*, 179–211. [CrossRef]
15. Ostrom, E. A general framework for analysing sustainability of Social-Ecological Systems. *Science* **2009**, *325*, 419–422. [CrossRef]
16. Kisingo, A.; Wilfred, P.; Magige, F.; Kayeye, H.; Nahonyo, C.; Milner-Gulland, E. Resource managers' and users' perspectives on factors contributing to unauthorised hunting in western Tanzania. *Afri. J. Ecol.* **2022**, *60*, 277–288. [CrossRef]
17. Ajzen, T. The Theory of Planned Behaviour. In *Handbook of Theories of Social Psychology*; Lange, P.A.M., Kruglanski, A.W., Higgins, E.T., Eds.; Sage: London, UK, 2012; Volume 1, pp. 438–459.
18. Geist, H.J.; Lambin, E.F. Proximate causes and underlying driving forces of tropical deforestation. *BioScience* **2002**, *52*, 143–150. [CrossRef]
19. Jellason, N.P.; Robinson, E.J.Z.; Chapman, A.S.A.; Neina, D.; Devenish, A.J.M.; Po, J.Y.T.; Adolph, B. A systematic review of drivers and constraints on agricultural expansion in sub-Saharan Africa. *Land* **2021**, *10*, 332. [CrossRef]
20. Moher, D.; Liberati, A.; Tetzlaff, J.; Altman, D.G.; PRISMA Group. Preferred reporting items for systematic reviews and meta-analyses: The PRISMA statement. *PLoS Med.* **2009**, *6*, e1000097. [CrossRef]
21. 't Sas-Rolfes, M.; Challender, D.W.S.; Hinsley, A.; Veríssimo, D.; Milner-Gulland, E.J. Illegal Wildlife Trade: Scale, Processes, and Governance. *Annu. Rev. Environ. Resour.* **2019**, *44*, 201–228. [CrossRef]
22. Steinmetz, H.; Knappstein, M.; Ajzen, I.; Schmidt, P.; Kabst, R. How Effective are Behavior Change Interventions Based on the Theory of Planned Behavior? *Z. Psychol.* **2016**, *224*, 216–233. [CrossRef]
23. Newth, J.L.; McDonald, R.A.; Wood, K.A.; Rees, E.C.; Semenov, I.; Chistyakov, A.; Mikhaylova, G.; Bearhop, S.; Cromie, R.L.; Belousova, A.; et al. Predicting intention to hunt protected wildlife: A case study of Bewick's swan in the European Russian Arctic. *Oryx* **2021**, *56*, 228–240. [CrossRef]
24. Konono, T. Understanding the Impact of Green Violence on Ndali Village, Zimbabwe. Master's Thesis, University of Cape Town, Cape Town, South Africa, 2021.
25. Witter, R. Why militarized conservation may be counter-productive: Illegal wildlife hunting as defiance. *J. Political Ecol.* **2021**, *28*, 175–192. [CrossRef]
26. Bennett, E.L.; Blencowe, E.; Brandon, K.; Brown, D.; Burn, R.W.; Cowlishaw, G.U.Y.; Davies, G.; Dublin, H.; Fa, J.E.; Milner-Gulland, E.J.; et al. Hunting for consensus: Reconciling bushmeat harvest, conservation and development policy in West and Central Africa. *Conserv. Biol.* **2007**, *21*, 884–887. [CrossRef] [PubMed]
27. Brown, T.; Marks, S.A. Livelihood, hunting and the game meat trade in Northern Zambia. In *Bushmeat and Livelihoods: Wildlife Management and Poverty Reduction*; Glyn, D., David, B., Eds.; Blackwell Publishing Ltd.: Oxford, UK, 2007; pp. 92–105.
28. Alexander, J.S.; McNamara, J.; Rowcliffe, J.M.; Oppong, J.; Milner-Gulland, E. The role of bushmeat in a West African agricultural landscape. *Oryx* **2014**, *49*, 643–651. [CrossRef]
29. Hariohay, K.; Ranke, P.S.; Fyumagwa, R.D.; Kideghesho, J.R.; Røskaft, E. Drivers of conservation crimes in the Rungwa-Kizigo-Muhesi Game Reserves, Central Tanzania. *Glob. Ecol. Conserv.* **2019**, *17*, e00522. [CrossRef]
30. Lubilo, R.; Hebinck, P. 'Local hunting' and community-based natural resource management in Namibia: Contestations and livelihoods. *Geoforum* **2019**, *101*, 62–75. [CrossRef]
31. Adams, W.M.; Aveling, R.; Brockington, D.; Dickson, B.; Elliott, J.; Hutton, J.; Roe, D.; Vira, B.; Wolmer, W. Biodiversity Conservation and the Eradication of Poverty. *Science* **2004**, *306*, 1146–1149. [CrossRef] [PubMed]
32. Knapp, E.J.; Peace, N.; Bechtel, L. Poachers and Poverty: Assessing Objective and Subjective Measures of Poverty among Illegal Hunters Outside Ruaha National Park, Tanzania. *Conserv. Soc.* **2017**, *15*, 24. [CrossRef]
33. Rogan, M.S.; Miller, J.R.B.; Lindsey, P.A.; McNutt, J.W. Socioeconomic drivers of illegal bushmeat hunting in a southern Africa savanna. *Biol. Conserv.* **2018**, *226*, 24–31. [CrossRef]
34. Lavadinović, V.M.; Islas, C.A.; Chatakonda, M.K.; Marković, N.; Mbiba, M. Mapping the Research Landscape on Poaching: A Decadal Systematic Review. *Front. Ecol. Evol.* **2021**, *9*, 630990. [CrossRef]
35. Kideghesho, J.R. The potentials of traditional African cultural practices in mitigating overexploitation of wildlife species and habitat loss: Experience of Tanzania. *Int. J. Biodivers. Sci. Manag.* **2009**, *5*, 83–94. [CrossRef]
36. Rizzolo, J.B.; Gore, M.; Ratsimbazafy, J.H.; Rajaonson, A. Cultural influences on attitudes about the causes and consequences of wildlife poaching. *Crime Law Soc. Chang.* **2016**, *67*, 415–437. [CrossRef]
37. Van Vliet, N. "Bushmeat Crisis" and "Cultural Imperialism" in Wildlife Management? Taking Value Orientations Into Account for a More Sustainable and Culturally Acceptable Wildmeat Sector. *Front. Ecol. Evol.* **2018**, *6*, 112. [CrossRef]
38. Kassa, S.; Baez-Camargo, C.; Costa, J.; Lugolobi, R. Determinants and Drivers of Wildlife Trafficking: A Qualitative Analysis in Uganda. *J. Int. Wildl. Law Policy* **2021**, *24*, 314–342. [CrossRef]
39. Lamarque, F.; Anderson, J.; Fergusson, R.; Lagrange, M.; Osei-Owusu, Y.; Bakker, L. *Human-Wildlife Conflict in Africa: Causes, Consequences and Management Strategies*; FAO Forestry Paper 157; FAO: Rome, Italy, 2009; p. 98.

40. Gross, E.; Jayasinghe, N.; Brooks, A.; Polet, G.; Wadhwa, R.; Hilderink-Koopmans, F. *A Future for All: The Need for Human-Wildlife Coexistence*; WWF: Gland, Switzerland, 2021; p. 103.
41. Cooney, R.; Roe, D.; Dublin, H.; Phelps, J.; Wilkie, D.; Keane, A.; Travers, H.; Skinner, D.; Challender, D.; Allan, J.R.; et al. From Poachers to Protectors: Engaging Local Communities in Solutions to Illegal Wildlife Trade. *Conserv. Lett.* **2016**, *10*, 367–374. [CrossRef]
42. Holden, M.H.; Biggs, D.; Brink, H.; Bal, P.; Rhodes, J.; McDonald-Madden, E. Increase anti-poaching law-enforcement or reduce demand for wildlife products? A framework to guide strategic conservation investments. *Conserv. Lett.* **2018**, *12*, e12618. [CrossRef]
43. Mfunda, I.M.; Holmern, I.M.; Røskaft, E. Benefits and access to natural resources influence conservation perceptions and relationship between local people and other stakeholder: The case of Serengeti, Tanzania. *Int. J. Biodivers. Conserv.* **2012**, *4*, 535–547.
44. Massé, F.; Gardiner, A.; Lubilo, R.; Themba, M. Inclusive Anti-poaching? Exploring the Potential and Challenges of Community-based Anti-Poaching. *S. Afr. Crime Q.* **2017**, *60*, 19–27. [CrossRef]
45. Gibson, C.C.; Marks, S.A. Transforming rural hunters into conservationists: An assessment of community-based wildlife management programs in Africa. *World Dev.* **1995**, *23*, 941–957. [CrossRef]
46. Marks, S.A. Back to the Future: Some Unintended Consequences of Zambia's Community-Based Wildlife Program (ADMADE). *Afr. Today* **2001**, *48*, 120–141. [CrossRef]
47. Nyhus, P.J. Human-wildlife conflict and coexistence. *Annu. Rev. Environ. Resour.* **2016**, *41*, 143–171. [CrossRef]
48. Haddaway, N.R.; Collins, A.; Coughlin, D.; Kirk, A.S. The Role of Google Scholar in Evidence Reviews and Its Applicability to Grey Literature Searching. *PLoS ONE* **2015**, *10*, e0138237. [CrossRef]
49. Carpenato, G.M.; Fusari, A. Subsistence hunting and bushmeat exploitation in central-western Tanzania. *Biodivers. Conserv.* **2000**, *9*, 1571–1585. [CrossRef]
50. Merson, S.D.; Dollar, L.J.; Johnson, P.J.; Macdonald, D.W. Retaliatory killing and human perceptions of Madagascar's largest carnivore and livestock predator, the fosa (Cryptoprocta ferox). *PLoS ONE* **2019**, *14*, e0213341. [CrossRef]
51. Knapp, E.J. Why poaching pays: A summary of risks and benefits illegal hunters face in Western Serengeti, Tanzania. *Trop. Conserv. Sci.* **2012**, *5*, 434–445. [CrossRef]
52. van Velden, J.L.; Wilson, K.; Lindsey, P.A.; McCallum, H.; Moyo, B.H.Z.; Biggs, D. Bushmeat hunting and consumption is a pervasive issue in Africa savannahs: Insights from four protected areas in Malawi. *Biodivers. Conserv.* **2020**, *29*, 1443–1464. [CrossRef]
53. King, E.C.P. Hunting for A Problem: An Investigation into Bushmeat Use around North Luangwa National Park, Zambia. Master's Thesis, Imperial College London, London, UK, 2014.
54. Loibooki, M.; Hofer, H.; Campbell, K.L.; East, M.L. Bushmeat hunting by communities adjacent to the Serengeti National Park, Tanzania: The importance of livestock ownership and alternative sources of protein and income. *Environ. Conserv.* **2002**, *29*, 391–398. [CrossRef]
55. Braga-Pereira, F.; Peres, C.A.; Campos-Silva, J.V.; Santos, C.V.-D.; Alves, R.R.N. Warfare-induced mammal population declines in Southwestern Africa are mediated by species life history, habitat type and hunter preferences. *Sci. Rep.* **2020**, *10*, 1–15. [CrossRef]
56. Lindsey, P.A.; Romañach, S.; Matema, S.; Matema, C.; Mupamhadzi, I.; Muvengwi, J. Dynamics and underlying causes of illegal bushmeat trade in Zimbabwe. *Oryx* **2011**, *45*, 84–95. [CrossRef]
57. De Merode, E.; Cowlishaw, G. Species protection, the changing informal economy, and the politics of access to the bushmeat trade in the Democratic Republic of Congo. *Conserv. Biol.* **2006**, *20*, 1262–1271. [CrossRef]
58. Golden, C.D. Bushmeat hunting and use in the Makira Forest, north-eastern Madagascar: A conservation and livelihood issue. *Oryx* **2009**, *43*, 386–392. [CrossRef]
59. Friant, S.; Paige, S.B.; Goldberg, T.L. Drivers of Bushmeat Hunting and Perceptions of Zoonoses in Nigerian Hunting Communities. *PLOS Negl. Trop. Dis.* **2015**, *9*, e0003792. [CrossRef] [PubMed]
60. Akinsorotan, O.; Olaniyi, O.; Oguntuase, B.; Raheem, T. Dynamics and Socioeconomic Drivers of Illegal Hunting of Wildlife Animal for Consumption in Oba Hills Forest Reserve in Southwest Nigeria. *J. Appl. Sci. Environ. Manag.* **2020**, *24*, 287–298. [CrossRef]
61. Ceppi, S.L.; Nielsen, M.R. A Comparative Study on Bushmeat Consumption Patterns in ten Tribes in Tanzania. *Trop. Conserv. Sci.* **2014**, *7*, 272–287. [CrossRef]
62. Abere, S.A.; Lateef, F.L.; Lameed, G.A. Assessment of Hunters and Other Rate of Illegal Activities in Afi-Mbe-Okwango Division, Cross River State, Nigeria. *Nat. Resour.* **2016**, *07*, 287–294. [CrossRef]
63. Manqele, N.S.; Selier, J.A.; Hill, T.R.; Downs, C.T. Drivers of the Illegal Hunting of Serval (*Leptailurus serval*) and Oribi (*Ourebiaourebi*) in the KwaZulu-Natal Midlands, South Africa. *Afr. J. Wildl. Res.* **2018**, *48*. [CrossRef]
64. Kümpel, N.F.; Milner-Gulland, E.; Cowlishaw, G.; Rowcliffe, M. Incentives for Hunting: The Role of Bushmeat in the Household Economy in Rural Equatorial Guinea. *Hum. Ecol.* **2010**, *38*, 251–264. [CrossRef]
65. Erena, M.G.; Bekele, A.; Debella, H.J. Bushmeat hunting practices in Jorgo-Wato Protected Forest, Western Ethiopia. *Int. J. Ecol. Environ. Sci.* **2020**, *46*, 61–72.
66. Nielsen, M.R. Importance, cause and effects of bushmeat hunting in the Udzungwa Mountains, Tanzania: Implications for community-based wildlife management. *Biol. Conserv.* **2006**, *128*, 509–516. [CrossRef]

67. Kiffner, C.; Peters, L.; Stroming, A.; Kioko, J. Bushmeat Consumption in the Tarangire-Manyara Ecosystem, Tanzania. *Trop. Conserv. Sci.* **2015**, *8*, 318–332. [CrossRef]
68. Nyaki, A.; Gray, S.A.; Lepczyk, C.A.; Skibins, J.C.; Rentsch, D. Local-Scale Dynamics and Local Drivers of Bushmeat Trade. *Conserv. Biol.* **2014**, *28*, 1403–1414. [CrossRef] [PubMed]
69. LeBreton, M.; Prosser, A.; Tamoufe, U.; Sateren, W.; Mpoudi-Ngole, E.; Diffo, J.L.D.; Burke, D.S.; Wolfe, N.D. Patterns of bushmeat hunting and perceptions of disease risk among central African communities. *Anim. Conserv.* **2006**, *9*, 357–363. [CrossRef]

sustainability

MDPI

Communication

Neighboring Green Network and Landscape Metrics Explain Biodiversity within Small Urban Green Areas—A Case Study on Birds

Marufa Sultana [1,*], Max Müller [2], Magdalena Meyer [2] and Ilse Storch [1]

[1] Wildlife Ecology and Management, Faculty of Environment and Natural Resources, University of Freiburg, D-79106 Freiburg, Germany; ilse.storch@wildlife.uni-freiburg.de

[2] Institute of Evolutionary Ecology and Conservation Genomics, University of Ulm, D-89081 Ulm, Germany; max.mueller@uni-ulm.de (M.M.); magdalena.meyer@uni-ulm.de (M.M.)

* Correspondence: marufa.sultana@wildlife.uni-freiburg.de or marufa_sultana@hotmail.com

Abstract: Cities' green areas are fragmented patches and are often confined to smaller sizes than the higher built-up proportions. Such small-sized green areas can be essential components of green infrastructure to compensate for biodiversity loss. As a proxy to biodiversity, we studied birds in nine small green area locations of Freiburg and eight area locations in Regensburg in Germany. We investigated the neighboring green networks (distance to the nearest water body and another green area) and landscape metrics (patch abundance and habitat heterogeneity at a 1 km radius) that might benefit and explain bird richness and composition in small green areas. We found that the variations in the observed species richness and composition at the surveyed locations were better explained solely by green networks in Freiburg and by green networks and landscape metrics in Regensburg. In general, it indicates that a small green area could be biodiverse if its spatial distribution considers a nearby water body and other green areas, allowing a higher abundance of similar patches and habitat heterogeneity in the neighborhood.

Keywords: park; urbanization; green infrastructure; Freiburg; Regensburg

Citation: Sultana, M.; Müller, M.; Meyer, M.; Storch, I. Neighboring Green Network and Landscape Metrics Explain Biodiversity within Small Urban Green Areas—A Case Study on Birds. *Sustainability* **2022**, *14*, 6394. https://doi.org/10.3390/su14116394

Academic Editor: Vasilios Liordos

Received: 4 April 2022
Accepted: 22 May 2022
Published: 24 May 2022

Publisher's Note: MDPI stays neutral with regard to jurisdictional claims in published maps and institutional affiliations.

Copyright: © 2022 by the authors. Licensee MDPI, Basel, Switzerland. This article is an open access article distributed under the terms and conditions of the Creative Commons Attribution (CC BY) license (https://creativecommons.org/licenses/by/4.0/).

1. Introduction

Urbanization is a long-standing phenomenon and can facilitate alteration in the natural landscape, threatening biodiversity in cities. Due to this concern, sustainable city planning integrating green infrastructure components is now an advancing concept [1]. Here, green areas, such as parks, street trees, campgrounds, and golf courses are essential components since they function as stepping stones and improve the connectivity of highly fragmented habitats in cities [2,3]. These features are also recognized as habitat islands [4], facilitate mobility and dispersal of urban species, and thereby benefit community composition. Thus, urban green areas provide ecological benefits by supporting biodiversity beyond their traditional functions for recreation and food sources for city residents.

It is well understood that larger green areas support greater species diversity in cities [2,4–6]. Therefore, a green area size threshold of circa 3.5 ha is suggested, the point below which species diversity declines [7,8]. Nevertheless, there is a usual trade-off in planning and allotment of green areas' extent in urban areas, which confines the size of such patches. Given the consequences, it is essential to understand whether and how a comparatively small-sized green area within an urban locality could still support biodiversity to some extent. In particular, studies are required to evaluate the role of the surrounding landscape characteristics in the effectiveness of small-sized urban green areas, such as community parks, street parks, and playgrounds, to inform urban biodiversity planning [2].

Several studies elucidated biodiversity patterns, namely species-specific occurrences, diversity, and composition patterns, within urban green areas [5,9,10]. It contributed to the understanding that species-specific responses may differ concerning urban environmental factors in small green areas [11,12]. However, overall species diversity is generally shaped by anthropogenic disturbances in the neighborhood. For example, species variety declines due to the adverse effects of noise level, car intensity, pedestrian movement, and built-up proportions, whereas it improves with the increasing amount of green area and native vegetation complexity [3,5,9,11–14]. In contrast, the understanding of how surrounding landscape factors might benefit biodiversity patterns in small urban green areas (i.e., are usually highly fragmented) is still unclear.

In this study, we focused on birds (as a proxy to biodiversity) in small urban green areas of two European cities, Freiburg and Regensburg, in Germany. Birds are one of the most widely studied taxa in urban ecology [15–17]. Since birds respond to any environmental change quickly and are easy to observe, avian diversity can be an excellent indicator of city biodiversity [18–20]. Furthermore, higher avian variety represents better urban habitat status and indicates greater recreational experiences for local people in green areas [21].

Here, we aimed to understand the contribution of neighboring landscape factors that likely benefit birds and explain variations in species richness and composition in small-sized green urban areas. Bird diversity has been shown to decline in highly built-up areas [10], however, increases in the presence of green areas and higher habitat diversity [5,9,22]. On this basis, we hypothesized that a network of ecologically important habitat patches (hereafter, a green network), such as other green areas and water bodies nearby, allows birds to make use of small green areas. Furthermore, Freiburg and Regensburg have a long history of spatial planning for greenery within cities [23,24]. This planning might have allowed a higher amount of well-planned small green patches (i.e., isolated units outside the forest and large green areas) and improved the quality of habitats for birds within cities. Therefore, we further hypothesized that the landscape metrics at the local scale, specifically a higher abundance of comparable patches (i.e., serving as stepping stones) and habitat heterogeneity (i.e., representing habitat quality) at a 1 km radius, enhance bird diversity in a small-sized urban green area.

2. Methods

2.1. Study Sites

In this study, we considered nine small green area locations (i.e., street parks, campus yards, and playgrounds) in Freiburg and eight in Regensburg in Germany (Figure 1). All were fragmented habitat patches (isolated from core green areas, i.e., forests) located in and near built-up areas. The selected green area locations were approximately > 700–1000 m far from each other in both cities. The average size of the green areas in Regensburg was 0.733 ± 0.51 ha, while Freiburg was 0.38 ± 0.16 ha.

Freiburg is located in the south of Baden-Württemberg and has approximately 230,000 inhabitants. The city area is 15,307 ha and comprises 6530 ha of forest and 664 ha of sports, leisure time, and recreational spaces [25]. Regensburg is in the east of Bavaria and has approximately 160,000 inhabitants. The city's 8070 ha area includes 467 ha of forest and 268 ha of recreational, sports, and leisure time places [26]. Both Freiburg and Regensburg are medium-sized (based on population estimate) and share a resemblance in the spatial configuration of different landscape features (i.e., crossed by a river). Moreover, both cities are unique, with more than 50% greenery within the city boundary [27,28] and are considered "green cities" with a background in spatial city planning towards sustainable growth [23,29].

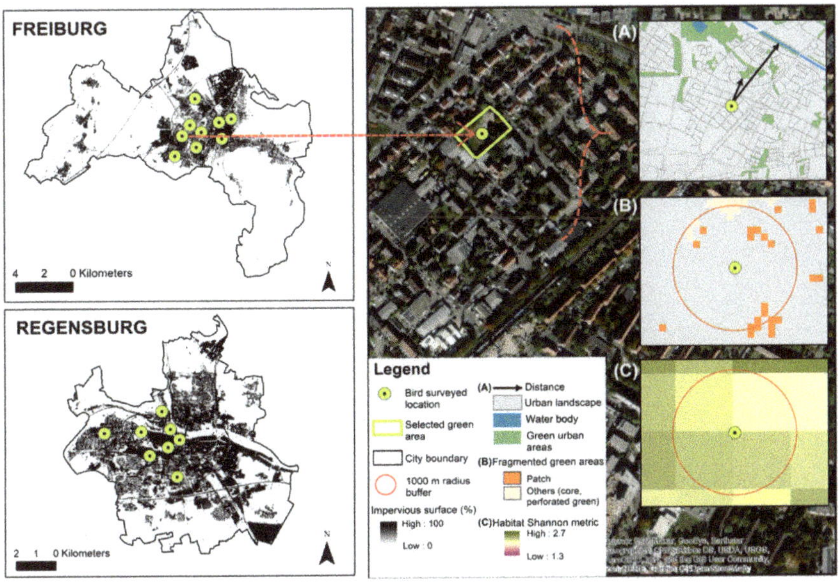

Figure 1. The map shows selected small green areas and bird survey locations in Freiburg and Regensburg. An example of one small green area is shown on the right side. Here, (**A**–**C**) displays the neighboring landscape factors of the small green area considered in this study. Image (**A**) shows the variables of green networks (i.e., distance to the nearest other urban green area and water body). The following images in (**B**,**C**) show variables of landscape metrics at a 1 km radius of the surveyed location. (**B**) shows the different types of green fragments (grid cells of 110 m × 110 m), from which abundance of small 'patch' units (i.e., comparable to studied small green areas) is considered, and (**C**) shows the measure of habitat heterogeneity (habitat Shannon metric). The projected coordinate systems used in the map were WGS_1984_UTM_Zone_32N for Freiburg and WGS_1984_UTM_Zone_33N for Regensburg.

2.2. Bird Survey

In both cities, we considered the centroid location of each selected small urban green area for a bird survey. We surveyed birds following a point count method [30] and documented all species spotted at the location within a 50 m radius for 10 min. In Regensburg, each location was visited four times during March–April 2021. In Freiburg, each site was visited twice during July–August 2018. All field visits were carried out during the early four hours in the morning by two observers on non-consecutive days.

From the bird survey data, we measured two response metrics for each of the green area locations in both cities: bird species richness (total number of observed species), species composition (sites-by-species data, i.e., number of individuals of each observed species at each specific surveyed location).

2.3. Explanatory Variables

We investigated bird species richness and composition in small green areas in relation to four explanatory variables representing landscape factors [31].

Within the urban matrix, urban green areas and water bodies are crucial habitat features. Thus, we measured two proximity variables representing the green network at the surveyed location—(i) distance to the nearest another urban green area (DUG) and (ii) distance to the nearest water body (DW) (Figure 1). We measured the distance from the centroid of the small green areas to the nearest urban green water features from the 'Urban Atlas 2018' vector map dataset [32,33] using the tool 'Near' in ArcMap 10.8.1.

Next, we measured two landscape metrics at a 1 km radius buffer scale extent of the surveyed location—(iv) habitat Shannon metric (HS) to represent habitat heterogeneity and quality at the bird surveyed location, and (v) the total number of 'patch' units (hereafter, patch abundance; PA) to represent the amount of isolated habitat parallel to our selected small green areas in the neighborhood [31] (Figure 1).

HS is the measure of diversity in the 'Enhanced Vegetation Index' (i.e., variety in vegetation concentration) and extracted from the "Global Habitat Heterogeneity" raster dataset (resolution ~1 km) [34]. PA represents the number of fragmented small patches that have no core green area (i.e., no small, medium, or large forest tracts along with an outside edge of 110 m from the nearest urban pixel) [35]. To identify such patches, we used the tool 'Landscape Fragmentation' (Center for Land use Education and Research; www.clear.uconn.edu) that is installed in ArcMap 10.8.1. We considered a classified binary image of forest-no forest developed from the 'Land Cover map' data of CGLS-LC100 [36]; we assumed an edge width of 110 m and urban areas as fragmenting land cover. It provided rasterized landscape fragmentation map data for each city with attributes of different types of forest fragments: patch, edge, perforated, core [35]. For our analysis, we extracted the total number of 'patch' grid cells (~110 m × 110 m) at a 1 km radius buffer of the surveyed locations. We considered only the data of the unit 'patch' [31] (Figure 1), since it was the most available type of green fragment within a 1 km radius of the surveyed locations. Moreover, we intended to understand if bird diversity and their composition observed at the small urban green area locations can be explained by a higher abundance of similar type patches in the neighborhood.

Further, it is well known that bird richness commonly declines with increasing imperviousness [17,37–39], however, higher bird richness is associated with larger-sized green areas in urban areas [4,9]. Therefore, besides the four explanatory variables, we used associated green area size and the proportion of impervious surface as the control factors during the model assessment of bird species richness.

2.4. *Analysis*

We investigated which variables of green network and landscape metrics in the neighborhood of small green areas better explained observed bird richness (number of species) through regression modelling and species composition (sites-by-species individual number) through variance partitioning assessment. We applied the analysis approach similarly to Regensburg and Freiburg. All data analyses were performed in the statistics program R, version 4.0.1 [40], and RStudio, version 1.2.5033 [41].

2.4.1. Modelling

We applied linear regression models to fit bird species richness (i.e., our response variable) with different explanatory variables. Bird species richness was log10-transformed to achieve normality in the case of Freiburg.

A preliminary investigation on modelling of bird richness with explanatory variables using the function 'lm' and 'glm' (with 'Poisson') (package MASS) [42] did not show a significant change in the model fits in the case of Regensburg. The model fit with 'lm' was also appropriate in the case of Freiburg. Hereafter, we used the 'lm' function in our model assessment for both cities. Further, we checked VIF (variance inflation factor) (package 'car') [43] to ensure no multicollinearity issue was present in the models (i.e., VIF < 0.4) [44].

In our study, bird richness at a survey location was strongly correlated (0.52) with green area size in Freiburg and with the built-up proportion (−0.59) in Regensburg. Therefore, our first generated model included only green area size (in the case of Freiburg)/built-up proportion (in the case of Regensburg), considering the relevant most correlated variable as the control factor. In the next models, we selected and added another variable distinctively from HS (habitat Shannon metric), PA (patch abundance), DW (distance to the nearest water body), and DUG (distance to the nearest urban green area). Here, we partially followed hierarchical regression and gradually entered a new variable into the first generated model to observe the change in explained variation. Due to the low sample size and to avoid multicollinearity, we allowed a maximum of two variables in each of the models and did not include any interactions of the predictors.

We majorly inspected if the addition of any variables of the green network and landscape metrics improved our first model's ability to explain bird richness at the surveyed locations within small green areas. Moreover, we contrasted all generated models using the function 'model.sel' (package 'MuMIn'; [45] and examined the coefficient parameters, change in the AICc value, and AICc weight.

2.4.2. Variation Partitioning

Our following analysis included an assessment of the partitioning of the effect of four selected variables on bird species composition—HS (habitat Shannon metric), PA (patch abundance), DW (distance to the nearest water body), and DUG (distance to the nearest urban green area). We performed this assessment on bird species composition for Freiburg and Regensburg individually. Specifically, we investigated how much variation in the bird species composition data observed at the small green area locations within the cities is explained by the specific or combined variables. For this, we conducted a 'variation partitioning' assessment [46] with redundancy analysis, using a site-by-species table (abundances of each observed species at a specific surveyed location) as the response matrix. In this assessment, we used the function 'varpart' of the R package 'vegan' [47].

3. Results

We documented a total of 329 individuals of 21 species of birds in Freiburg and 539 individuals of 28 species in Regensburg during our survey (Supplementary Materials—Tables S1 and S2). In both cities, the majority of the birds observed at the locations within small green areas were medium to highly common species (i.e., following [48,49]).

3.1. *Modelling*

Our generated models of bird species richness did not include any issue of multicollinearity. The first generated model, which included the variable GA in the case of Freiburg and IS in the case of Regensburg, contained the lowest AICc value and highest AICc weight (Table 1).

In the case of Freiburg, the explained variation (R-squared) in the first model was 23% which increased by 4% when DW was added. The changes in the R-squared value were negligible in other models. No variables with significant values were retained. Although, the coefficient parameters indicated a negative effect of the distance to water body on bird richness at the observed sites in small green areas (Table 1).

In the case of Regensburg, the explained variation in the first model was 34% which noticeably improved by 22% when DG was added and by 14% when PA was added. Again, no variables with significant values were retained. However, the coefficient parameters indicated a negative effect of DG and a positive effect of PA on observed bird richness at a 1 km radius at the surveyed locations within small green areas (Table 1).

3.2. *Variation Partitioning*

Our further assessment, through variation partitioning with redundancy analysis, presented a partition of the influences of a green network (DW and DUG) and landscape metrics (PA and HS) on bird species composition. The results showed that the variation in bird species assemblage at the surveyed location explained by variables of a green network (46% by DUG and 40% by DW) was higher than any other single or combined variables in the case of Freiburg (Figure 2). While in the case of Regensburg, the variations in bird species composition were marginally explained by the sole effect of each variable. Only the combined effects of different variables were visible. Variation in species composition in small green areas was better explained by PA (14%) in combination with HS (i.e., habitat heterogeneity at a 1 km radius). Moreover, the explained variations in bird species composition by DW (12%) and by DUG (12%) in combination with HS were noticeable (Figure 2).

Table 1. A comparison of models' coefficients and explained variations in bird species richness in the case of Freiburg (A) and Regensburg (B). In both cases, the first generated model contained the lowest AICc and is marked in italics. Here, Δ_i is the change in AICc and w_i is the AICc weight. The model showing the highest explained variation based on R-squared (R^2) value is marked in bold. Bird species richness was log10-transformed during modelling in the case of Freiburg. Here, GA = green areas size, IS = impervious surface, HS = habitat Shannon metric, PA = patch abundance, DW = distance to the nearest water body, DUG = distance to the nearest other urban green areas.

(A). Case of Freiburg										
Models	(Int)	GA	DG	DW	HS	PA	AICc	Δ_i	w_i	R^2
~GA	*0.76*	*0.06*					*−6.4*	*0.00*	*0.892*	*0.23*
~GA + DW	**0.76**	**0.05**	**−0.02**				**0.3**	**6.76**	**0.030**	**0.27**
~GA + DG	0.76	0.05		−0.01			0.6	7.04	0.026	0.25
~GA + HS	0.76	0.06			0.01		0.7	7.07	0.026	0.25
~GA + PA	0.76	0.06				0.01	0.7	7.13	0.025	0.24
(B). Case of Regensburg										
Models	(Int)	IS	DW	DG	HS	PA	AICc	Δ_i	w_i	R^2
~IS	*12.37*	*−1.5*					*45.2*	*0.00*	*0.91*	*0.35*
~IS + PA	**12.37**	**−1.08**				**1.28**	**51.3**	**6.05**	**0.04**	**0.57**
~IS + DG	12.37	−1.23		−0.99			52.7	7.43	0.02	0.49
~IS + DW	12.37	−1.40	−0.33				54.4	9.15	0.01	0.36
~IS + HS	12.37	−1.56			−0.09		54.6	9.32	0.01	0.35

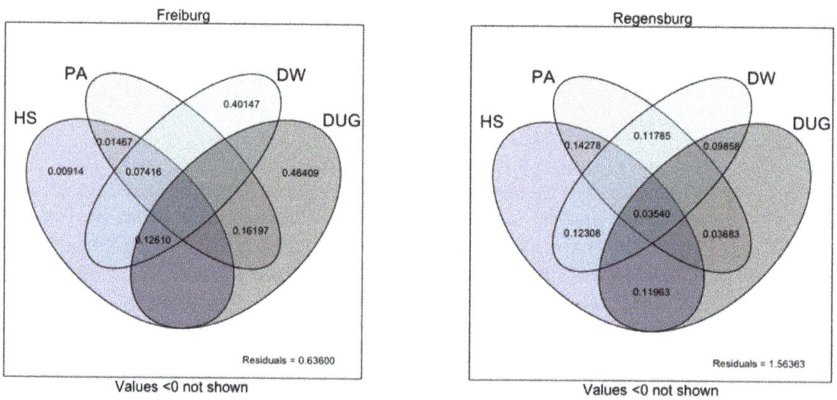

Figure 2. Partitioning of the effects of explanatory variables on bird species composition. The four variables used are HS (habitat Shannon metric), PA (patch abundance), DW (distance to the nearest water body), and DUG (distance to the nearest urban green area). HS and PA represent landscape metrics, and DW and DUG represent green networks in the vicinity of a small green area. The rectangular boundary represents the total variation in the avian species composition data. The numerals and their placement within the circles represent the portion of variation (i.e., the value of 'adjusted R.square') explained by specific explanatory variables or combined variables.

4. Discussion

In this study, we inspected how the neighboring green network and landscape metrics explain species richness and composition of birds (as a proxy for biodiversity) in small urban green areas of two green cities in Germany—Freiburg and Regensburg. Our model assessment in Freiburg suggested that the proximity (distance) to the water body along with the usual positive effect of the associated green area size better explained bird species richness at the surveyed locations. It indicated that small green areas that were spatially located nearby a water body consisted of higher bird richness in Freiburg. In the case of Regensburg, patch abundance at a 1 km radius along with the usual greater negative effect of imperviousness explained bird species richness better at the surveyed locations. It pointed out that small green areas which contained a higher abundance of similar green 'patches' at a 1 km radius supported bird richness in Regensburg.

Our further assessment of the partitioning of the explained variation in bird species composition by the green network and landscape metrics variables indicated two alternate neighborhood settings of small green areas. In Freiburg, variables of the green network were influential; thus, variations in bird species composition in small-sized green areas were better explained by the sole effects of proximity to other ecologically important habitat areas (i.e., another green area and water body). In Regensburg, however, the highest explained variation in bird species composition was by the combined effect of landscape metrics (i.e., habitat heterogeneity and patch abundance) at a 1 km radius of small green areas.

Within cities, small-sized green areas, such as pocket parks, gardens, or playgrounds, are ecologically different components of green infrastructures since vegetation cover within these areas is entirely altered by human interventions. These green areas are spatially separated habitat patches and are often compared with islands since they are isolated from the core green area/forest area. Such patches can be crucial for animals, such as birds that are highly mobile [11]. Here, our results in Freiburg indicated that the presence of a nearby other urban green area and water body might improve the variety of birds and their composition. This is reasonable, however, since these features serve as refugia by providing shelter, nesting sites, and food [50,51]. In Regensburg, our results supported the fact that the presence of higher habitat heterogeneity (i.e., variety in complex vegetation structures) and an increasing quantity of green patches—as far as possible on a local scale (1 km radius)—might improve neighborhood habitat quality for bird species richness and composition [52,53]. The green patches may also vary in accessibility, depending on the width of surrounding roads as well as traffic volumes, which have been shown to influence the number and species composition of birds crossing the roads [54,55]. Thus, neighboring landscape configuration may help to improve the functionality and connectivity to promote the existence of different species in fragmented small green areas [56–59].

In our study, rare bird encounters were underrepresented at the surveyed small green area locations. This might be due to the fact that generalist bird species already replaced specialist birds in green areas within cities (i.e., due to avian homogenization) [52,60,61]. However, our study of limited duration (i.e., only one seasonal survey) cannot provide a solid understanding of this aspect. The purpose of our study was to assess patterns in species richness and composition in small green areas in relation to neighboring landscape factors. It did not aim to obtain a complete species checklist, which requires multiple years of consistent surveying.

Our study provides insights into the effects of only local scale neighboring landscape factors, which are fundamental and influential in enhancing birds' mobility and presence in an urban area [17,21]. However, local scale avian diversity can also be associated with green networks and landscape metrics at a broader spatial scale since birds are highly mobile [16,62]. In addition, existing studies have found that anthropogenic disturbances such as daytime noise levels and pedestrian movement can also influence species occurrence and composition, as well as the dominance of native versus invasive species in urban areas [12,62,63].

The high conservation value of small green patches is only recently recognized for species diversity conservation [64]. In this paper, we targeted to support that neighboring landscape factors explain species richness and composition in small-sized green areas within cities. A preliminary investigation was first conducted in Freiburg and later replicated in Regensburg. We presented both case assessments to explain the individual city's scenario better. However, the interpretation of our results requires some caution, since our analysis was based on limited-scale studies in two cities. Further, we could not perform the assessment considering a combined dataset due to the cautiousness of sampling effort variability between the cities. Birds were surveyed during the breeding season in Regensburg and during post-breeding into the migration season in Freiburg. The influence of green networks and landscape metrics on bird diversity and species composition could be different at different times of the year, as shown in other studies [65,66]. A clear understanding of how species-specific use and preference for green infrastructure features differ between seasons, and how such circumstances change throughout the year, would be worth exploring in the future.

'Conservation' usually targets threatened animals and natural habitats, such as forest areas or large-sized green areas [67]. In contrast, small-sized urban green areas within cities are mainly considered for recreational use by humans with little or no clear target for biodiversity. However, the green area management planning in Freiburg and Regensburg is exceptional. Freiburg's effort to maintain its green spaces concerning natural and ecological principles trace back more than 20 years [24]. Regensburg has also been implementing spatial development planning to improve the urban greenery system concerning its suburban areas and natural and cultural heritage [23]. At the regional level, there is again a long history of bird conservation action networks as part of the European Union [68]. Despite a remarkable similarity in urban greenery and its management between the two cities, we observed that different variables of green networks and landscape metrics and their composition effectively explained variations in species composition in small green areas of Freiburg and Regensburg. A broad-scale study (i.e., with a higher amount of sampling sites) would be necessary to assess if the difference in the landscape factors' effects is valid and whether a variation in variables' effect on species composition in small green areas actually correlates with city-specific planning.

Nevertheless, our study indicates that landscape metrics of habitat quality (i.e., the higher number of patches and heterogeneity in complex vegetation structure) and a green network (i.e., the proximity of other green areas and water bodies) in the neighborhood can be fundamental to maintain species richness and composition in small green areas. It might ultimately contribute to strengthening the green infrastructure system in urban localities to support overall biodiversity within a city.

Supplementary Materials: The following supporting information can be downloaded at: https://www.mdpi.com/article/10.3390/su14116394/s1.

Author Contributions: M.S. and I.S. designed the study; M.S. compiled the data, conducted the analysis and wrote the manuscript; M.M. (Max Müller) and M.M. (Magdalena Meyer) conducted the bird survey in Regensburg and contributed to revising the manuscript; I.S. contributed to the revising of the manuscript. All authors have read and agreed to the published version of the manuscript.

Funding: This research received no external funding.

Institutional Review Board Statement: Not applicable.

Informed Consent Statement: Not applicable.

Data Availability Statement: All data used in this assessment are available in the citation [31].

Acknowledgments: We thank Sueon Ahn for her contribution during the bird survey in Freiburg and Marius Huber for assisting during the preliminary assessment of the data. Marufa Sultana was supported by the LGFG (Landesgraduiertenförderungsgesetz) scholarship from the International Graduate Academy (IGA) of the University of Freiburg during the research period. Finally, we acknowledge the support of the Open Access Publication Fund of the University of Freiburg.

Conflicts of Interest: The authors declare no conflict of interest.

References

1. Benton-Short, L.; Keeley, M.; Rowland, J. Green infrastructure, green space, and sustainable urbanism: Geography's important role. *Urban Geogr.* **2017**, *40*, 330–351. [CrossRef]
2. Lepczyk, C.A.; Aronson, M.F.J.; Evans, K.L.; Goddard, M.; Lerman, S.B.; MacIvor, J.S. Biodiversity in the City: Fundamental Questions for Understanding the Ecology of Urban Green Spaces for Biodiversity Conservation. *BioScience* **2017**, *67*, 799–807. [CrossRef]
3. Khera, N.; Mehta, V.; Sabata, B. Interrelationship of birds and habitat features in urban greenspaces in Delhi, India. *Urban For. Urban Green.* **2009**, *8*, 187–196. [CrossRef]
4. Leveau, L.M.; Ruggiero, A.; Matthews, T.J.; Bellocq, M.I. A global consistent positive effect of urban green area size on bird richness. *Avian Res.* **2019**, *10*, 30. [CrossRef]
5. Callaghan, C.T.; Major, R.E.; Lyons, M.B.; Martin, J.; Kingsford, R. The effects of local and landscape habitat attributes on bird diversity in urban greenspaces. *Ecosphere* **2018**, *9*, e02347. [CrossRef]
6. Biroli, A.P.; Van Doren, B.M.; Grabowska-Zhang, A. Drivers of avian species richness and community structure in urban courtyard gardens. *J. Urban Ecol.* **2020**, *6*, juz026. [CrossRef]
7. Drinnan, I.N. The search for fragmentation thresholds in a Southern Sydney Suburb. *Biol. Conserv.* **2005**, *124*, 339–349. [CrossRef]
8. Kang, W.; Minor, E.; Park, C.-R.; Lee, D. Effects of habitat structure, human disturbance, and habitat connectivity on urban forest bird communities. *Urban Ecosyst.* **2015**, *18*, 857–870. [CrossRef]
9. Strohbach, M.; Lerman, S.B.; Warren, P.S. Are small greening areas enhancing bird diversity? Insights from community-driven greening projects in Boston. *Landsc. Urban Plan.* **2013**, *114*, 69–79. [CrossRef]
10. Schütz, C.; Schulze, C.H. Functional diversity of urban bird communities: Effects of landscape composition, green space area and vegetation cover. *Ecol. Evol.* **2015**, *5*, 5230–5239. [CrossRef]
11. Liordos, V.; Jokimäki, J.; Kaisanlahti-Jokimäki, M.-L.; Valsamidis, E.; Kontsiotis, V. Niche Analysis and Conservation of Bird Species Using Urban Core Areas. *Sustainability* **2021**, *13*, 6327. [CrossRef]
12. Liordos, V.; Jokimäki, J.; Kaisanlahti-Jokimäki, M.-L.; Valsamidis, E.; Kontsiotis, V.J. Patch, matrix and disturbance variables negatively influence bird community structure in small-sized managed green spaces located in urban core areas. *Sci. Total Environ.* **2021**, *801*, 149617. [CrossRef] [PubMed]
13. Chace, J.F.; Walsh, J.J. Urban effects on native avifauna: A review. *Landsc. Urban Plan.* **2006**, *74*, 46–69. [CrossRef]
14. Threlfall, C.G.; Mata, L.; Mackie, J.A.; Hahs, A.K.; Stork, N.E.; Williams, N.S.G.; Livesley, S.J. Increasing biodiversity in urban green spaces through simple vegetation interventions. *J. Appl. Ecol.* **2017**, *54*, 1874–1883. [CrossRef]
15. Aronson, M.F.J.; La Sorte, F.A.; Nilon, C.H.; Katti, M.; Goddard, M.A.; Lepczyk, C.A.; Warren, P.S.; Williams, N.S.G.; Clilliers, S.; Clarkson, B.; et al. A global analysis of the impacts of urbanization on bird and plant diversity reveals key anthropogenic drivers. *Proc. R. Soc. B Biol. Sci.* **2014**, *281*, 20133330. [CrossRef]
16. Lepczyk, C.A.; Sorte, F.A.L.; Aronson, M.F.; Goddard, M.A.; MacGregor-Fors, I.; Nilon, C.H.; Warren, P.S. Global patterns and drivers of urban bird diversity. In *Ecology and Conservation of Birds in Urban Environments*; Springer International Publishing: Cham, Switzerland, 2017; pp. 13–33.
17. Sultana, M.; Corlatti, L.; Storch, I. The interaction of imperviousness and habitat heterogeneity drives bird richness patterns in south Asian cities. *Urban Ecosyst.* **2020**, *24*, 335–344. [CrossRef]
18. Strohbach, M.W.; Haase, D.; Kabisch, N. Birds and the City: Urban Biodiversity, Land Use, and Socioeconomics. *Ecol. Soc.* **2009**, *14*, 31. [CrossRef]
19. Trimble, M.J.; van Aarde, R.J. Geographical and taxonomic biases in research on biodiversity in human-modified landscapes. *Ecosphere* **2012**, *3*, 1–16. [CrossRef]
20. Li, E.; Parker, S.S.; Pauly, G.B.; Randall, J.M.; Brown, B.V.; Cohen, B.S. An Urban Biodiversity Assessment Framework That Combines an Urban Habitat Classification Scheme and Citizen Science Data. *Front. Ecol. Evol.* **2019**, *7*, 277. [CrossRef]
21. Estevo, C.A.; Nagy-Reis, M.B.; Silva, W.R. Urban parks can maintain minimal resilience for Neotropical bird communities. *Urban For. Urban Green.* **2017**, *27*, 84–89. [CrossRef]
22. Yang, X.; Tan, X.; Chen, C.; Wang, Y. The influence of urban park characteristics on bird diversity in Nanjing, China. *Avian Res.* **2020**, *11*, 45. [CrossRef]
23. Wojtyszyn, B. Spatial Development of Regensburg/Ratyzbona Towards the City of "Green" Housing Estates. *Civ. Environ. Eng. Rep.* **2020**, *30*, 185–195. [CrossRef]
24. FWTM. Approaches to Sustainability, Green City Freiburg. Freiburg. 2017. Available online: https://www.awec2017.com/images/GreenCity_English.pdf (accessed on 1 April 2022).
25. Stadt Freiburg. FR.ITZ Online Das Statistik Portal, FR.ITZ Informationsportal. 2021. Available online: https://fritz.freiburg.de/Informationsportal/#app/startpage (accessed on 28 November 2021).
26. Stadt Regensburg. Informationen und Zahlen. 2021. Available online: http://www.statistik.regensburg.de/menue/informationen_u_zahlen.php (accessed on 5 August 2021).
27. Taubenböck, H.; Reiter, M.; Dosch, F.; Leichtle, T.; Weigand, M.; Wurm, M. Which city is the greenest? A multi-dimensional deconstruction of city rankings. *Comput. Environ. Urban Syst.* **2021**, *89*, 101687. [CrossRef]
28. City Mayors. City Mayors: Germany's Greenest Cities. 2017. Available online: http://www.citymayors.com/environment/german-green-cities.html (accessed on 30 March 2022).

29. Buehler, R.; Jungjohann, A.; Keeley, M.; Mehling, M. How Germany Became Europe's Green Leader: A Look at Four Decades of Sustainable Policymaking. *Solutions* **2011**, *2*, 51–63.
30. Bibby, C.J.; Hill, D.A.; Burgess, N.D.; Mustoe, S. *Bird Census Techniques*, 2nd ed.; Academic Press: London, UK, 2000.
31. Sultana, M.; Müller, M.; Magdalena, M.; Storch, I. OSF | Data of the Study on Small Green Areas. Available online: https://osf.io/9jsav/?view_only=c77fd1b5d8224146a058c70c82415600 (accessed on 6 March 2022).
32. Copernicus Land Monitoring Service. Urban Atlas 2018—Copernicus Land Monitoring Service. 2022. Available online: https://land.copernicus.eu/local/urban-atlas/urban-atlas-2018?tab=download (accessed on 30 March 2022).
33. Ares, R. Mapping Guide for a European Urban Atlas. 2012. Available online: http://ec.europa.eu/regional_policy/sources/tender/pdf/2012066/annexe2.pdf (accessed on 6 March 2022).
34. Tuanmu, M.-N.; Jetz, W. A global, remote sensing-based characterization of terrestrial habitat heterogeneity for biodiversity and ecosystem modelling. *Glob. Ecol. Biogeogr.* **2015**, *24*, 1329–1339. [CrossRef]
35. Vogt, P.; Riitters, K.H.; Estreguil, C.; Kozak, J.; Wade, T.G.; Wickham, J.D. Mapping Spatial Patterns with Morphological Image Processing. *Landsc. Ecol.* **2006**, *22*, 171–177. [CrossRef]
36. Buchhorn, M.; Lesiv, M.; Tsendbazar, N.-E.; Herold, M.; Bertels, L.; Smets, B. Copernicus Global Land Cover Layers—Collection 2. *Remote Sens.* **2020**, *12*, 1044. [CrossRef]
37. Silva, C.P.; García, C.E.; Estay, S.; Barbosa, O. Bird Richness and Abundance in Response to Urban Form in a Latin American City: Valdivia, Chile as a Case Study. *PLoS ONE* **2015**, *10*, e0138120. [CrossRef]
38. Carvajal-Castro, J.D.; Ospina-L, A.M.; Toro-López, Y.; Pulido-G, A.; Casas, L.C.; Guerrero-Peláez, S.; García-Merchán, V.H.; Vargas-Salinas, F. Birds vs bricks: Patterns of species diversity in response to urbanization in a Neotropical Andean city. *PLoS ONE* **2019**, *14*, e0218775. [CrossRef]
39. Marcacci, G.; Westphal, C.; Wenzel, A.; Raj, V.; Nölke, N.; Tscharntke, T.; Grass, I. Taxonomic and functional homogenization of farmland birds along an urbanization gradient in a tropical megacity. *Glob. Chang. Biol.* **2021**, *27*, 4980–4994. [CrossRef]
40. R Development Core Team. *R: A Language and Environment for Statistical Computing;* R Foundation for Statistical Computing: Vienna, Austria, 2013. Available online: https://www.R-project.org/ (accessed on 10 April 2021).
41. RStudio Team. *RStudio: Integrated Development Environment for R;* RStudio: Boston, MA, USA, 2021. Available online: https://www.rstudio.com/ (accessed on 10 April 2021).
42. Venables, W.N.; Ripley, B.D. *Modern Applied Statistics with S*, 4th ed.; Springer: New York, NY, USA, 2002.
43. Fox, J.; Weisberg, S. *An R Companion to Applied Regression*, 3rd ed.; Sage: Thousand Oaks, CA, USA, 2019.
44. Akinwande, M.O.; Dikko, H.G.; Samson, A. Variance Inflation Factor: As a Condition for the Inclusion of Suppressor Variable(s) in Regression Analysis. *Open J. Stat.* **2015**, *5*, 754–767. [CrossRef]
45. Bartoń, K. Package 'MuMIn' Title Multi-Model Inference. 2022. Available online: https://cran.r-project.org/web/packages/MuMIn/MuMIn.pdf (accessed on 2 April 2022).
46. Borcard, D.; Legendre, P.; Drapeau, P. Partialling out the Spatial Component of Ecological Variation. *Ecology* **1992**, *73*, 1045–1055. [CrossRef]
47. Oksanen, J.; Blanchet, G.F.; Kindt, R.; Legendre, P.; Minchin, R.P.; O'hara, B.R.; Simpson, L.G.; Solymos, P.; Stevens, H.M.H.; Wagner, H.; et al. Package 'vegan'. Community Ecology Package. *Community Ecol. Package* **2019**, 2, 1–297. Available online: https://cran.r-project.org/web/packages/vegan/vegan.pdf (accessed on 10 April 2021).
48. Bauer, H.-G.; Boschert, M.; Förschler, M.I.; Hölzinger, J.; Kramer, M.; Mahler, U. *Rote Liste und Kommentiertes Verzeichnis der Brutvogelarten Baden-Württembergs;* LUBW Landesanstalt für Umwelt, Messungen und Naturschutz Baden-Württemberg: Stuttgart, Germany, 2016; Volume 6.
49. Bayerisches Landesamt für Umwelt (LfU). Rote Liste und Liste der Brutvögel Bayerns. 2016. Available online: https://www.lfu.bayern.de/natur/rote_liste_tiere/2016/index.htm (accessed on 30 March 2022).
50. Sorace, A.; Gustin, M. Species Richness and Species of Conservation Concern in Parks of Italian Towns. In *Ecology and Conservation of Birds in Urban Environments;* Springer: Cham, Switzerland, 2017; pp. 425–448. [CrossRef]
51. Tryjanowski, P.; Morelli, F.; Mikula, P.; Kristin, A.; Indykiewicz, P.; Grzywaczewski, G.; Kronenberg, J.; Jerzak, L. Bird diversity in urban green space: A large-scale analysis of differences between parks and cemeteries in Central Europe. *Urban For. Urban Green.* **2017**, *27*, 264–271. [CrossRef]
52. Ferenc, M.; Sedláček, O.; Fuchs, R.; Dinetti, M.; Fraissinet, M.; Storch, D. Are cities different? Patterns of species richness and beta diversity of urban bird communities and regional species assemblages in E urope. *Glob. Ecol. Biogeogr.* **2013**, *23*, 479–489. [CrossRef]
53. van Heezik, Y.; Freeman, C.; Porter, S.; Dickinson, K.J.M. Garden Size, Householder Knowledge, and Socio-Economic Status Influence Plant and Bird Diversity at the Scale of Individual Gardens. *Ecosystems* **2013**, *16*, 1442–1454. [CrossRef]
54. Johnson, C.D.; Evans, D.; Jones, D. Birds and Roads: Reduced Transit for Smaller Species over Roads within an Urban Environment. *Front. Ecol. Evol.* **2017**, *5*, 36. [CrossRef]
55. Rao, S.; Koli, V.K. Edge Effect of Busy High Traffic Roads on the Nest Site Selection of Birds Inside the City Area: Guild Response. *Transp. Res. Part D Transp. Environ.* **2017**, *51*, 94–101. Available online: https://www.infona.pl//resource/bwmeta1.element.elsevier-7668b2fe-4df1-35a6-a667-4314cd70d841 (accessed on 13 May 2022). [CrossRef]
56. Donnelly, R.; Marzluff, J.M. Relative importance of habitat quantity, structure, and spatial pattern to birds in urbanizing environments. *Urban Ecosyst.* **2006**, *9*, 99–117. [CrossRef]

57. Evans, K.L.; Newson, S.E.; Gaston, K.J. Habitat influences on urban avian assemblages. *Ibis* **2008**, *151*, 19–39. [CrossRef]
58. Faeth, S.H.; Saari, S.; Bang, C. Urban Biodiversity: Patterns, Processes and Implications for Conservation. *eLS* **2012**. [CrossRef]
59. Beninde, J.; Veith, M.; Hochkirch, A. Biodiversity in cities needs space: A meta-analysis of factors determining intra-urban biodiversity variation. *Ecol. Lett.* **2015**, *18*, 581–592. [CrossRef] [PubMed]
60. Clergeau, P.; Croci, S.; Jokimäki, J.; Kaisanlahti-Jokimäki, M.-L.; Dinetti, M. Avifauna homogenisation by urbanisation: Analysis at different European latitudes. *Biol. Conserv.* **2006**, *127*, 336–344. [CrossRef]
61. Morelli, F.; Benedetti, Y.; Ibáñez-Álamo, J.D.; Jokimäki, J.; Mänd, R.; Tryjanowski, P.; Møller, A.P. Evidence of evolutionary homogenization of bird communities in urban environments across Europe. *Glob. Ecol. Biogeogr.* **2016**, *25*, 1284–1293. [CrossRef]
62. Heggie-Gracie, S.D.; Krull, C.R.; Stanley, M.C. Urban divide: Predictors of bird communities in forest fragments and the surrounding urban matrix. *Emu-Austral Ornithol.* **2020**, *120*, 333–342. [CrossRef]
63. da Silva, B.F.; Pena, J.C.; Viana-Junior, A.B.; Vergne, M.; Pizo, M.A. Noise and tree species richness modulate the bird community inhabiting small public urban green spaces of a Neotropical city. *Urban Ecosyst.* **2020**, *24*, 71–81. [CrossRef]
64. Riva, F.; Fahrig, L. The Disproportionately High Value of Small Patches for Biodiversity Conservation. *Conserv. Lett.* **2022**, e12881. Available online: https://onlinelibrary.wiley.com/doi/full/10.1111/conl.12881 (accessed on 13 May 2022).
65. Leveau, L.M.; Leveau, C.M. Street design in suburban areas and its impact on bird communities: Considering different diversity facets over the year. *Urban For. Urban Green.* **2019**, *48*, 126578. [CrossRef]
66. Cox, D.T.C.; Inger, R.; Hancock, S.; Anderson, K.; Gaston, K.J. Movement of feeder-using songbirds: The influence of urban features. *Sci. Rep.* **2016**, *6*, 37669. [CrossRef]
67. Elmqvist, T.; Zipperer, W.; Güneralp, B. Urbanization, habitat loss, biodiversity decline: Solution pathways to break the cycle | Treesearch. In *Routledge Handbook of Urbanization and Global Environmental Change*; Seta, K., Solecki, W.D., Griffith, C.A., Eds.; Routledge: London, UK; New York, NY, USA, 2016.
68. European Union. *The Birds Directive: 40 Years Conserving Our Shared Natural Heritage*; Publications Office of the European Union (EU): Luxembourg, 2019.

sustainability

MDPI

Article

Utility of Human Footprint Pressure Mapping for Large Carnivore Conservation: The Kafue-Zambezi Interface

Robin Lines [1], Dimitrios Bormpoudakis [1], Panteleimon Xofis [2,*], Douglas C. MacMillan [1], Lucy Pieterse [3] and Joseph Tzanopoulos [1,4]

[1] Durrell Institute of Conservation and Ecology, School of Anthropology and Conservation, University of Kent, Canterbury CT2 7NZ, UK; robin.lines@gmail.com (R.L.); bormpd@gmail.com (D.B.); d.c.macmillan@kent.ac.uk (D.C.M.); J.Tzanopoulos@kent.ac.uk (J.T.)
[2] Department of Forestry and Natural Environment, International Hellenic University, 66100 Drama, Greece
[3] BioCarbon Partners (BCP), Unit 3 Leopards Hill Business Park, Leopards Hill Road, Lusaka 50830, Zambia; lucypieterse@gmail.com
[4] Kent Interdisciplinary Centre for Spatial Studies, School of Anthropology and Conservation, University of Kent, Canterbury CT2 7NZ, UK
* Correspondence: pxofis@for.ihu.gr; Tel.: +30-6973035416 or +30-25210-60430

Abstract: Proxies and indicators to monitor cumulative human pressures provide useful tools to model change and understanding threshold pressures at which species can persist, are extirpated, or might recolonize human-impacted landscapes. We integrated modelling and field observations of human pressure variables to generate a site-specific, fine scale Human Footprint Pressure map for 39,000 km^2 of rangelands at the Kafue–Zambezi interface—a key linkage in the Kavango-Zambezi Transfrontier Conservation Area. We then modelled Human Footprint Pressure against empirically derived occurrence data for lion (*Panthera leo*), leopard (*Panthera pardus*), and spotted hyena (*Crocuta crocuta*) to generate Human Footprint Pressure threshold ranges at which each species were persisting or extirpated within ten wildlife managed areas linking Kafue National Park to the Zambezi River. Results overcame many limitations inherent in existing large-scale Human Footprint Pressure models, providing encouraging direction for this approach. Human Footprint Pressure thresholds were broadly similar to existing studies, indicating this approach is valid for site- and species-specific modelling. Model performance would improve as additional datasets become available and with improved understanding of how asymmetrical and nonlinear threshold responses to footprint pressure change across spatial-temporal scales. However, our approach has broader utility for local and region-wide conservation planning where mapping and managing human disturbance will help in managing carnivore species within and without protected area networks.

Keywords: carnivores; Kavango-Zambezi Transfrontier Conservation Area; connectivity

Citation: Lines, R.; Bormpoudakis, D.; Xofis, P.; MacMillan, D.C.; Pieterse, L.; Tzanopoulos, J. Utility of Human Footprint Pressure Mapping for Large Carnivore Conservation: The Kafue-Zambezi Interface. *Sustainability* **2022**, *14*, 116. https://doi.org/10.3390/su14010116

Academic Editors: Vasilios Liordos and Panayiotis Dimitrakopoulos

Received: 22 October 2021
Accepted: 22 December 2021
Published: 23 December 2021

Publisher's Note: MDPI stays neutral with regard to jurisdictional claims in published maps and institutional affiliations.

Copyright: © 2021 by the authors. Licensee MDPI, Basel, Switzerland. This article is an open access article distributed under the terms and conditions of the Creative Commons Attribution (CC BY) license (https://creativecommons.org/licenses/by/4.0/).

1. Introduction

Humanity's impact on the planet stretches from the deep ocean to mountaintops, manifesting through direct demands on natural resources and indirect effects of these demands on wider global systems [1,2]. The wide-ranging implications of increasing spatio-temporal resource demands lead to loss and fragmentation of key wildlife habitats [3,4], constraining species movement [5] and resulting to the reduction and extinction of wildlife populations at multiple scales [6–8]. While the decline in human pressures on natural system is presenting new opportunities for rewilding and carnivore conservation throughout much of continental Europe [9], many of the world's developing regions supporting large tracts of existing wildlife habitat and high levels of biodiversity [10] are experiencing intensifying spatio-temporal human pressures in and around protected areas [11,12]. Increased human resource demands in these areas are also impacting conservation efforts and political support for the maintenance and expansion of wildlife-based land uses and wildlife economies at regional, national, and transboundary scales [13,14].

Attempts to capture these anthropogenic pressures include the Human Footprint Pressure which takes into account inter alia population growth, the expansion of built areas and settlement, transport infrastructure and linkages, agropastoralism, and extractive industries [15]. Significant Human Footprint Pressure often results in profound and complex effects impacting the structure and function of ecosystems, including changes to key resources driving socioecological system productivity and resilience [16] and livelihood opportunities for communities residing within them [17]. Furthermore, elevated Human Footprint Pressure decreases structural and functional connectivity between wildlife managed areas for many species of conservation concern [18].

Existing Human Footprint Pressure analyses have traditionally been generated at relatively low resolution to provide overviews and indicators of Human Footprint Pressure at global scales [17,19,20]. Increasingly, the focus of Human Footprint Pressure is shifting to consider its utility as a proxy or predictive indicator for measuring and understanding finer scale impacts on species and processes, including studies on species movement [5], behaviour [21], extinction risk [22], range use [23], and more broadly as a conservation planning tool [24]. These approaches seek to overcome many of the questions and limitations surrounding data availability, accuracy, and resolution posed by conventional coarser scale multivariate models.

Generating site- and species-specific Human Footprint Pressure models that can be used as a proxy or indicator of species-level habitat suitability and sensitivity to human pressure can aid our understanding of thresholds at which species persist, are extirpated, or are likely to recolonize both protected and non-protected areas, leading to improved application of conservation science in management [25]. However, beyond large scale assessments [22], these tools are poorly understood and developed owing chiefly to an absence of integrated fine-scale remote sensing and in situ data, precluding appropriate accuracy and resolution [26].

The Kavango-Zambezi Transfrontier Conservation Area (KAZA TFCA, hereafter KAZA) in central Southern Africa seeks to promote connectivity between clusters of wildlife managed areas at the interface of five neighboring countries. Connectivity at the species and scale of interest are poorly studied within and between many of the proposed landscape-scale linkages in KAZA [27], but with human pressure increasing throughout the region [11,28], there is a need to understand how Human Footprint Pressures are impacting connectivity for key species of interest throughout core linkages. The large carnivores exert significant top-down influence on ecosystems, imparting strong regulatory pressures driving ecosystem structure and function [29,30]. They are highly susceptible to direct and indirect human activities including (legal and illegal) hunting, reduction of wild prey, and habitat fragmentation and loss [6,31]. Large carnivores are also a key asset for the development of wildlife economies [32], and have been identified by the KAZA programme as target species for conservation action, including the stabilization and growth of populations in key habitats, and maintenance of secure and active connectivity pathways between core wildlife managed areas [33]. In concert, these factors highlight large carnivores as appropriate target species against which to model Human Footprint Pressure.

The current study examines the effect of Human Footprint Pressures on the distribution of three emblematic carnivores, namely: lion (*Panthera leo*), leopard (*Panthera pardus*), and spotted hyena (*Crocuta crocuta*). It aims to generate site-specific, fine scale maps of Human Footprint Pressure to (1) test the validity of this approach for predicting species occurrence and (2) explore if this approach can determine discernible Human Footprint Pressure thresholds at which target species persist or are extirpated at the wildlife managed area scale.

2. Materials and Methods

The study area, which is the Kafue–Zambezi interface, covers central part of KAZA. The KAZA extends over c. 520,000 km^2 of central Southern Africa, spanning the borders of Angola, Botswana, Namibia, Zambia, and Zimbabwe, centered around the Kavango and Zambezi River basins (Figure 1) [34]. The KAZA landscape incorporates a network of ~70 protected areas in accordance to the International Union for Conservation of Nature (IUCN) in the categories I–VI and Not Reported categories [35]. These protected areas are characterized by a wide spectrum of investment and management effectiveness [36]. Spatial, connectivity between these protected areas has been identified as one of KAZA's central objectives [34].

Kafue National Park and surrounding protected areas, collectively known as the Greater Kafue System, represents KAZA's major northern cluster (Figure 1) and Zambia's majority contribution to the KAZA Programme [34]. Connectivity between Kafue National Park and adjacent protected areas, centered on Chobe National Park and East Zambezi Region in Namibia, is contingent on movement across eight partially and nominally protected areas plus an adjacent open Communal Areas identified by Lines et al. [37], as potentially important for corridor planning. In concert, these areas span ~13,000 km^2, extending 140–170 km from the Kafue National Park border south-southwest towards the Zambezi River at the confluence of Zambia, Namibia, and Botswana [38].

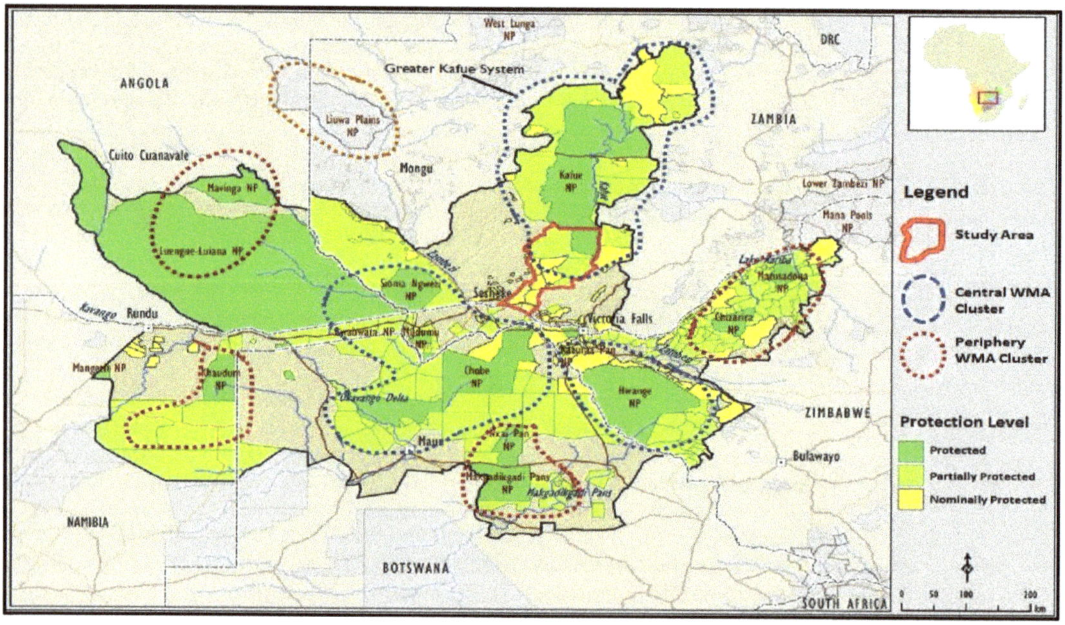

Figure 1. The Kavango-Zambezi Transfrontier Conservation Area landscape, indicating study area, clusters of wildlife managed areas (WMAs) and their degrees of protection. Protected = National Parks; IUCN II; Partially Protected = IUCN III-VI; Nominally Protected = IUCN Not Reported (adapted from [39]).

The landscape is historically, and still remains, characterized by dynamic spatiotemporal human pressures, though few data on the areas' wildlife and human population are available prior to the 1960s [38,40]. Much of the study area was sparsely settled until the development of a railway from Livingstone to Mulobezi from 1923 to 1924 to exploit the region extensive tracts of Zambezi teak forest (*Baikiaea plurijuga*). Access to formerly remote areas had profound impacts on its people and wildlife [41]. Southern areas around Simalaha, bordering the Zambezi River, were heavily depopulated during the 1966–1990 Angolan War, and thereafter increasing numbers of agro-pastoralists have settled this landscape (Yeta, pers comms), significantly increasing human pressures [38]. Systematic censuses from 2000 onward indicate Districts with boundaries intersecting the study area have experienced annualized population growth of ~2.8%, with an average population density of ~4.5 people/km^2 [40]. However, these larger scale surveys hide significant finer scale variation.

2.1. Generating Human Footprint Pressure Maps

Early Geographical Information System-based versions of the Human Footprint Pressure sought to build on the concept of Ecological Footprint mapping [42], utilizing availability of new Earth observation data sets and advances in satellite imagery capabilities covering human activities and the physical world, including land use and cover, transport linkages and human population density. This increase in resolution facilitated the development of geographical proxies for inferring variation in global human influences believed to have the most important direct pressure on wildlife [15]. Based on previous efforts, Venter et al. [17] extended the methodology of aggregating pressure scores at a global 1 km^2 resolution, based on long-term datasets, to generate updated Human Footprint Pressures and trends over time. Sanderson et al. [15] and Venter et al. [17] assign pressure scores to anthropogenic land-uses or activities. They integrate individual human pressure layers into a GIS for a composite human footprint pressure layer. While the Kafue–Zambezi landscape lacks long term datasets from which to derive trend data, our reworking of the Sanderson et al. [15] and Venter et al. [17] methodology sought to integrate the highest resolution data sets currently available for the landscape to generate outputs at two orders of magnitude finer scale. Details on the layers employed, spatial resolution and pressure scores adopted in the current study are shown in Table 1 and in the paragraphs below. Due to lack of pastureland data availability, this particular human pressure was omitted from this study.

Table 1. Human Pressure Variables and Scoring used in the current study.

Variable	Pressure Score	Source	Spatial Resolution	Details
Settlement	0, 10	[43]	30 m	All settled areas mapped given score of 10
Population Density	0–10 Continuous	[43]	30 m	Pressure score = 3.333 × log (population density + 1)
Roads	8 Direct impacts 0–4 Indirect impacts	PPF	10 m	Direct pressure score of 8 for 500 m either side of road, exponentially decaying out to 4 at 15 km
Railways	0–8	PPF	10 m	Direct pressure score of 8 for 500 m either side
Navigable Water	0–4	PPF	10 m	Pressure score of 4 exponentially decaying out to 15 km
Arable	0, 7	[37]	10 m	All areas mapped as crops given score of 7
Night Lights/NTL	0–9	[44]	100 m	Pressure score = 3.333 × log (NTL + 1)

Scoring of individual human pressure variables follows the same approach of Sanderson et al. [15]. Since more than one pressure variable may be present in a particular location, the maximum score, when all variables are present to their maximum scores, results in a pressure score of 43.8.

1. Settlement data was derived from Bonafilia et. al. [43] at 30 m resolution. All pixels overlapping settlement areas were given a pressure score of 10 representing the highest level of direct pressure (implying settled area were unsuitable for wildlife), with all other pixels given a score of 0.
2. Human Population Density data was unavailable at sufficiently fine scale for the landscape to include as a stand-alone data layer. Given the largely homogenous nature of settlement throughout the area (an absence of large multi-story buildings and dense conurbations versus ubiquitous single-story concrete block and tin buildings with scattered adobe and grass huts throughout rural area (Lines, pers obs)), we calculated average population density for the study area from the district scale data using 2019 population projected data [40]. Assuming that the total population of a district exists within the area of settlements, the total population was divided with the total area occupied by settlements, to provide their average population density. We then applied

the log formula employed by Venter et al. [17] for scarcely populated areas (Table 1). The calculated score was applied to all pixels that constitute part of a settlement.

3. Roads, acquired by the Peace Parks Foundation (PPF), constitute both a direct and an indirect human pressure. They reduce the extent of suitable habitats and the degree of habitat fragmentation while at the same time they are associated with increased traffic-induced mortality [45]. The indirect impacts are associated with the increased accessibility to wild areas ensured by a dense road network.
4. The same approach adopted by Venter et al. [17] for scoring the human pressure associated with roads was adopted in the current study. A pressure score of 8 was assigned to all pixels in a distance of 0.5 km either side of roads, indicating high direct human pressure. A pressure score of 4, exponentially decaying out to 15 km, was assigned to pixels in a distance longer from 0.5 km either side of the roads up to 15 km away, indicating lower indirect pressures as distance from the roads increases. The threshold of 15 km was set as it represents the approximate distance a person might reasonably access on foot within a day. A vector roads layer, provided by Peace Parks Foundation (unpublished data), including major tar and secondary dirt roads linking settlements, formed the baseline for the generation of the roads raster layer at a spatial resolution of 10 m. Tertiary dirt tracks were omitted from the analysis due to their dynamic nature and inconsistent mapping. The range of pressure scores varied between 0.25 at a distance of 15 km from roads to 8 for pixels next to roads.
5. Railways, acquired by the PPF, represent direct drivers of habitat conversion and conduits of human access into wildlife areas similar to roads. Since passengers cannot commonly disembark at will, indirect effects away from the railway line are considered minimal. Following Venter et al. [17], we gave railways a direct pressure score of 8 for a distance of 0.5 km either side of the railway using the same method as for roads.
6. Navigable Waterways, acquired by the PPF, like roads, provide direct access to wildlife habitats along the waterway, and indirect access in periphery areas. The Zambezi River is the only permanent navigable waterway in our study area, and following Venter et al. [17], we assigned a pressure score of 4 to pixels adjacent to the river, exponentially decaying out to 15 km.
7. Arable land throughout the Kafue–Zambezi interface is characterized by majority maize and pulses cultivated using the traditional Chitemene low input, rain fed, slash and burn farming method [46]. Arable land cover classifications are considered by Venter et al. [17] to provide intermediate disturbance to wildlife though direct reduction of wildlife habitat.
8. The arable land was extracted by a land cover map produced by Lines et al. [37] using a mosaic of 24, geometrically and atmospherically corrected, Sentinel 2 images in an Object Oriented Image Analysis environment. The land cover map had a spatial resolution of 10 m and an overall classification accuracy of 91.6% [37]. A pressure score of 7 was assigned to all pixels covered by arable land and 0 to all other pixels [17].
9. Night-time light infrastructure, while sparse and of low intensity throughout much of our study area, is considered a direct human pressure limiting wildlife through a range of negative impacts [47].
10. The "vcm-orm-ntl" (VIIRS Cloud Mask—Outlier Removed—Night-time Lights) annual average layer was used [44], for generating the respective pressure layer. Pixels with a value of 0 (no light) were assigned the value of 0 in the generated layer. For all other pixels, following Venter et al. [17], we applied the same log formula used for pop density (Table 1) resulting in pixels with scores ranging from 0 to 8.971.
11. Aggregating the layers: All generated layers were added to generate an aggregate layer indicating for each pixel the total Human Footprint Pressure. Before aggregation all layers generated at a spatial resolution coarser than 10 m were resampled to a spatial resolution of 10 m. The resolution of 10 m was adopted for this analysis, because it corresponds to the resolution of the land cover data and it is the finest

among all data. While the resampling to a finer resolution does not affect the quality of data provided at coarser resolutions, a resampling of fine resolution data to a coarser resolution would probably result to information loss. The aggregated Human Footprint Pressure layer was used as the single explanatory variable in a habitat suitability modelling analysis.

2.2. *MaxEnt Habitat Suitability Modelling*

Following the methods described in Lines et al. [37], habitat suitability maps for lion, leopard, and spotted hyaena were generated using MaxEnt [48] which performs well compared to other modelling techniques using presence only data [49], and has been repeatedly used to model large carnivores distribution [50–53]. We present the modelling briefly below, and refer to Lines et al. [37] for more details.

We incorporated empirically generated occurrence data from Lines et al. [38]. In total, 102 × 4 km transects, optimized for site conditions, were surveyed on foot three times by the author and two experienced local trackers from the safari hunting industry, amounting to 1224 km of spoor transects during the dry season of May–October 2015, based on a pilot study to determine optimal sampling effort to detect target species and cover the landscape in a single field season. To account for sampling bias, we spatially rarefied occurrence records for all species by thinning (using a 500 × 500 m pixel-size grid of the area). In total, 43 occurrence records were used for lions, 84 for leopards, and 78 for spotted hyenas. Data we split into two sets, a training (70%) and a testing (30%) set for all species, 10,000 thousands background points were randomly selected as pseudoabsence data, and 50 iterations were run for all species. We used receiver operating characteristic area under the curve for evaluating the models' efficiency (ROC AUC). While we sought to incorporate occurrence data for the entire extant large carnivore guild known from the Greater Kafue System, sample sizes were too small for cheetah (*Acinonyx jubatus*) and African wild dog (*Lycaon pictus*) to include in final analyses. The predictor variable modelled against single occurrence was the aggregated human footprint.

An additional analysis was undertaken to investigate Species Sensitivity to Human Footprint Pressure. The relationship of large carnivores to changing Human Footprint Pressure is well established at the global scale [23,54]. However, application of this relationship towards an understanding of thresholds at which species occur, locally extirpate, or might recolonize is poorly developed, irrespective of its clear utility as a conservation tool [24]. In order to identify the thresholds of Human Footprint Pressure, at the wildlife management area scale, above which the species do not occur we calculated the mean Human Footprint Pressure for each protected area. The resulted scores were examined against the derived occurrence data for lion, leopard, and spotted hyena, then compared outputs against species-scale relative sensitivities to extinction from Di Marco et al. [22], and the ranking of sensitivities to localized extirpation following Riggio et al. [31].

3. Results

Figure 2 indicates areas of high to low human pressure, with notable areas of highest pressure around Sesheke/Katima Mulilo in the southwest, along the Zambia/Namibia border following the east-west tar road, along much of the Zambezi River and in the central/eastern areas dominated by access roads, settlements, and agricultural development. Broadly, settlement and agricultural development is widespread throughout the landscape, concurrent with the formal and informal road network. Areas of low apparent human pressure include Kafue National Park (where settlement and agriculture are illegal and non-existent), and adjacent areas of northern Mulobezi, Sichifulo Game Management Areas, Nachitwe and Martin Forests (Figure 2).

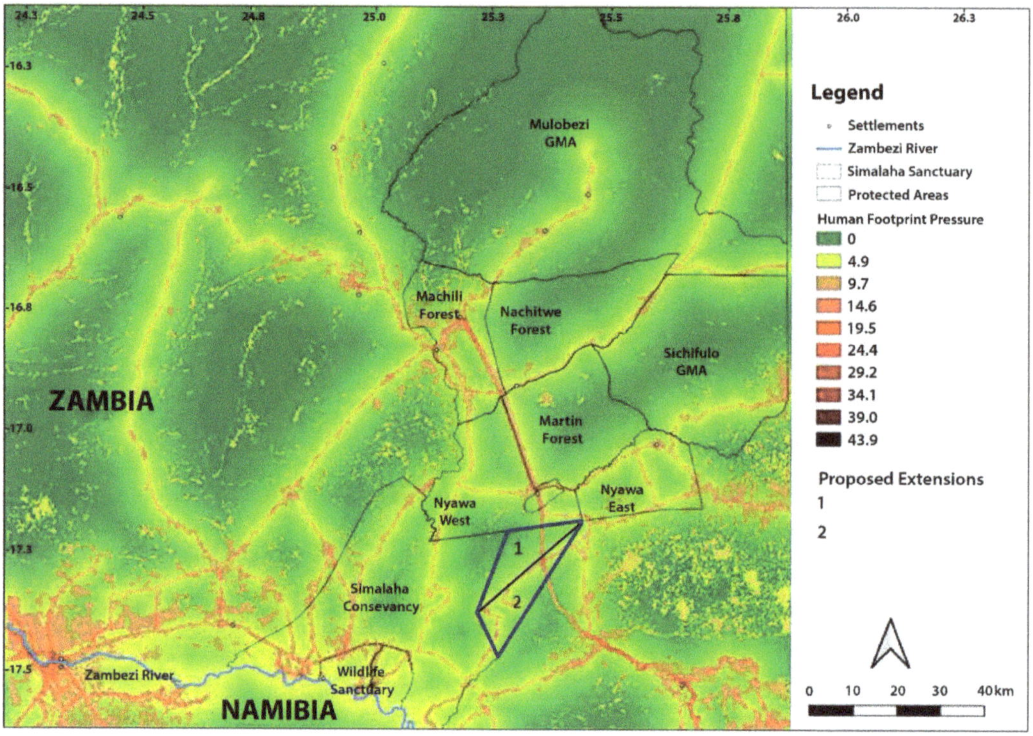

Figure 2. Human Footprint Pressure, Kafue–Zambezi Interface, 10 m resolution.

3.1. MaxEnt Habitat Suitability Modelling Outputs

The best performing model was for lion (ROC AUC = 0.72), then leopard (ROC AUC = 0.65), and finally spotted hyena (ROC AUC = 0.61), indicating strong to moderate model performance, considering these are single-variable MaxEnt models. As expected, we found a negative correlation between aggregated human pressure and species occurrence (Figure 3). Human pressure had the clearer impact on lion, then leopard then hyena, as shown by the sharper drop of suitability as the Human Footprint Pressure increases as well as by the thresholds presented in Table 2. Significant unsuitable areas for all species in central-southern areas, and especially along the Zambezi river parallel to the main tar road where settlements and agriculture mainly occur, were identified by the analysis. Another significant linear feature of human pressure affecting all species but predominantly lions followed the railway line and parallel roads, interspersed with settlements and arable land, demonstrating the strong relationship between access infrastructure, settlement and agricultural development driving the human footprint throughout this landscape.

We should note that single-variable models with a limited number of presences are likely to result in models with high uncertainty, especially for species like hyenas or leopards that exhibit behavioral plasticity vis-à-vis the presence of humans. In Figure 3 below, we can see that for the response curves of leopards and hyenas as Human Footprint increases, so does model uncertainty as evidenced by the large difference in habitat suitability for each run. Therefore, the accuracy and predictive ability of the models for levels of human footprint exceeding the point where the response curve minimizes, is very low and does not merit any ecological or behavioral interpretation.

At the wildlife management area scale, Human Footprint Pressure was lowest in Kafue National Park, Mulobezi and western parts of Sichifulo Game Management Areas. Nachitwe and Martin Forest Reserves appear relatively intact with significant pressure on their western boundaries. Machili Forest Reserve is heavily impacted throughout by human pressure. There are still areas within Nyawa communal lands with relatively low human pressure and again to the northeast and eastern sections of Simalaha Conservancy, extending into the adjacent unprotected areas. Extensive pressure exists around the settlement of Bombwe, formerly a registered Forest Reserve. Simalaha Wildlife Recovery Sanctuary, sandwiched between the Zambezi River and main Tar road, is subject to significant human pressure, including settlement and agriculture both within the Sanctuary and on its borders.

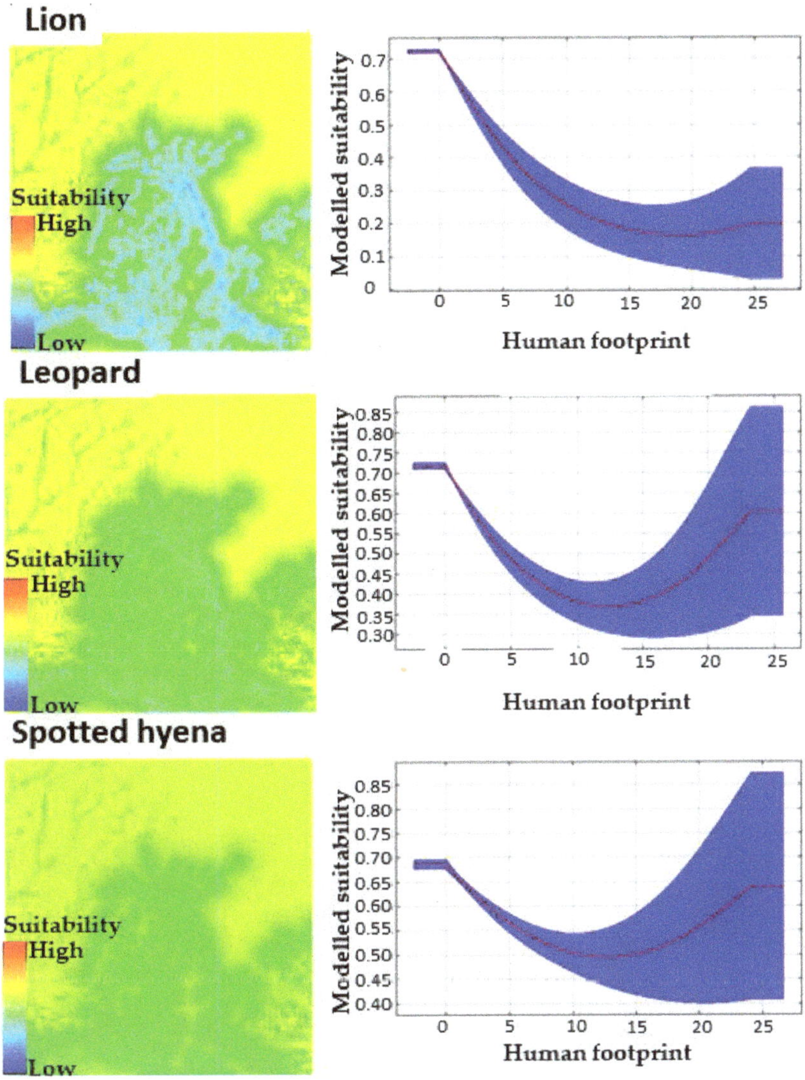

Figure 3. A visual presentation of the results of the human footprint habitat suitability model. Left: habitat suitability maps of the three species; colors on the map indicate habitat suitability, going from blue (low suitability) to green (medium suitability) to orange (high suitability). Right: the response curves between habitat suitability and human footprint scores for each species; the red line indicates the mean of the 50 runs for each model, and the blue surface the deviation from the mean.

Table 2. Mean Human Footprint Pressures within Wildlife Managed Areas against Species Occurrence.

Area	IUCN	Area Ha	Mean HFP	Occurrence		
				Leopard	Lion	Hyena
Kafue National Park	II	206,314	1.1	Yes	Yes	Yes
Mulobezi GMA	VI	347,481	1.2	Yes	Yes	Yes
Sichifulo GMA	VI	133,734	2.2	Yes	Yes	Yes
Nachitwe Forest Reserve	unreported	71,075	2.3	Yes	Yes	Yes
Martin Forest Reserve	unreported	62,948	2.4	Yes	Yes	Yes
Nyawa West	unreported	57,439	3.7	Yes	No	Yes
Simalaha Conservancy	unreported	181,936	4.6	No	No	No
Open Area extension 1 *	unreported	32,712	3.9	Yes	No	Yes
Open Area extension 1 + 2 *	unreported	73,660	3.9	Yes	No	Yes
Nyawa East	unreported	39,565	4.6	No	No	No
Machili Forest Reserve	unreported	49,269	5.6	Yes	No	Yes
Simalaha Sanctuary E	unreported	11,020	5.8	No	No	No
Simalaha Sanctuary W	unreported	11,282	6.8	No	No	No

* Proposed extensions to protected area network within HFP thresholds limits for selected species.

3.2. Human Footprint Pressure Thresholds and Species Persistence

The means Human Footprint Pressure at the wildlife managed area scale varied from 1.1 in Kafue National Park to 6.8 in the western section of the Simalaha Sanctuary, with IUCN categorized protected areas experiencing lowest mean Human Footprint Pressure (Table 2). With the exception of Machili Forest (Human Footprint Pressure 5.6) there was a steady increase in Human Footprint Pressure moving south away from Kafue National Park towards the Zambezi River.

The Human Footprint Pressure threshold (Table 2) at which each species occurs in each wildlife management area revealed broadly similar high sensitivities. Lion exhibited highest sensitivity to Human Footprint Pressure with an occurrence threshold between 2.4–3.7, followed by hyena and leopard, with threshold values between 3.7–4.6, mirroring species sensitivity presented by Riggio et al. [31]. An apparent anomaly is Machili Forest Reserve, with extensive settlement, agriculture, and transport infrastructure, having a mean Human Footprint Pressure score of 5.6, and with both leopard and spotted hyena occurring. The proposed supplemental addition to the protected area network identified in Lines et al. [37], Open Area extension 1 (and 2), has a mean Human Footprint Pressure of 3.9, within threshold limits presented here, which indicates that both leopard and spotted hyena could inhabit these protected areas.

4. Discussion

Human Footprint Pressure modelling has traditionally been undertaken at global scales, and typically at low spatial resolution, mainly due to lack of availability of high resolution global datasets [15,55]. There is a constant attempt to overcome resolution constraints which could facilitate improved accuracy and applications for conservation planning [26], including deriving impacts of human pressure at more appropriate site- and species-specific scales where the utility of proxies such as Human Footprint maps might be most valuable as conservation tools [24]. Our study successfully overcomes limitations to existing models by generating and integrating site-specific, multiple high-resolution data sets at two orders of magnitude finer scale, then applying it directly to key questions surrounding the impacts of Human Footprint Pressure on large carnivores throughout a network of wildlife managed areas under varying degrees of Human Footprint Pressure at the Kafue–Zambezi landscape, a key proposed corridor in the KAZA.

Model output performed best for lion, a species exhibiting very high sensitivity to human disturbance [31], which we would expect to capture in multi-variate model analyses. Both leopard and spotted hyena are species known for intrinsic ecological traits and behavioral plasticity. These characteristics facilitate greater coexistence with humans in landscapes with increased human pressure, and our model output for both these species captured this, presenting lower predictive power as expected by such characteristics. Therefore, species and site sensitivity to complex, interrelated human disturbance variables explains why the best performing model output is represented by the species with the highest sensitivity to human pressure, and less so for species with increased tolerance to human pressures.

While there are limits to the predictive power of single variable models, the predictive power of this model would likely benefit from supplemental data layers, when available, notably, interference and exploitative pressures of pastoralism and (legal and illegal) wildlife consumption [56], and synergistic effects between human behaviour and climate change [57]. A multi-variate exploration of the same species in the same landscape using a similar methodology wielded better predictive power, but was significantly more laborious in terms of creating the necessary data layer for analysis. There is also debate over the appropriate scale or extent at which to measure Human Footprint Pressure, whether that be at the population level, proportion of species total range, home range, or other scales [54]. Additionally, there is scope for a greater understanding of site- and species-specific pressure score calibrations, including impacts of formal and informal road linkages [26].

Model response curves and secondary explorative analyses of thresholds at which species are extirpated at the wildlife management area scale closely match global mammalian Human Footprint Pressure extinction thresholds [22], while they are also in accordance with the results presented by Riggio et al. [31] who conducted a sensitivity analysis of African large mammals with high susceptibility to human disturbance. Collectively these data provide compelling evidence that Human Footprint Pressure scores ranging between 2.4 and 4.6 represent a threshold limit for these three species of large carnivores, beyond which they are unlikely to persist in human-dominated landscapes using Venter et al.'s [17] existing pressure score methodology. The persistence of leopard and hyena in Machili Forest, with a mean Human Footprint Pressure score of 5.6, is likely explained by the proximity of this area to extensive lower Human Footprint Pressure areas closer to Kafue National Park, a carnivore guild core habitat with highest level of wildlife management support and protection in the long term. In this regard, Machili Forest could be characterized as a threshold area or an attractive sink (areas of relatively pure habitat quality where species tend to inhabit in cases of rapid environmental change), limiting the range expansion of these species to broader areas with lower Human Footprint Pressure.

The identification of potential additions to the protected area network in Open Areas east of the Simalaha, first suggested by Lines et al. [38], and further posited here, serves a two-fold purpose: (a) the possible increase in wildlife habitat for a range of species and (b) the likely increase in connectivity between Kafue National Park and the Zambezi River for both leopard and spotted hyena in areas of low human habitation and agricultural development, limiting the scope for human–wildlife conflict in the otherwise increasingly human-dominated landscapes at the central-southern extents of the Kafue–Zambezi interface.

These site-specific, high-resolution maps have broad utility as a baseline against which subsequent changes to human footprint pressure can be mapped and modelled over time as more data sets come available to refine this iterative process. Pressure score calibration merits more explicit treatment to improve model response given species and/or processes of interest. The value and applicability of generating standardized approaches to mapping Human Footprint Pressure underlines their use as a proxy or indicator of broader drivers impacting habitat degradation, ecosystem function, species loss, or potential for species recovery. Progress with Human Footprint Pressure modelling depends in part on understanding and addressing limitations and assumptions of model development [58] and recognition of the dynamic nature of human pressure in terms of asymmetrical and nonlinear threshold responses to total footprint pressure changes across spatial-temporal scales [19,59].

5. Conclusions

We have demonstrated that Human Footprint Pressure analyses can be utilized as indicators of habitat suitability for a suit of large carnivores of conservation value, including predicting species persistence, extirpation, and potential for recolonization, even at this preliminary, proof-of-concept stage of model development.

Model output broadly follows existing data in support of understanding human pressure impacts on landscape-level connectivity at the Kafue–Zambezi interface, providing a valuable additional tool in conservation planning for this landscape, the broader KAZA region, and beyond. As additional data layers become available and the pressure score calibration process evolves, site-specific human pressure maps can be expanded to the broader Zambian and KAZA landscape to model how spatiotemporal human pressure impacts species and processes of interest to key conservation and human–wildlife management objectives.

Author Contributions: Conceptualization, R.L., J.T. and D.C.M.; methodology, R.L., D.B., P.X., J.T. and D.C.M.; validation, R.L.; formal analysis, R.L., D.B., P.X. and J.T.; validation, R.L.; investigation, R.L.; data curation, R.L., D.B., P.X. and L.P.; writing—original draft preparation, R.L., D.B., P.X. and J.T.; writing—review and editing, P.X., D.B., R.L. and J.T.; visualization, R.L., D.B. and P.X.; supervision, J.T. All authors have read and agreed to the published version of the manuscript.

Funding: This research was partially funded by: (1) University of Kent 50th Anniversary Scholarship, (2) WWF Namibia, (3) Humane Society International, (4) Westwood.

Institutional Review Board Statement: The study was conducted according to the guidelines of the Declaration of Helsinki, and approved by the Ethics Committee of the School of Anthropology and Conservation, University of Kent. Permission to contact research in Zambia was provided by the Zambia Wildlife Authority.

Informed Consent Statement: Not applicable.

Data Availability Statement: Not applicable.

Acknowledgments: We thank A. Nambota for study development; P. Moss, J. Hanks, and M. Musgrave for insights into Kafue National Park and surrounding Game Management Areas; and the Chilanga, Ngoma, Mulobezi, and Mulanga offices of the Department of Parks and Wildlife Management (formerly the Zambia Wildlife Authority) for permissions and field support. We also thank HRH Chief Inyambo Yeta of the Lozi and the late HRH Chief Moomba of the Nkoya, for granting smooth passage through their Chiefdoms.

Conflicts of Interest: The authors declare no conflict of interest.

References

1. Vitousek, P.; Mooney, H.; Lubchenco, J.; Melillo, J. Human domination of Earth's ecosystems. *Science* **1997**, *277*, 494–499. [CrossRef]
2. Wackernagel, M.; Schulz, N.B.; Deumling, D.; Linares, C.A.; Jenkins, M.; Kapos, V.; Monfreda, C.; Loh, J.; Myers, N.; Norgaard, R.; et al. Tracking the ecological overshoot of the human economy. *Proc. Natl. Acad. Sci. USA* **2002**, *99*, 9266–9271. [CrossRef] [PubMed]
3. Riitters, K.; Wickham, J.; O'Neill, R.; Jones, B.; Smith, E. Global-scale patterns of forest fragmentation. *Conserv. Ecol.* **2000**, *4*, 3. Available online: http://www.consecol.org/vol4/iss2/art3/ (accessed on 20 October 2021). [CrossRef]
4. Foley, J.A.; Defries, R.; Asner, G.P.; Barford, C.; Bonan, G.; Coe, M.T.; Daily, G.C.; Gibbs, H.K.; Carpenter, S.R.; Chapin, F.S.; et al. Global consequences of land use. *Science* **2005**, *333*, 301–306. [CrossRef]
5. Tucker, M.A.; Fagan, W.F.; Fryxell, J.M.; Van Bram, M.; Susan, C.A.; Abdullahi, H.A.; Andrew, M.A.; Attias, N.; Tal, A.; Brooks, B.; et al. Moving in the Anthropocene: Global reductions in terrestrial mammalian movements. *Science* **2018**, *359*, 466–469. [CrossRef] [PubMed]
6. Maxwell, S.L.; Fuller, R.A.; Brooks, T.M.; Watson, J.E. Biodiversity: The ravages of guns, nets and bulldozers. *Nat. News* **2016**, *546*, 143. [CrossRef]
7. Ceballos, G.; Ehrlich, P.R. Mammal population losses and the extinction crisis. *Science* **2002**, *296*, 904–907. [CrossRef]
8. Ceballos, G.; Ehrlich, P.R.; Barnosky, A.D.; Garcia, A.; Pringle, R.M.; Palmer, T.M. Accelerated modern human–induced species losses: Entering the sixth mass extinction. *Sci. Adv.* **2015**, *1*, e1400253. [CrossRef]
9. Navarro, L.M.; Pereira, H.M. Rewilding abandoned landscapes in Europe. In *Rewilding European Landscapes*; Springer: Berlin/Heidelberg, Germany, 2015; pp. 3–23.

10. Brooks, T.M.; Mittermeier, R.A.; Fonseca, G.A.; Gerlach, J.; Hoffmann, M.; Lamoreux, J.F.; Mittermeier, C.G.; Pilgrim, J.D.; Rodrigues, A.S. Global biodiversity conservation priorities. *Science* **2006**, *313*, 58–61. [CrossRef] [PubMed]
11. Newmark, W.D. Isolation of African protected areas. *Front. Ecol. Environ.* **2008**, *6*, 321–328. [CrossRef]
12. Geldmann, J.; Joppa, L.N.; Burgess, N.D. Mapping change in human pressure globally on land and within protected areas. *Conserv. Biol.* **2014**, *28*, 1604–1616. [CrossRef] [PubMed]
13. Duffy, R. Global governance and environmental management: The politics of transfrontier conservation areas in Southern Africa. *Political Geogr.* **2006**, *25*, 89–112. [CrossRef]
14. Gren, M.; Svensson, T.; Elofsson, K.; Engelmann, M. Economics of wildlife management—An overview. *Eur. J. Wildl. Res.* **2018**, *64*, 22. [CrossRef]
15. Sanderson, E.W.; Jaiteh, M.; Levy, M.A.; Redford, K.H.; Wannebo, A.V.; Woolmer, G. The human footprint and the last of the wild: The human footprint is a global map of human influence on the land surface, which suggests that human beings are stewards of nature, whether we like it or not. *Bioscience* **2002**, *52*, 891–904. [CrossRef]
16. Walker, B.; Salt, D. *Resilience Thinking: Sustaining Ecosystems and People in a Changing World*; Island Press: Washington, DC, USA, 2006; 192p, ISBN 978-1-59726-622-2.
17. Venter, O.; Sanderson, E.W.; Magrach, A.; Allan, J.; Beher, J.; Jones, K.R.; Possingham, H.P.; Laurance, W.F.; Wood, P.; Fekete, B.M.; et al. Global terrestrial Human Footprint maps for 1993 and 2009. *Sci. Data* **2016**, *3*, 160067. [CrossRef]
18. Ayram, C.A.C.; Mendoza, M.E.; Etter, A.; Salicrup, D.R.P. Anthropogenic impact on habitat connectivity: A multidimensional human footprint index evaluated in a highly biodiverse landscape of Mexico. *Ecol. Indic.* **2017**, 895–909. [CrossRef]
19. Watson, J.E.M.; Shanahan, D.F.; Di Marco, M.; Allan, J.; Laurance, W.F.; Sanderson, E.W.; Mackey, B.; Venter, O. Catastrophic declines in wilderness areas undermine global environment targets. *Curr. Biol.* **2016**, *26*, 2929–2934. [CrossRef]
20. Jones, K.R.; Venter, O.; Fuller, R.A.; Allan, J.; Maxwell, S.L.; Negret, P.J.; Watson, J.E. One-third of global protected land is under intense human pressure. *Science* **2018**, *360*, 788–791. [CrossRef]
21. Gaynor, K.M.; Hojnowski, C.E.; Brashares, J.S. The influence of human disturbance on wildlife nocturnality. *Science* **2018**, *360*, 1232–1235. [CrossRef] [PubMed]
22. Di Marco, M.; Venter, O.; Possingham, H.P.; Watson, J.E. Changes in human footprint drive changes in species extinction risk. *Nat. Commun.* **2018**, *9*, 4621. [CrossRef]
23. Di Marco, M.; Santini, L. Human pressures predict species' geographic range size better than biological traits. *Glob. Chang. Biol.* **2015**, *21*, 2169–2178. [CrossRef]
24. Trombulak, S.C.; Baldwin, R.F.; Woolmer, G. *The Human Footprint as a Conservation Planning Tool in Landscape-Scale Conservation Planning*; Trombulak, S.C., Baldwin, R.F., Eds.; Springer: Dordrecht, The Netherlands, 2010; pp. 281–302.
25. Sutherland, W.J.; Pullin, A.S.; Dolman, P.M.; Knight, T.M. The need for evidence-based conservation. *Trends Ecol. Evol.* **2004**, *19*, 305–308. [CrossRef]
26. Woolmer, G.; Trombulak, S.C.; Ray, J.C.; Doran, P.J.; Anderson, M.G.; Baldwin, R.F.; Morgan, A.; Sanderson, E.W. Rescaling the human footprint: A tool for conservation planning at an ecoregional scale. *Landsc. Urban Plan.* **2008**, *87*, 42–53. [CrossRef]
27. Cumming, D.H. Constraints to conservation and development success at the wildlife-livestock-human interface in southern African transfrontier conservation areas: A preliminary review. *Wildl. Conserv. Soc.* **2011**. Available online: https://www.povertyandconservation.info/en/biblio/B1785 (accessed on 20 October 2021).
28. Wittemyer, G.; Elsen, P.; Bean, W.T.; Burton, A.C.; Brashares, J.S. Accelerated human population growth at protected area edges. *Science* **2008**, *321*, 123–126. [CrossRef]
29. Ripple, W.J.; Estes, J.A.; Beschta, R.L.; Wilmers, C.C.; Ritchie, E.G.; Hebblewhite, M.; Berger, J.; Elmhagen, B.; Letnic, M.; Nelson, M.P. Status and ecological effects of the world's largest carnivores. *Science* **2014**, *343*, 124–148. [CrossRef]
30. Estes, J.A.; Terborgh, J.; Brashares, J.S.; Power, M.E.; Berger, J.; Bond, W.J.; Carpenter, S.R.; Essington, T.E.; Holt, R.D.; Jackson, B.C.J.; et al. Trophic downgrading of planet Earth. *Science* **2011**, *333*, 301–306. [CrossRef] [PubMed]
31. Riggio, J.; Kija, H.; Masenga, E.; Mbwilo, F.; Van de Perre, F.; Caro, T. Sensitivity of Africa's larger mammals to humans. *J. Nat. Conserv.* **2018**, *43*, 136–145. [CrossRef]
32. Funston, P.J.; Groom, R.J.; Lindsey, P.A. Insights into the management of large carnivores for profitable wildlife-based land uses in African savannas. *PLoS ONE* **2013**, *8*, e59044. [CrossRef]
33. KAZA. *The Kavango Zambezi Transfrontier Conservation Area Carnivore Conservation Strategy*; KAZA: Lusaka, Zambia, 2018.
34. KAZA. *Master Integrated Development Plan*; KAZA: Lusaka, Zambia, 2014.
35. UNEP-WCMC. United Nations Environment Programmes World Conservation Monitoring Centre. Protected Planet Database. Available online: https://www.protectedplanet.net/ (accessed on 1 May 2015).
36. Lindsey, P.; Nyirenda, V.; Barnes, J.; Becker, M.S.; McRoob, R. Underperformance of the Zambian protected area network: Steps needed to improve functionality and effectiveness. *PLoS ONE* **2014**, *9*, e94109.
37. Lines, R.; Bormpoudakis, D.; Xofis, P.; Tzanopoulos, J. Modelling multi-species connectivity at the Kafue-Zambezi Interface: Implications for Transboundary Carnivore Conservation. *Sustainability* **2021**, *13*, 12886. [CrossRef]
38. Lines, R.; Tzanopoulos, J.; MacMillan, D. Status of terrestrial mammals at the Kafue–Zambezi interface: Implications for transboundary connectivity. *Oryx* **2018**, *53*, 764–773. [CrossRef]
39. KAZA. Treaty Between the Governments of Angola, Botswana, Namibia, Zambia and Zimbabwe on the Establishment of the Kavango-Zambezi Transfrontier Conservation Area. *J. Environ. Dev.* **2008**, *17*, 99–117.

40. CSO. Central Statistics Office of Zambia. Available online: http://www.zamstats.gov.zm/ (accessed on 23 September 2019).
41. Calvert, G.M. *Sitimela: A History of the Zambesi Sawmills Logging Railway, 1911–1972*; Barotse Development Trust: Mongu, Zambia, 2005.
42. Rees, W.; Wackernagel, M. *Our Ecological Footprint: Reducing Human Impact on the Earth*; New Society Publishers: Gabriola Island, BC, Canada, 1996.
43. Bonafilia, D.; Gill, J.; Basu, S.; Yang, D. Building High Resolution Maps for Humanitarian Aid and Development with Weakly- and Semi-Supervised Learning. In Proceedings of the IEEE Conference on Computer Vision and Pattern Recognition, CVPR Workshops 2019, Long Beach, CA, USA, 16–20 June 2019; pp. 1–9.
44. NOAA. *National Oceanic and Atmospheric Administration. National Geophysical Data Center. DMSP Data Collected by US Air Force Weather Agency*; NOAA: Washington, DC, USA, 2019.
45. Woodroffe, R.; Ginsberg, J.R. Edge Effects and the Extinction of Populations Inside Protected Areas. *Science* **1998**, *280*, 2126–2128. [CrossRef] [PubMed]
46. Musgrave, M. *Scale, Governance and Change in Zambezi Teak Forests: Sustainable Development for Commodity and Community*; Cambridge Scholars Publishing: Newcastle upon Tyne, UK, 2016.
47. Rich, C.; Longcore, T. *Ecological Consequences of Artificial Night Lighting*; Island Press: Washington, DC, USA, 2013.
48. Phillips, S.J.; Dudík, M. Modelling of species distributions with Maxent: New extensions and a comprehensive evaluation. *Ecography* **2008**, *31*, 161–175. [CrossRef]
49. Elith, J.; Phillips, S.J.; Hastie, T.; Dudík, M.; Chee, Y.E.; Yates, C.J. A statistical explanation of MaxEnt for ecologists. *Divers. Distrib.* **2011**, *17*, 43–57. [CrossRef]
50. Jackson, C.R.; Marnewick, K.; Lindsey, P.; Røskaft, E.; Robertson, M.P. Evaluating habitat connectivity methodologies: A case study with endangered African wild dogs in South Africa. *Landsc. Ecol.* **2016**, *312*, 1433–1447. [CrossRef]
51. Di Minin, E.; Macmillan, D.C.; Goodman, P.S.; Escott, B.; Slotow, R.; Moilanen, A. Conservation businesses and conservation planning in a biological diversity hotspot. *Conserv. Biol.* **2013**, *27*, 808–820. [CrossRef]
52. Ahmadi, M.; Nezami Balouchi, B.; Jowkar, H.; Hemami, M.; Fadakar, D.; Malakouti-Khah, S.; Ostrowski, S. Combining landscape suitability and habitat connectivity to conserve the last surviving population of cheetah in Asia. *Divers. Distrib.* **2017**, *23*, 592–603. [CrossRef]
53. Angelieri, C.C.S.; Adams-Hosking, C.; de Barros Ferraz, K.M.P.M.; de Souza, M.P.; McAlpine, C.A. Using species distribution models to predict potential landscape restoration effects on puma conservation. *PLoS ONE* **2016**, *11*, e0145232.
54. Di Marco, M.; Rondinini, C.; Boitani, L.; Murray, K. Comparing multiple species distribution proxies and different quantifications of the human footprint map, implications for conservation. *Biol. Conserv.* **2013**, *165*, 203–211. [CrossRef]
55. Leu, M.; Hanser, S.E.; Knick, S.T. The human footprint in the west: A large-scale analysis of anthropogenic impacts. *Ecol. Appl.* **2008**, *18*, 1119–1139. [CrossRef]
56. Everatt, K.T.; Moore, J.; Kerley, G.I. Africa's apex predator, the lion, is limited by interference and exploitive competition with humans. *Glob. Ecol. Conserv.* **2019**, *20*, e00758. [CrossRef]
57. Brodie, J.F. Synergistic effects of climate change and agricultural land use on mammals. *Front. Ecol. Environ.* **2016**, *14*, 20–26. [CrossRef]
58. Halpern, B.S.; Fujita, R. Assumptions, challenges, and future directions in cumulative impact analysis. *Ecosphere* **2013**, *4*, 1–11. [CrossRef]
59. Toews, M. *Managing Human Footprint with Respect to its Effects on Large Mammals: Implications of Spatial Scale, Divergent Responses and Ecological Thresholds*; University of Victoria: Victoria, BC, Canada, 2016.

sustainability

MDPI

Article

Fish Hobbyists' Willingness to Donate for Wild Fighting Fish (*Betta livida*) Conservation in Klang Valley

Muhammad Fairus Abd Rahman [1] and Nitanan Koshy Matthew [1,2,*]

1 Department of Environment, Faculty of Forestry and Environment, Universiti Putra Malaysia, Serdang 43400, Malaysia; abdr.fairus@gmail.com
2 Institute of Tropical Agriculture and Food Security (ITAFoS), Universiti Putra Malaysia, Serdang 43400, Malaysia
* Correspondence: nitanankoshy@upm.edu.my

Abstract: *Betta livida* is an endangered endemic species of wild fighting fish affected by habitat degradation and exploitation. Despite this concern, the literature on the conservation of wild betta is negligible. Conservation is a non-use value, whereas the species itself is a use-value because they are sought after in the ornamental fish trade business. Therefore, the contingent valuation method (CVM) was applied in this study to establish the monetary value for species conservation by determining hobbyists' willingness to donate (WTD) for conservation. Fish hobbyists are the most prominent backbone of the industry and are the most acquainted with the targeted species. Hence, hobbyists' knowledge, perceptions and attitude towards species conservation have also been explored and weighed against the WTD. Purposive sampling was employed with a total of 150 respondents in Klang Valley. The findings show that the WTD was influenced by double-bound CVM, age (AGE) and hobbyists who owned the species (OWNB). In contrast, knowledge, perceptions and attitudes were not significant. Using probit regression analysis, hobbyists' WTD for species conservation was MYR 9.04 annually. The survey also revealed concern for species that are wild-caught by hobbyists. Hence, the results of this study offer preliminary insights into the WTD for wild betta and local freshwater fish conservation in Malaysia.

Keywords: attitude; contingent valuation method; knowledge; perception; wild betta; willingness to donate

Citation: Abd Rahman, M.F.; Matthew, N.K. Fish Hobbyists' Willingness to Donate for Wild Fighting Fish (*Betta livida*) Conservation in Klang Valley. *Sustainability* **2021**, *13*, 10754. https://doi.org/10.3390/su131910754

Academic Editor: Vasilios Liordos

Received: 31 July 2021
Accepted: 2 September 2021
Published: 28 September 2021

Publisher's Note: MDPI stays neutral with regard to jurisdictional claims in published maps and institutional affiliations.

Copyright: © 2021 by the authors. Licensee MDPI, Basel, Switzerland. This article is an open access article distributed under the terms and conditions of the Creative Commons Attribution (CC BY) license (https://creativecommons.org/licenses/by/4.0/).

1. Introduction

Fighting fish (betta) have always been a popular preference for ornamental fish due to their vivid colour schemes, minimal set-up requirement and the unique showcase of the male betta displaying dominance by flaring their operculum and finnage when threatened. However, the most common betta is a captive crossbreed with particular traits of the *Betta splenden*, which are common and not under threat of endangerment. Wild betta fish are found in some endemic regions throughout Malaysia, Indonesia and Thailand. Tan and Ng (2005) [1] described numerous wild betta species found in Malaysia, including *B. coccina*, *B. persephone* and *B. livida*, the latter of which are the focus of this study (see Table 1).

Tan and Ng [1] described adult *B. livida* as having small, unique bodies (the standard length is less than 40 mm), with a uniform deep red (maroon) body colour. They have mid-lateral bodies and both genders have an iridescent green blotch that often fades throughout adulthood. They have a parallel vertical iridescent gold bar on the operculum and a falcate pelvic fin with an iridescent green tip (Figure 1). The species may also be mistaken for another species, *B. coccina*. They are almost identical, except that the tip of the pelvic fin of *B. coccina* is black and they are found in different geographic areas. This classical look of wild betta and their rareness has made this species the most sought after in the ornamental fish trade. A pair might cost more than MYR 100 from private sellers and collectors.

Table 1. Taxonomy of *B. livida.*

Kingdom	Phylum	Class	Order	Family	Genus	Species
Animalia	Chordata	Actinopterygii	Perciformes	Osphronemidae	*Betta*	*livida*

Source: Ng and Kotellat (1992).

Figure 1. Photos of Betta livida (Photo credit: Muhammad Ilham Norhakim).

Research on the conservation of wild betta is still inadequate and requires further study. Keat-Chuan et al. [2] noted that the challenges in freshwater fish conservation were due to a lack of interest, funding and problems with unresolved taxonomic classification. The only established conservation research of wild betta was for *B. persephone*, a critically endangered and endemic species found in Johor [3]. Moreover, economic valuation for the conservation of local freshwater fish species in Malaysia has yet to be established. Eliciting monetary value would indicate how much the public is willing to pay (WTP) to protect ornamental, native fish and eventually urge authorities to act to conserve the species.

Developments in peat swamp forest areas have been detrimental to the environment; most ecosystems have been destroyed, resulting in the depletion of flora and fauna biodiversity. *Betta livida* is an example of an endemic species of the peat swamp forest in Selangor, which is in the Klang Valley (KV) area, a hub for rapid modern development. The recent issuance of a local state government proposing the degazetting of the North Kuala Langat Reserve Forest, which is home to *B. livida*, became a testament to a lack of conservation mindedness. Giam et al. [4] projected the extinction of *B. livida* and other swamp fish species by 2050 if environmental management and policies fail to protect their ecosystem. *B. livida* has also been classified as an endangered (EN) species in the Red List's International Union of Conservation of Nature (IUCN) [5]. Overexploitation in the ornamental fish trade will further endanger their status [2,6]. Ng [6] raised concern that the wild betta was excessively caught in wild habitats. Establishing the conservation of the species with a monetary value would provide critical information and enable the initiation of conservation and sustainable development with sound environmental management and policy formulation. Afterall, it would be a loss to Malaysia's biodiversity heritage if this native and endemic species only existed in history books.

Hobbyists are the primary consumers for this industry and key stakeholders in conservation. Therefore, it would be interesting to probe their response to the WTP for conservation of these and other endemic species. Hobbyists would contribute to species conservation via donation. Haefele et al. [7] noted that the use of taxes as elicitation might not appeal to the public due to tax burdens and scepticism about how the money would be used for species conservation. Willingness to pay (WTP) typically dictates that the fund be allocated to specific conservation funds or facilities. As no specific trust fund has been established for this species, a hypothetical trust fund was established. This trust aims to develop conservation efforts that encourage breeding and awareness programs and habitat gazetting for the *B. livida* species. Therefore, employing willingness to donate (WTD) instead of WTP throughout this study was more fitting.

The contingent valuation method (CVM) has been widely used in environmental studies as an economic tool. Environmentalists prefer it for its compatibilities with appraising the value and non-value environmental commodities [8,9]. For example, Greece's public WTP mean for edible sea urchins (*Paracentrotus lividus*) was EUR 56, which greatly influenced the perception of the species [10]. The perception revolves primarily around who is responsible for conserving the species, which falls to the government. In Sri Lanka, the reputation of Asian elephants (*Elephas maximus*) as agricultural pests was evaluated [11]. However, the public, especially those residing in urban areas, still favoured species conservation with a WTP per month of LKR 110.16 for five years, which could compensate for the farmers' losses in rural areas. In contrast, Asmamaw et al. [12] identified that the WTP for *Labeobarbus* fish in Lake Tana using labour days was 48.48 per household, which is equivalent to USD 4,422,792.4 per year. The WTP was significantly influenced by the respondents' age, sex and economic activities (i.e., fishery). This study found that species-level valuation focused on iconic species or species with a socio-economic effect. Because wild *B. livida* act as value and non-value commodities, CVM is an appropriate economic tool for evaluating the conservation of the species in monetary value.

Numerous economic valuations for Malaysia's environmental commodities revolve around WTP valuation for eco-tourism, such as reserved forests and recreational nature parks. For example, Shahwahid et al. [13] conducted a study on the perception of wild elephants in Pahang. They found a WTP mean of MYR 5.86 and a median of MYR 5.00 per person for species conservation. The amount implied the results, the age gap between the respondents and the older generation's enthusiasm for preserving biodiversity heritage for subsequent generations. In Semporna, Sabah, attention was centred on the total environmental value (TEV) of marine turtle conservation [14]. The estimated TEV utilising the CVM was USD 23 million per year from tourists and locals, indicating that the non-use value of the species surpassed the consumption use-value.

On the other hand, Vianna et al. [15] assessed shark-diving tourism as a funding mechanism for shark conservation, with a WTP mean of MYR 35.73 from tourists and guides. Marine Park authorities also generated approximately USD 2 million annually for environmental management and enforcement. These are significant indicators demonstrating the importance of non-use environmental commodities supporting society and ensuring that anthropocentric characteristics heavily influence WTP. At the same time, other factors, such as biocentrism, aesthetics and existence, are valued less.

Therefore, this study acts as preliminary research for the WTP for conserving wild betta and native freshwater fish in Malaysia. To better comprehend the matter, the study probed hobbyists' knowledge regarding the species and measured their perceptions and attitudes towards species conservation. Lastly, the estimation of the WTD for the species conservation was determined using double-bound CVM.

The results of this study aid in discussing and addressing the relevance of using WTD for *B. livida* wild fighting fish conservation among hobbyists in Klang Valley with possible inheriting factors, as follows:

Hypothesis 1 (H1). *The mean WTD per hobbyist's income (INC) for species conservation is statically greater than 0.*

$$(H_{01}: WTP_{INC} > 0, H_{A1}: WTP_{INC} < 0).$$

Hypothesis 2 (H2). *The mean WTD per hobbyist's perception towards the conservation of the species (PERC) is statistically greater than 0.*

$$(H_{02}: WTP_{PERC} > 0, H_{A2}: WTP_{PERC} < 0).$$

Hypothesis 3 (H3). *The mean WTD per hobbyist's attitude towards the conservation of the species (ATT) is statistically greater than 0.*

(H_{03}: $WTP_{ATT} > 0$, H_{A4}: $WTP_{ATT} < 0$).

2. Materials and Methods

2.1. Study Area, Sampling and Target Population

Klang Valley (KV), with an unknown hobbyist population, was selected to be a study area based on the endemic distribution of *B. livida* [5]. KV, which comprises several state districts and federations, is the central hub for modernisation, transportation and industry. Although the description of KV differs in literature, eight (8) districts were selected as study areas. Figure 2 shows the geo-referencing map of *B. livida* distribution coupled with the study area's scope border. 'Resident' refers to the original distribution of the species within the established habitat, while 'extent' indicates species reported outside the habitat area due to the floodplain.

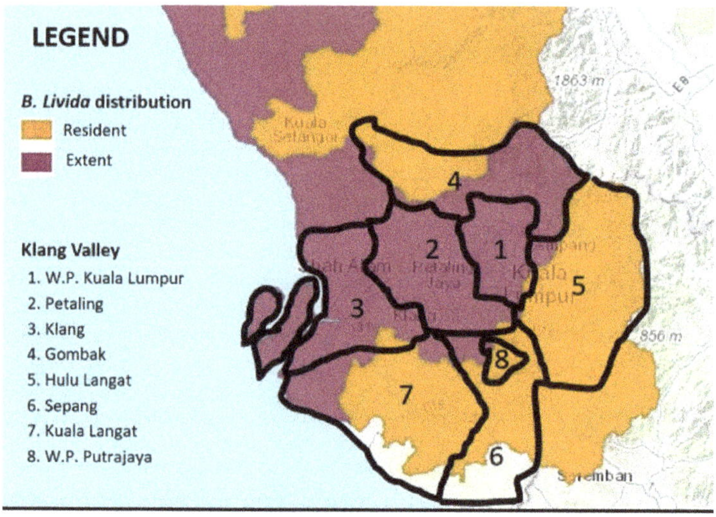

Figure 2. Study Area and *B. livida* hotspots [5].

An online survey was administered using a 'Google Form' circulated to the fish hobbyist groups on social media. Engaging with members ensured the platform algorithm because the survey posted on their social media feed was challenging. This was an issue expected by Ronald [16] when using the internet-based survey to increase response rates. Moreover, each hobbyist group is bombarded with over 20 new posts every day, reducing the chance of the survey appearing on their social media feeds. Support from admin moderators and group members were sought to distribute and circulate the survey post. However, the social group member seeing the survey post provides no guarantees that they will participate. Therefore, incentives might be a consideration in future studies to mitigate the issue of low participation.

A purposive sampling method was used in this study targeting fish hobbyists living in KV. A total of $N = 150$ respondents were classified with a 90% confidence interval and a 6.6% margin of error. According to a conventional sample size formula like Cochrane [17], Yamane [18] and Zikmund [19], a sample size of at least $N = 300$ respondents is required for a survey to represent the actual population. However, due to the lack of hobbyists in Klang Valley, lack of engagement in surveys and time constraints, this study used a smaller population sample size with less representation accuracy. Purposive sampling is a non-probability technique in which respondents are chosen based on qualities or expertise they possess [20]. In this case, the expertise is the hobbyists' familiarity with the target species.

Stebbins [21] argued that there was a distinction between amateurs, hobbyists and experts in leisure and recreational activities. Hobbyists spend a tremendous amount of time on leisure activities. In this study, 'hobbyist' was purposely characterised as public group members of fish hobbyist groups on social media, including betta breeders, goldfish keepers and aquascapers. Apart from these criteria, hobbyists also had to be residents of KV. The hobbyist group was selected because they are the primary stakeholders of the ornamental fish trade and are familiar with *B. livida*. Therefore, they are the group most likely to donate to *B. livida* conservation.

2.2. *Survey Instrument*

The survey started with general icebreakers regarding hobbyists' perceptions towards the species' value. Then, additional information concerning whether they had collected the species as a hobby and the mode and collection purpose was recorded as supportive facts. The survey continued with knowledge, perceptions and attitude towards *B. livida* conservation. Finally, the WTD section was introduced. The survey ended by eliciting the demographic background of respondents.

In the knowledge section, multiple-choice questions were developed to gauge their knowledge of the species. Study of the Golden Warbler [22] established that general knowledge of the species significantly affected the respondent's WTP for species conservation. Respondents were firstly shown pictures portraying various betta species and they had to choose the correct image of *B. livida*. Furthermore, they were questioned whether they knew of *B. livida* in KV, the species habitats and further analysis of conservation and IUCN status. The mean score represents the hobbyists' knowledge regarding the conservation of *B. livida*.

Next, hobbyists responded to Likert-type scaled questions with a score for a series of perception statements contributing to WTD for the species conservation. The questions ollowed the example provided by Shahwahid et al. [13]. Statements covered perceptions of ecology, biodiversity and heritage niche of the species were included. Then, the best possible ways for species conservation were covered. Protection of the species habitat should be prioritised, but translocation to new habitats could also serve as an alternative to conservation. Some betta breeders have done captive breeding and released them back to the wild to foster natural stocks. Respondents were also questioned regarding the effort of authorities and the existing legislation to protect the species. The last statement sets in the survey addressed the possible reasons for species endangerment.

Likert scaled questions were also applied in the attitude section, where hobbyists responded to possible attitude statements. Their views towards nearby wildlife indicated that they were affected by experience and had preferences for particular species [23]. Hence, it was prudent to ask hobbyists if they had dedicated themselves to species conservation by researching their willingness to collect wild *B. livida* or purchase captive breeds. Catching the species is one of the factors threatening the species; the tendency of refraining and preventing others from catching the species in the wild was also included in survey questions. This section also measured the hobbyist's concerns regarding the dwindling wild species population and their habitat degradation. The following two sets identify their potential to contribute to the monetary fund or expertise in conservation. The final question sets dealt with controversial measures in protecting wildlife related to regulations and the feasibility of getting a licence to own the species.

A manuscript called 'Wild Red Fighting Fish, a Klang Valley Natural Heritage' was initially presented to hobbyists in the WTD section. The text discussed the unique characteristics of *B. livida* and its market in the ornamental fish trade. The North Kuala Langat Forest Reserve was an example of the degradation of the species' habitat, which has led to the endangerment of the wild species population. Readers were then told about monetary funds that were crucial in developing conservation plans for the species, such as gazetting habitats, breeding and awareness programmes. With all the details collected, questions emerged as to whether they were willing to donate (WTD) a set amount annually for *B. livida* conservation.

Double bound dichotomous key questions were developed to probe hobbyists' WTD for species conservation [24]. A total of five (5) sets of different bids were presented to respondents starting from RM1, RM3, RM5, RM7 and RM9, respectively, based on the pilot test. The WTD questions were double-bounded; when respondents accepted the first bid (BID1), a second bid (BID2) with a higher amount was introduced and they had to accept or reject the bid. If the respondent refused BID1, BID2 was presented with a lower amount and had to accept or reject the offer. Respondents were free to reject the entire bid. Possible reasons for each outcome were probed.

The last section of the survey gathered social background data from the respondents, including age, gender, marital status, race, education level, occupation and household income. These data were critical to compare with population census data. KV districts were also noted as within the targeted sample population. Out-group respondents were removed from this sample size.

2.3. Validity and Reliability

The questionnaire was validated by a panel of three experts with expertise in wildlife valuation and perception studies. A four-point scale rating with 1 = Not Relevant, 2 = Somewhat relevant, 3 = Relevant, 4 = Very relevant was used for expert scoring. Only items that scored more than two were retained. Furthermore, the comments and suggestions by the reviewers were taken into consideration to further improve the questionnaire. Next, a reliability test was done on hobbyists' perception and attitude towards species conservation involving $N = 30$ pilot respondents. Validation was assessed based on the survey format and contents to verify external and internal inconsistencies. The reliability of the survey was based on the Kuder–Richardson coefficient of reliability (K-R 20) test for the knowledge section and the Cronbach's alpha test for the perceptions and attitudes section to ensure that item questions on each section represent the best score for each attribute. The reliability of knowledge component was K-R 20 = 0.46, which was the rule for moderate agreement [25]. Perception was $\alpha = 0.54$, which was lower than the average standard. However, reviews from various papers showed that an $\alpha = 0.5$ can be satisfactory and sufficient (see Taber, 2018); thus, the score of $\alpha = 0.54$ was considered acceptable. In comparison, the attitude section was $\alpha = 0.78$, which was universally used in most studies.

2.4. Data Analysis and Model Specification

Descriptive analysis was conducted on general questions, perception and attitude towards species conservation and the respondents' social demographics using SPSS v23. Frequency means and percentages were tabulated to reflect the attributes.

WTD questions were analysed using STATA v15. Probit regression was conducted using WTP11 and WTP2 data. This model was chosen because of its well-established theoretical background and its compatibilities with dichotomous variables. The model Equation (1) is illustrated as follows:

$$\mathrm{WTPi} = \ln\left(\frac{P}{1-P}\right) = \beta 0 + \beta 1\mathrm{X}1 + \beta 2\mathrm{X}2 \ldots \beta n\mathrm{X}n \tag{1}$$

where P = the probabilities, Y = 1, X is a set of independent variables, β is the coefficient to be estimated corresponding to the logistic distribution. Thus, WTP depends on various factors, including bid price and unobservable characteristics to be captured in error (εi). Hobbyists were then given a follow-up WTP question:

$$\mathrm{WTPi} = \mathrm{Xi}\beta + \varepsilon\mathrm{i} \tag{2}$$

$$\mathrm{WTP2i} = (1 - \mathrm{y})\,\mathrm{WTPi} + \mathrm{Y}\beta + \delta$$

In which Y is the parameter on the starting bid βi, while δ is a shifting parameter. Therefore, hobbyists were questioned for double intervals. WTP = β2 accepts both starting bid β1 and a follow-up bid (β2), β2 ≤ WTP < β1 rejects the starting bid (β1) and accepts the follow-up bid (β2), or WTP < β2 rejects both bids (β1), followed by (β2). From the possible choices, the sequences are as follows (3):

$$Li\left(\frac{WTPi}{\beta 1}\right) = Pr\ (WTP1i + \varepsilon ij > \beta i\ WTP2 + \varepsilon 2j \geq \beta 2)\ YY \tag{3}$$

$$Pr\ (WTP1i + \varepsilon ij > \beta i\ WTP2 + \varepsilon 2j \geq \beta 2)\ YN$$

$$Pr\ (WTP1i + \varepsilon ij > \beta i\ WTP2 + \varepsilon 2j \geq \beta 2)\ NY$$

$$Pr\ (WTP1i + \varepsilon ij > \beta i\ WTP2 + \varepsilon 2j \geq \beta 2)\ NN$$

WTPi and WTP2 are the means for the first and latter bid response, while YY = yes-yes and YN = I for a yes-no answer and NY for a no-yes answer, NN = 0 for no-no answer. This likelihood function is then estimated using the following model (4):

$$Li\left(\frac{WTPi}{\beta\beta 1}\right) = \varnothing\ \frac{\varepsilon}{\varepsilon 2}\left(d1j\left(\frac{\beta 1 - WTP1}{\sigma 1}\right)d2j\left(\frac{\beta 2 - WTP2}{\sigma 2}\right), d1jd2j\rho\right) \tag{4}$$

In which WTP1J = 1 if the hobbyist answered the first question yes or WTP2J if the response to the second question was yes or otherwise, dij = 2WTP1j = 1 and 2WTP2j = 1. Thus, the mean (5) and median (6) are derived as follow:

$$\text{Mean WTP} = \exp\left(\frac{X\hat{\beta}'}{\hat{\beta}\,0} + 0.5\hat{\sigma}2\right) \tag{5}$$

$$\text{Median WTP} = \left(\frac{\overline{X}\beta'}{\hat{\beta}\,0}\right) \tag{6}$$

where X is a k + 1 row vector of the mean value of the possible variable, including 1 for constant term, $\hat{\beta}'$ is a k + 1 column vector of estimated coefficient and $\hat{\sigma}$ is the estimated variance. Alejandro (2012) [24] argued that using this double-bound method enables the efficient utilisation of data to estimate WTP (or, in this case, WTD).

The model below shows factors influencing the WTD of hobbyists to conserve wild *B. livida* in Klang Valley (see Table 2). The WTP elicitation method to be used for dependent variables is consistent with single-bound CVM and double-bound CVM.

$$WTD = \beta 0 + \beta 1BID + \beta 2AGE + \beta 3INC + \beta 4KNOW + \beta 5PERC + \beta 6ATT + \varepsilon \tag{7}$$

Table 2. Variables used in the WTD model.

WTD	Willingness to Donate (MYR) (1: Yes, 0: No)	Dependent Variable
β0	Constant Coefficient	
BID 1	Initial bids	Independent variable
BID 2	Follow up bids	Independent variable
AGE	Age of hobbyist	Independent variable
INC	Household Income of hobbyist	Independent variable
KNOW	Knowledge of the *B. livida* conservation	Independent variable
PERC	Perception towards the species conservation	Independent variable
ATT	Attitude towards the species conservation	Independent variable
OWNB	Owned *B. livida* (1: Yes. 0: No)	Independent variable
ε	Random error	

3. Results and Discussion

3.1. Social Demographics and General Question

Social demographics data of hobbyists were collected as the foundation of this study. Table 3 below displays the demographics summary of the respondents:

Table 3. Hobbyists' social demographics background.

	Item	Frequency	%
F1	**Gender**		
	Male	142	94.7
	Female	8	5.3
F2	**Age**		
	21–30	60	40.0
	31–40	72	48.0
	41–50	17	11.3
	>51	1	0.7
F3	**Marital Status**		
	Married	104	69.3
	Single	45	30.0
	Others	1	0.7
F4	**Race/Ethnicity**		
	Malay	124	82.7
	Chinese	19	12.7
	Indian	7	4.7
F5	**The district within Klang Valley**		
	Petaling	28	18.7
	Hulu Langat	27	18.0
	Klang	26	17.3
	Kuala Lumpur	22	14.7
	Gombak	16	10.7
	Sepang	14	9.3
	Kuala Langat	9	6.0
	Putrajaya	8	5.3
F6	**Education level**		
	Bachelor's Degree	57	38.0
	Diploma and Equivalent	45	30.0
	High school	34	22.7
	Master's Degree	13	8.7
	PhD	1	0.7
F7	**Occupation**		
	Private	81	54.0
	Government	38	25.3
	Self-Employed	28	18.7
	Student	3	2.0
F8	**Household Income Category**		
	B40 (lower than RM4500/Month)	84	56.0
	M20 (between RM4501–RM10,000/month)	55	36.7
	T20 (higher than RM10,001/month)	10	6.7
	No Income	3	0.7

Most (94.7%) respondents were male, indicating that men conduct this hobby in higher numbers. Almost half of respondents were between 31 and 40 years of age (48%) and most were married (69.3) middle-aged men. Race composition reveals that Malay was the majority (82.7%), followed by Chinese (12.7%) and Indian (7%). The highest education level reported was a bachelor's degree at 38%. Interestingly, diploma and high school graduates exhibited weights of 30% and 22.7%, respectively. This explains why most respondents were in the B40 income category (56%). In terms of employment, most worked in the private sector (54%), followed by the government sector (25%) and self-employed sector (18%). The hobbyist distribution was almost evenly distributed within Klang Valley. The Petaling district (18.7%) contained marginally more than Hulu Langat district (18%). However, 33 respondents were removed from the total sample because they were not within the study area. Feedback from the out-group willing to participate in this online survey was also encouraging, intending to broaden the scope of future study.

The initial section of the survey stipulated several general questions regarding hobbyists' leisure activities. This information may be crucial for building a fundamental understanding of the nature and essence of this study. Therefore, the results in Table 4 below were tabulated so that the mode percentage frequency is of the upper-class.

Table 4. Result for general questions.

	Item	Frequency	%
A1	**Owned *B. livida* (*N* = 150)**		
	NO	85	56.70
	YES	65	43.30
A2	**Mode of Obtaining *B. livida* (*N* = 65)**		
	Self-Caught	37	46.3
	Private Seller	32	40.0
	Aquarium/Pet shop	11	13.8
A3	**Reason for Owning (*N* = 65)**		
	Hobby	44	47.3
	Breeding	39	41.9
	Selling	10	10.8
A4	**No. of *B. livida* owned (*N* = 65)**		
	1–10 owned	46	70.7
	11–50 owned	9	13.8
	>51 owned	10	14.6
A5	**A reason not owning *B. livida* (*N* = 85)**		
	Difficult to Obtain	45	52.9
	Not Interested	28	32.7
	Very Expensive	11	12.8
	Others	1	1.7
A6	**Value Perceived towards the species (*N* = 150)**		
	Bequest Value	90	60.0
	Biodiversity Value	45	30.0
	Economic Value	14	9.3
	Other	1	0.7

Almost half (56.7%) of respondents (*N* = 150) owned *B. livida* and most caught the species themselves in the wild (46.3%), followed by purchasing from private sellers (40%) and pet shops (13.8%). This trend has highlighted concerns for an unregulated and excessive catch of wild betta by Ng [6]. Overexploitation could further endanger and jeopardise the conservation status of the species. This survey could not determine the hobbyist's 'hotspot' for the species' wild catch, which could provide crucial insights in monitoring the wild catch activities, especially the species distribution in forest reserves and private lands. It must be noted, however, that 41.9% of the species owners breed them. Most wild bubble-nester betta like *B. livida* already have established breeding programs, producing offspring like the common colourful Siamese betta. However, the success of reproducing the species may vary based on the number of *B. livida* they own. The majority own 1–10 betta fishes (70.7%), although this figure does not distinguish among the numbers of *B. livida* caught in the wild and the ones that are bred or purchased. On the other hand, hobbyists who did not own the species explained why they (52.9%) have difficulty obtaining the *B. livida*, followed by a lack of interest in the species and the high cost. Hobbyists also responded concerning the value of species and over half said it was of bequest value (60%) and that it should be protected for future generations.

3.2. Knowledge

The knowledge section assessed the respondent's knowledge of species conservation based on their mean score for each item question, as shown in Table 5. Thus, the knowledge in this survey did not reflect the actual knowledge of respondents but rather represents their familiarity with the subject.

Table 5. Hobbyist's knowledge of *B. livida* Conservation.

	Item	Frequency %		Mean	Sd	Variance
		1	0			
B1	Identified B. livida	94.0	6.00	0.94	0.24	0.06
B2	Existence of the Species in KV	68.0	32.0	0.68	0.47	0.22
B4	Meaning of Endemic	22.0	78.0	0.78	0.42	0.17
B5	Species Habitat	94.70	5.30	0.95	0.23	0.05
B6	IUCN Status	48.70	51.30	0.51	0.5	0.25
B7	Possible Reason for IUCN Status	22.70	77.30	0.77	0.42	0.18
B8	Meaning of Species Conservation	15.30	84.70	0.85	0.36	0.13
B9	Authorities	31.30	68.70	0.69	0.47	0.22
	Average total mean knowledge			0.77		

1 = correct answer, 0 = others.

The results indicate that most hobbyists can identify the picture of *B. livida*, among other betta species, with a mean of 0.94 and a slight variance of 0.06. In addition, the betta species have distinguishing features, which was evident in that respondents were not confused with almost identical looking species, such as *B. coccina* and *B. brownorum*.

Only 68% of respondents knew that *B. livida* existed in KV areas. However, they have an average comprehension of what endemic species means and recognised the species natural habitat. Hobbyists could also grasp the meaning of species conservation and acknowledge species protection by an authority like the Department of Fisheries (DoF).

Half of the respondents were uncertain about the IUCN status for the species (51.3%) with a significant standard deviation and variance. However, they can guess that conservation is due to the species' endemic nature, habitat degradation and exploitation. However, the conservation status of this species is one of the main problem statements for this study that is supposed to influence the WTD. It must be noted as well the mode in which the hobbyists gathered information regarding the species in Table 6, with social media being the most frequent source followed by websites. This study, therefore, justified the purposive sampling technique on hobbyist groups on social media.

Table 6. Information obtaining modes.

B3	Mode	Frequency	%
	Media Social	79	30.74
	Website/Internet	64	24.90
	Friends/Relatives	53	20.62
	Books	34	13.23
	Exhibition	27	10.51

3.3. Perception

Perception is a means of interpreting, understanding or regarding something. Frequency in Table 7 below shows the hobbyists' response to the perception of wild *B. livida* conservation in Klang Valley:

Table 7. Hobbyist's perception towards *B. livida* conservation.

	N = 150	Frequency %					Mean	Score
		1	2	3	4	5		
C1	Ecological Niche	1.3	4.0	19.3	20.0	55.3	4.24	4
C2	Biodiversity Role	0	0	2.7	19.3	78.0	4.75	4
C3	Natural Heritage	0.7	0.7	2.7	18.0	78.0	4.72	4
C4	Preserving Habitat	0.7	0	2.0	16.7	80.7	4.77	4
C5	Translocation	33.3	14.0	24.0	12.0	16.7	2.65	3
C6	Captive Breeding	7.3	11.3	35.3	27.3	18.7	3.39	3
C7	Authority Effort	30.7	24.7	29.3	5.3	10.0	2.39	3
C8	Wildlife Law	16.0	16.0	36.0	17.3	14.7	2.99	3
C9	Habitat Destruction	0.0	0.7	4.0	15.3	80.0	4.75	4
C10	Exploitation	11.3	7.3	19.3	20.7	41.3	3.73	3

Note: <2.00 = strongly disagree (1), 2.01–3.00 = disagree (2), 3.01–4.00 = agree (3), 4.01–5.00 = strongly agree (4) (adapted from Pallant, 2007).

Hobbyists firmly agreed with the perception statements that *B. livida* habitat was the leading cause of the species endangerment (mean = 4.75), followed by the importance of preserving the species' habitat (mean = 4.24). They also strongly agreed that it was essential to protect the species because of its role in ecological niches, biodiversity enrichment, in addition to being part of the natural heritage of the state of Selangor. This way of thinking has to do with the fact that they are aware of the species' habitat. The notion of captive breeding was agreeable (mean = 3.39), indicating that hobbyists were perceived as viable conservation options. Captive breeding increases survival for the species population but opens the door to the commercial market. In-depth research is needed to characterise the hobbyists' perception of captive breeding. However, respondents disagreed with the notion of transferring the species to new safe habitats (mean = 2.65), which shows intense perception about the importance of the existing species habitat. The high variance means of 2.149 on this item should also be noted among other perception components. Respondents also argued that they did not have faith in current wildlife law to protect the species or in efforts by the authorities to conserve the wild *B. livida* population, with mean values of 2.39 and 2.99, respectively. The survey shows that overall, the hobbyist showed an agreeable perception about the species conservation (total average mean= 3.83). However, they are hesitant about how management of the conservation will be conducted.

3.4. Attitude

Attitude is a mental state of mind that characterises an individual towards a subject. Table 8 shows the hobbyist's attitudes towards the *B. livida* conservation.

Table 8. Hobbyist's attitude towards *B. livida* conservation.

	N = 150	Frequency %					Mean	Score
		1	2	3	4	5		
D1	Stop buying Wild	8.0	6.7	29.3	13.3	42.7	3.76	3
D2	Only buy Captive Breed	4.7	2.7	22.0	22.0	48.7	4.07	4
D3	Refrain Wild Catch	2.0	3.3	22.0	20.7	52.0	4.17	4
D4	Advise People	3.3	2.7	24.7	26.7	42.7	4.03	4
D5	Concern (population loss)	0.7	0.0	8.7	20.7	70.0	4.59	4
D6	Concern (Habitat loss)	0.7	0.7	6.0	16.0	76.7	4.67	4
D7	Contribute Financially	7.3	1.3	34.7	26.7	30.0	3.71	3
D8	Helping Researcher	3.3	1.3	25.3	28.7	41.3	4.03	4
D9	Rules and Regulation	6.0	3.3	18.0	17.3	55.3	4.13	4
D10	Ownership License	23.3	8.0	16.0	19.3	33.3	3.31	3

Note: <2.00 = strongly disagree (1), 2.01–3.00 = disagree (2), 3.01–4.00 = agree (3), 4.01–5.00 = strongly agree (4) (adapted from Pallant, 2007).

The hobbyists were firm on their commitment to purchase wild *B. livida* (mean = 3.76). However, they appeared to be firmly in agreement to buy a captive-bred fish instead of wild caught fish (mean = 4.07). This corresponds to the initial understanding of the captive breeding species. The response also showed that they firmly decided to withdraw from capturing betta from the wild and were happy to advise others to join or support the initiative. This is a strong indication for hobbyists to withdraw from a wild catch. The initial results data indicated that most of the respondents that owned *B. livida* had caught the betta from the wild. This may also be the explanation of why the hobbyists tended to purchase a captive-bred fish. Hobbyists demonstrated considerable concern (strongly agreed) for both biodiversity losses (mean = 4.67) and the dwindling population of species (mean = 4.59) compatible with previous views of habitat and exploitation. Consensus respondents on the willingness to make a charitable contribution (mean = 3.71) will possibly affect the WTD section. In addition, they have strongly agreed to support biodiversity protection by enabling the researchers to obtain valuable data (mean = 4.03). This shows a positive attitude by the hobbyist in conserving the species.

Concerning governance from authorities, the hobbyists must comply with the laws and regulations laid down for the species. Therefore, they consented to comply with the ownership licence provision for *B. livida*. It should be noted, based on the feedback, that some respondents strongly object to the requirement for a species ownership licence, which was confirmed by the significant variance mean for this item (2.458) relative to other attitude items. However, the survey could not differentiate between those who accepted or opposed this notion: hobbyists, breeders, or sellers. Therefore, further in-depth research is required to validate this inference and probe why they are against licencing.

3.5. Willingness to Donate

The survey documented the hobbyist's willingness to donate to wild *B. livida* conservation in Klang Valley by utilising closed-ended dichotomous key questions. Table 9 indicates the frequency of those who are eager or unwilling to do so:

Table 9. Hobbyist WTD for Wild *B. livida* Conservation.

	Item	Frequency	%
E1	Willingness to Donate (*N* = 150)		
	YES	99	66
	NO	51	44
E5	Reason for WTD (*N* = 99)		
	Bequest	59	59.5
	Conservation	38	38.4
	Others	2	2.1
E6	Reason NOT WTD (*N* = 51)		
	Want to Contribute with Other Mean	27	52.9
	Could not Afford	18	35.3
	It is unnecessary	6	11.8

The majority of the hobbyists who participated in this survey were willing to contribute (66%) to conserve the species and vice versa (44%). Furthermore, they are ready to donate because of the importance of the species' perceived bequest value (59.5%). This corresponded to the earlier finding that hobbyists preferred this species to be a natural heritage and want future generations to appreciate it. Therefore, the donation should be used solely for the conservation of the species (38.4%). In addition, hobbyists were willing to contribute because of the species' aesthetic value (2.1%). On the other hand, a significant number of hobbyists opted not to donate because they wanted to participate by other methods (52.9%), which, backed by the mindset of most respondents, firmly consent to aid the researcher with conservation efforts. Other factors include the possibility that they cannot afford a contribution or consider a *B. livida* reservation unnecessary.

Probit regression analysis of single bound CVM for initial bid (BID1) against WTD was performed and tabulated, as shown in Table 10.

Table 10. Single Bound CVM.

Variables	Coefficient	Std Error	Significant
Constant	0.73339	0.21872	0.00 ***
Bid1	−0.0839	0.03729	0.024 **
	Number of obs = 150		
	Prob > chi^2 = 0.0235		
	Pseudo R^2 = 0.0257		
	*** Significant at 1% ** significant at 5% level		

Stata command: probit WTP1 BID1.

The results revealed that the initial bid negatively impacted WTD, indicating that BID1 has an inverse relationship with WTD. According to demand theory, the value was expected to be significant at a 5% level (0.024). R^2 showed that 2.5% of the initial bid as a dependent variable influences the WTD. Further Probit analysis with independent variables of age and income of the respondents, along with the hobbyists' mean for knowledge, perception and attitude towards the species conservation, was conducted and shown in Table 11:

Table 11. Single-bound CVM modified.

Variables	Coefficient	Std Error	Significant
Constant	−1.06396	0.6215333	0.087
BID1	−0.0731149	0.0383354	0.05 **
AGE	0.0643578	0.0200888	0.001 ***
INC	−0.0000623	0.0000321	0.05 **
		Number of obs = 150	
	Prob > chi^2 = 0.0010		
	Pseudo R^2 = 0.0816		*** significant at 1% ** significant at 5%

Stata Command: probit WTP1 BID1 AGE INC.

Independent variables indicated that they influenced the result of WTD by 8.2%. At Single bound CVM, all independent variables of respondents were not significant, except for age and income. However, the respondents' age was significant at 1% level. In comparison, income was significant at 5%, indicating that the older the respondents were, the higher the possibility to donate. Significances from these variables, compute for WTD value for initial bid, was estimated using the WTD estimation model, as shown in Table 12.

Table 12. Estimation of WTD using Single-Bound CVM.

	Coefficient	Std Error	Significant
WTD1	9.351782	2.682928	0.000 ***

*** Significant at 1%. Stata Command: nlcom(WTP1:-(_b[_cons] + AGE_m*_b[AGE] + INC_m*_b[INC])/_b[BID1]), noheader.

The WTD single bound for the survival of wild *B. livida* was estimated at a value of RM9.35 per year. This figure will function as a conservation work fund for the species. For the second bid submitted to the respondent, a double-bound CVM was used to evaluate both the original and the second bid for WTD species, as shown in Table 13:

Table 13. Double-Bound CVM.

Variables	Coefficient	Std Error	Significant
Beta Constant	8.947126	1.41945	0.000 ***
Sigma Constants	14.4819	2.117302	0.000 ***
Number of obs = 150			
Prob > chi2 = 0.0000			

Stata command: doubleb BID1 BID2 WTP1 WTP2; *** significant at 1%.

The table revealed that the initial and the second bids were relevant at 1%, with a favourable relationship with WTD. The discrepancy between Beta and Sigma constant (bids) was RM5.53. Independent variables were then applied to the model to evaluate some critical relationships with the WTD, as shown in Table 14:

Table 14. Double-bound CVM modified.

Variables	Coefficient	Std Error	Significant
Constant	−15.84793	7.400414	0.032
AGE	0.6987584	0.2270002	0.002 ***
OWNB	5.400544	2.912554	0.064 *
Sigma Constant	14.02604	2.048678	0.000
Number of obs = 150			
	Wald chi^2 (5) = 12.64		
	Prob > chi^2 = 0.0018		
	* Significant at 10% level *** Significant at 1% level.		

Stata command: doubleb BID1 BID2 WTP1 WTP2 AGE OWNB.

In contrast to a single bound CVM earlier, the double-bound CVM has no significant effect on the WTD. However, species-owned hobbyists (OWNB) have a considerable influence at a 10% level. This was focused on the notion that those who possessed *B. livida* were more inclined to donate and vice versa. The respondent's age was the only reliably significant 1% positive relationship with WTD using a double-bound model. Subsequently, significant variables were estimated in the WTD estimation model using a double-bound CVM, as shown in Table 15:

Table 15. Estimation of WTD using Double-Bound CVM.

	Coefficient	Std Error	Significant
WTD	9.04	1.408674	0.000 ***

*** Significant at 1% level, Stata command: nlcom(WTP:(_b[_cons] + AGE_m*_b[AGE] + OWNB_m*_b[OWNB])), noheader.

Hobbyist's WTD for the *B. livida* conservation in Klang Valley was MYR 9.04 per year, estimated using double-bound CVM with independent attributes. This estimation comprises both initial and second bids to value a mean WTD. Hence, approximately MYR 1356 from 150 hobbyists in Klang Valley is expected to conserve wild *B. livida* per year. However, the annual funds for a conservation fund are often minimal. Thus, it must be spent effectively to ensure the survivability of the species.

In this analysis, local WTD for species-level was calculated to be higher than the WTP for elephants performed by Shahwahid et al. [13]. Using closed-ended bids combined with double-bound dichotomous bid keys makes it possible for the respondent to be impartial when stating a monetary value. Moreover, as this study only specified the conservation *B. livida*, not the other wild betta species, the economic value elicited was lower than the survey by Teh et al. [14] and Vianna et al. [15]. However, their WTP was elicited towards various marine turtles and shark species and they gave more weight for the respondent to consider their willingness to pay for conservation. The mean WTD could also have been influenced by the number of respondents in this study. Moreover, those residing outside KV who could be informative were also excluded. Studies by Haefele et al. [7] and Michalski et al. [26] showed significance between two regions willing to conserve species in other areas.

Hobbyist's knowledge, perception and attitude did not significantly influence the WTD. Results for the hypotheses H_{A1}, H_{A2} and H_{A3}, assuming knowledge, perception and attitude of hobbyists influenced the WTD were rejected. Thus, the null hypothesis was accepted. This result contrasted with other established studies that independent variables influenced the WTD [22,27,28]. However, in the knowledge section, an item regarding the IUCN status of *B. livida*, despite only half of the respondents knew its endangered status, hobbyists were willing to donate for species conservation. This corresponded with Colléony et al. [29] that the respondents chose to donate not because of the endangerment status but rather because of the species' charisma and aesthetic. Age of hobbyist was the only variable that consistently and significantly influenced the WTD for conservation. According to Asmamaw et al. [12], age greatly affected the affordability to pay for protection. Older people tended to have more stable employment and income and are possibly more familiar with the theme. The income variable was significant for the single bound WTD but failed to influence double-bound bids significantly. Regression was computed for the age of hobbyists against knowledge perception and attitude and income, but they were also found not to be significant. However, hobbyists who owned *B. livida* (OWNB) were more likely to donate for species conservation than those who did not (significant). This was parallel with familiarity with the targeted species being able to influence WTP, as shown in the study by Lundberg et al. [28]. It suggests older hobbyists who are more familiar with *B. livida* are most likely to donate for species conservation for the sake of heritage value.

4. Conclusions

Degradation of natural ecosystems and over-exploitation of wild species in the commercial market have raised concerns regarding the endangerment and survival of the flora and fauna heritages of Klang Valley. Via this preliminary study, the hobbyist's knowledge, perception, attitude and willingness to donate to protect wild *B. livida* in Klang Valley was addressed using an online survey questionnaire. This study served as a baseline for WTD towards conserving wild betta and freshwater fish in Malaysia. This study also inferred that the Klang Valley hobbyists' WTD for wild *B. livida* conservation using double-bound CVM was MYR 9.05 per year. The hobbyist's knowledge, perception and attitude toward *B. livida* conservation have no significant relationship with the WTD.

In contrast, the respondent's age indicated a significant connection concerning the WTD, attributed to the species familiarity among older citizens. The possibility that owners of the species (OWNB) would contribute to its survival was also significant. It was also found that the number of wild species caught by the hobbyist appeared to be substantial, but the implications could not be established.

As a recommendation, the decision-makers should consider the amount of donation that the hobbyists are willing to contribute and set up a trust fund for habitat conservation for the species. Online donation collecting through social media has been proven a successful fundraising platform. Furthermore, the survey responses suggested that individual hobbyists could contribute to the parent-pair breeding programme for *B. livida*. For example, on transferring them for ex-situ conservation, i.e., captive breeding, some successfully bred and released the species back into their natural habitat. Therefore, the decision-makers should reach out to hobbyist groups and collaborate to preserve this valuable fauna.

This research has some weaknesses; future study could involve more respondents representing the hobbyist sample population. Next, the examination must be inclusive for respondents outside the species distribution region to determine any substantial variation. In addition, more exploration of the target species' charisma and aesthetic appeal must be included to assess respondent attraction to the species when it comes to donating for conservation. Lastly, it is recommended that this type of study be conducted on other significant local freshwater fish.

Author Contributions: Conceptualisation, M.F.A.R. and N.K.M.; Data curation, M.F.A.R. and N.K.M.; Formal analysis, M.F.A.R.; Investigation, M.F.A.R. and N.K.M.; Methodology, M.F.A.R.; Supervision, N.K.M.; Project administration, M.F.A.R.; Resources, M.F.A.R. and N.K.M.; Software, N.K.M.; Validation, N.K.M.; Writing—original draft, M.F.A.R.; Writing—review and editing, M.F.A.R. and N.K.M. All authors have read and agreed to the published version of the manuscript.

Funding: This research received no external funding; however, the Research Management Centre of Universiti Putra Malaysia will fund the publication fee.

Data Availability Statement: Not applicable.

Acknowledgments: We would like to thank those who have participated in this study, especially admin moderators of hobbyist groups, who have helped us to distribute this survey.

Conflicts of Interest: The authors declare no conflict of interest.

References

1. Tan, H.H.; Ng, P.K.L. The fighting fishes (Teleostei: Osphronemidae: Genus Betta) of Singapore, of Malaysia and Brunei. *Raffles Bull. Zool.* **2005**, *13*, 43–99.
2. Keat-Chuan, N.C.; Aun-Chuan, O.P.; Wong, W.L.; Khoo, G. An overview of the status, trends and challenges of freshwater fish research and conservation in Malaysia. *J. Surv. Fish. Sci.* **2017**, *3*, 7–21. [CrossRef]
3. Chan, K.-G. Conservation of the critically endangered endemic Malaysian black fighting fish Betta persephone Schaller (Teleostei: Osphronemidae): A brief review. *PeerJ. Prepr.* **2015**, *3*, e1048v1. [CrossRef]
4. Giam, X.; Koh, L.P.; Tan, H.H.; Miettinen, J.; Tan, H.T.W.; Ng, P.K.L. Global extinctions of freshwater fishes follow peatland conversion in Sundaland. *Front. Ecol. Environ.* **2012**, *10*, 465–470. [CrossRef]
5. Low, B.W. Betta livida. *IUCN Red List. Threat. Species* **2019**, *8235*. [CrossRef]
6. Ng, C. The ornamental freshwater fish trade in Malaysia. *UTAR Agric. Sci. J.* **2016**, *2*, 7–18.
7. Haefele, M.A.; Loomis, J.B.; Merideth, R.; Lien, A.; Semmens, D.J.; Dubovsky, J.; Wiederholt, R.; Thogmartin, W.E.; Gary, T.H.; Rodrigo, M.; et al. Willingness to Pay for Conservation of Transborder Migratory Species: A Case Study of the Mexican Free-Tailed Bat in the United States and Mexico. *Environ. Manag.* **2018**, *62*, 229–240. [CrossRef]
8. Martín-López, B.; Montes, C.; Benayas, J. Economic valuation of biodiversity conservation: The meaning of numbers. *Conserv. Biol.* **2008**, *22*, 624–635. [CrossRef]
9. Nijkamp, P.; Vindigni, G.; Nunes, P.A.L.D. Economic valuation of biodiversity: A comparative study. *Ecol. Econ.* **2008**, *67*, 217–231. [CrossRef]
10. Matsiori, S.; Aggelopoulos, S.; Soutsas, K. Economic value of conservation. The case of the edible sea urchin Paracentrotus lividus. *J. Environ. Prot. Ecol.* **2012**, *12*, 269–274.
11. Bandara, R.; Tisdell, C. The economic value of conserving the Asian elephant: Contingent valuation estimates for Sri Lanka. *GAJAH J. Asian Elephant Spec. Gr.* **2003**, *22*, 22–29.
12. Asmamaw, B.; Beyene, B.; Tessema, M.; Kara, A.; Goshu, B.; Assefa, A. Estimating Willingness to Pay for Labeobarbus fish Species Conservation in Lake Tana, Ethiopia: A Contingent Valuation Study. *Int. J. Nat. Resour. Ecol. Manag.* **2016**, *1*, 155–161. [CrossRef]
13. Shahwahid, M.; Yong, P.L.; Sius, T. Perceptions on wild elephants among the public in Pahang, Malaysia. *Malays. For.* **2009**, *72*, 37–44.
14. Teh, L.S.L.; Teh, L.C.L.; Jolis, G. An economic approach to marine megafauna conservation in the coral triangle: Marine turtles in Sabah, Malaysia. *Mar. Policy* **2017**, *89*, 1–10. [CrossRef]
15. Vianna, G.M.S.; Meekan, M.G.; Rogers, A.A.; Kragt, M.E.; Alin, J.M. Shark-diving tourism as a financing mechanism for shark conservation strategies in Malaysia. *Mar. Policy* **2018**, *94*, 220–226. [CrossRef]
16. Ronald, D. Fricker, J. The SAGE Handbook of Online Research Methods. *SAGE Handb. Online Res. Methods* **2012**, 195–216. [CrossRef]
17. Cochran, W.G. *Sampling Techniques*, 2nd ed.; John Wiley & Sons, Inc.: New York, NY, USA, 1963.
18. Yamane, T. *Statistics an Introductory Analysis*, 2nd ed.; Harper & Row: New York, NY, USA, 1967.
19. Zikmund, W.G. *Business Research Methods*, 7th ed.; South-Western: New York, NY, USA, 2002; ISBN 0030350840.
20. Etikan, I. Comparison of Convenience Sampling and Purposive Sampling. *Am. J. Theor. Appl. Stat.* **2016**, *5*, 1–4. [CrossRef]
21. Stebbins, R.A. "Amateur" and "Hobbyist" as Concepts for the Study of Leisure Problems. *Soc. Probl.* **1980**, *27*, 413–417. [CrossRef]
22. Ferrato, J.R.; Brown, D.J.; McKinney, A. Assessment of Public Knowledge and Willingness to Pay for Recovery of an Endangered Songbird, the Golden-Cheeked Warbler. *Hum. Dimens. Wildl.* **2016**, *21*, 86–94. [CrossRef]
23. Mohamad, N.H.N. Urban residents' attitudes toward wildlife in their neighbourhoods: The case study of Klang Valley, Malaysia. *Plan. Malays.* **2011**, *9*, 19–36. [CrossRef]
24. Alejandro, L.-F. Introduction to Contingent Valuation Using Stata. 2012. Available online: https://mpra.ub.uni-muenchen.de/41018/2/MPRA_paper_41018.pdf (accessed on 30 May 2020).

25. Kuder, G.F.; Richardson, M.W. The Theory of the Estimation of Test Reliability. *Psychometrica* **1937**, *2*, 151–160. [CrossRef]
26. Michalski, F.; Boulhosa, R.L.P.; Nascimento, Y.N.D.; Norris, D. Rural Wage-Earners' Attitudes Towards Diverse Wildlife Groups Differ Between Tropical Ecoregions: Implications for Forest and Savanna Conservation in the Brazilian Amazon. *Trop. Conserv. Sci.* **2020**, *13*, 1940082920971747. [CrossRef]
27. Cárdenas, S.A.; Lew, D.K. Factors influencing willingness to donate to marine endangered species recovery in the Galapagos National Park, Ecuador. *Front. Mar. Sci.* **2016**, *3*, 60. [CrossRef]
28. Lundberg, P.; Vainio, A.; MacMillan, D.C.; Smith, R.J.; Veríssimo, D.; Arponen, A. The effect of knowledge, species aesthetic appeal, familiarity and conservation need on willingness to donate. *Anim. Conserv.* **2019**, *22*, 432–443. [CrossRef]
29. Colléony, A.; Clayton, S.; Couvet, D.; Saint Jalme, M.; Prévot, A.C. Human preferences for species conservation: Animal charisma trumps endangered status. *Biol. Conserv.* **2017**, *206*, 263–269. [CrossRef]

sustainability

MDPI

Article

Niche Analysis and Conservation of Bird Species Using Urban Core Areas

Vasilios Liordos [1,*], Jukka Jokimäki [2], Marja-Liisa Kaisanlahti-Jokimäki [2], Evangelos Valsamidis [1] and Vasileios J. Kontsiotis [1]

1 Department of Forest and Natural Environment Sciences, International Hellenic University, 66100 Drama, Greece; vagvals@yahoo.gr (E.V.); kontsiotisv@yahoo.gr (V.J.K.)

2 Arctic Centre, University of Lapland, 96101 Rovaniemi, Finland; jukka.jokimaki@ulapland.fi (J.J.); marja-liisa.kaisanlahti@ulapland.fi (M.-L.K.-J.)

* Correspondence: liordos@for.ihu.gr

Abstract: Knowing the ecological requirements of bird species is essential for their successful conservation. We studied the niche characteristics of birds in managed small-sized green spaces in the urban core areas of southern (Kavala, Greece) and northern Europe (Rovaniemi, Finland), during the breeding season, based on a set of 16 environmental variables and using Outlying Mean Index, a multivariate ordination technique. Overall, 26 bird species in Kavala and 15 in Rovaniemi were recorded in more than 5% of the green spaces and were used in detailed analyses. In both areas, bird species occupied different niches of varying marginality and breadth, indicating varying responses to urban environmental conditions. Birds showed high specialization in niche position, with 12 species in Kavala (46.2%) and six species in Rovaniemi (40.0%) having marginal niches. Niche breadth was narrower in Rovaniemi than in Kavala. Species in both communities were more strongly associated either with large green spaces located further away from the city center and having a high vegetation cover (urban adapters; e.g., Common Chaffinch (*Fringilla coelebs*), European Greenfinch (*Chloris chloris*), Eurasian Blue Tit (*Cyanistes caeruleus*)) or with green spaces located closer to the city center and having high gray area cover and anthropogenic disturbance level (urban exploiters; e.g., Western Jackdaw (*Corvus monedula*), House Sparrow (*Passer domesticus*), Eurasian Magpie (*Pica pica*)). The eleven species that were common to both study areas similarly used the environmental variables and had similar niches, indicating that birds respond similarly to urbanization irrespective of latitude. Sixteen species in Kavala and eleven species in Rovaniemi were identified as conservation priority species, based on their niche specialization level and conservation status. The management actions proposed for the conservation of priority species will also benefit other species with similar ecological requirements and ultimately help maintain diverse bird communities in small-sized green spaces in urban core areas.

Keywords: generalists; specialists; niche breadth; marginality; urban core areas; small green spaces; Mediterranean; Fennoscandia

Citation: Liordos, V.; Jokimäki, J.; Kaisanlahti-Jokimäki, M.-L.; Valsamidis, E.; Kontsiotis, V.J. Niche Analysis and Conservation of Bird Species Using Urban Core Areas. *Sustainability* **2021**, *13*, 6327. https://doi.org/10.3390/su13116327

Academic Editor: Richard A. Loyn

Received: 25 April 2021
Accepted: 1 June 2021
Published: 2 June 2021

Publisher's Note: MDPI stays neutral with regard to jurisdictional claims in published maps and institutional affiliations.

cc BY

Copyright: © 2021 by the authors. Licensee MDPI, Basel, Switzerland. This article is an open access article distributed under the terms and conditions of the Creative Commons Attribution (CC BY) license (https://creativecommons.org/licenses/by/4.0/).

1. Introduction

Hutchinson's [1] concept of the realized niche refers to the range of environmental conditions in which a species can survive, grow, reproduce and maintain a viable population, even in the presence of competitors and predators. Hutchinson [1] and Whittaker et al. [2] defined the ecological niche as an n-dimensional hypervolume (with 'n' being the number of environmental conditions examined) that determines species distribution. According to niche theory, species can be assigned along a narrow-broad niche continuum, from specialists that use a narrow range of environmental conditions to generalists that use a broad range of environmental conditions.

Birds are important components of urban landscapes, promoting ecosystem health [3]. They are also good indicators of the diversity of other animal groups and of habitat condition [4,5] because they are conspicuous, easy to quantify and quickly respond to habitat changes caused by urbanization [6,7]. The maintenance of a diverse bird community in cities provides important ecosystem services, such as pest control [8], plant pollination [9], art inspiration [10] and improvement of human well-being by increasing vitality and happiness and reducing stress and anxiety in urban residents [11,12]. Furthermore, the importance of urban green spaces has greatly increased during the COVID-19 pandemic. Birdwatchers from 97 countries reported that they primarily visit local small urban green spaces due to strict lockdown regulations [13]. However, birds face novel conditions in cities, mostly due to the destruction and fragmentation of natural habitats, especially forests, which are replaced by open habitats and impervious surfaces [14]. Moreover, birds in cities are generally more susceptible to predation, by both natural (e.g., corvids, hawks) and domestic (e.g., cats, dogs) predators [15,16], and disease due to anthropogenic pollution (e.g., organic, air, noise [17,18]). Such novel conditions induced by urbanization act as ecological filters, altering bird species distribution and community composition along the urbanization gradient [19–23]. Birds that occupy urban habitats are usually either exploiters—which often utilize human subsidized resources, such as artificial nesting and feeding sites as well as food waste, and their abundance is usually not dependent on vegetation—or adapters, which require considerable vegetation for shelter and food and utilize fewer anthropogenic resources [24,25].

Urban green spaces act as island habitats, hosting diverse and abundant bird communities [26,27]. In addition, urban areas are locally important for the conservation of birds, as they host many threatened species because they often occupy biodiversity hotspots [28,29]. As urban birds and their habitats also have great social and educational value, conservation efforts should increase in areas where people live and work [30]. Knowing the ecological requirements of bird species and their degree of specialization is essential for setting conservation priorities that would allow for successful conservation management for maintaining diverse urban bird communities.

Environmental variables might vary and differently affect bird species distribution, between green spaces located in urban core areas that are usually intensively managed, small-sized and heavily used by local residents, and green spaces located at the periphery of cities that are usually less managed, larger in size and less visited by people [31,32]. However, managed, small-sized green spaces in urban core areas have received relatively little attention, although they host diverse and important bird communities (e.g., [33–37]). In addition, environmental variables at the local scale are often more important than regional ones in determining species distribution and community composition in urban landscapes [38–41]. Such variables include patch level variables (e.g., green space size, amount of gray, tree, shrub and bare ground cover [6,26,35,42]), matrix level variables (e.g., distance from the city edge and center, building height [37,43,44]) and disturbance variables (e.g., anthropogenic noise, pedestrian and car traffic [45–47]).

Earlier studies considered the species richness, diversity and organization of urban bird communities at various spatial scales (e.g., [7,22,23,48–50]). Measures of niche characteristics have been mostly used to explain why some species adapt to the novel urban conditions while others do not, generally comparing species' ecological requirements with habitat conditions in cities and in natural habitats [51–55]. Studies comparing the niches of bird species in urban core areas are lacking. We used niche theory concepts to describe, at the local scale, the structure and niche characteristics of the bird communities in two similar-sized European cities, Kavala (Greece) in southern Europe and Rovaniemi (Finland) in northern Europe. We described species niches using niche position, marginality and breadth. Niche position shows how typical the environmental conditions used by a species are relative to those that are available in the region [56]. Niche position is measured by niche marginality: species with marginal niches use atypical or uncommon conditions within a region and species with non-marginal niches use typical or common conditions within a region. Niche breadth measures a species' tolerance to contrasting environmental conditions. Low species tolerance means use of a limited range of environmental conditions (specialist species), while high tolerance means the use of a wide range of environmental conditions (generalist species). Such measures can be used to identify the degree of specialization of bird species breeding in urban core areas and as such they could have important implications for their conservation management. Many studies have found that niche position is a strong predictor of ecological patterns and proposed that it can be a better predictor of specialization than niche breadth [53,57–60]).

We measured the abundance of breeding bird species and local environmental variables to describe the distribution and niche characteristics of urban bird species at managed small-sized green spaces located in the most urbanized core areas of Kavala and Rovaniemi. We also set conservation priorities, depending on niche specialization and conservation status, and proposed suitable management actions. We predicted that (1) niche characteristics will differ between different species, (2) environmental variables will varyingly influence species distribution, allowing for their classification either as adapters or as exploiters, (3) species with similar niche characteristics will be similarly influenced by environmental variables in both bird communities irrespective of the latitudinal location and (4) both bird communities will be important for conservation of priority species.

2. Materials and Methods

2.1. Study Area

The study was carried out in Kavala, Greece (40°56′ N, 24°24′ E), located in southern Europe, and Rovaniemi, Finland (66°30′ N, 25°44′ E), located in northern Europe, in their most urbanized core areas, i.e., the central part of the municipality that is covered by more than 50% impervious surface area, containing large buildings, primarily stores, offices and dense residential areas (Figure 1). In these areas, all the green spaces are man-made and managed and anthropogenic impacts on birds are supposed to be maximal.

The Kavala municipality has 70,501 inhabitants [61], covering approximately 351.35 km^2. Field work was performed in the core urban area of Kavala, were most inhabitants of the municipality live (56,371 [61]). It covers approximately 8.0 km^2 (about 7050 inhabitants/km^2) and is delimited by the sea to the south and by a Turkish Pine (*Pinus brutia*) dominated peri-urban forest to the north. Two Turkish Pine-dominated woodlands, the Panagiouda (17.0 ha) to the west and the Pentakosion (1.3 ha) to the east, are among the most important green spaces of Kavala. However, the most common green spaces are small square gardens (<3 ha), which are dispersed throughout the city and are usually partly covered by planted trees and shrubs, and gray infrastructure such as paved walks, playgrounds, cafés and restaurants.

Rovaniemi municipality has 63,631 inhabitants [62], covering approximately 8016 km^2 (7601 km^2 by land). It is located near the Arctic Circle and daylight duration is very high in summer since the sun does not set between 7 June and 6 July. Forests (61.4%) and mires with trees (25.1%) cover most of the land. The proportion of urban areas of the municipality is only about 0.3% of the land area. Most of the inhabitants (50,000; about 4000 inhabitants/km^2) of the municipality live in the core city area, where the field work was performed. Although the forests surrounding the city are pine-dominated (Scots Pine (*Pinus sylvestris*)), deciduous trees (e.g., Silver Birch (*Betula pendula*), Rowan (*Sorbus aucuparia*)) are favored in urban park planning.

2.2. Bird Surveys

We selected 19 green spaces from each city, representing all available managed green spaces in the core areas of both cities, and established one survey station at the center of each green space. Green spaces and their corresponding survey stations were located at least 250 m apart to avoid the double counting of individuals. The single-visit fixed-radius ($\leq$50 m) point count method was employed to assess the abundance and diversity of bird species [63]. Birds were surveyed very early in the morning, during the period of maximum bird activity. All surveys were conducted before 1030 h. They were conducted following the recommendations of Bibby et al. [63], with a 5 min silent period before starting the

5 min bird surveys. Surveys were conducted during fine weather by the same observer (E.V. in Kavala, J.J. in Rovaniemi) to avoid introducing an observer effect. Counts were carried out during the peak breeding season, 9–13 May 2016 in Kavala and 1–15 June 2020 in Rovaniemi.

The use of point counts is appropriate in urban areas due to built-up structures [39,64]. Several studies have validated the efficiency of single-year, single-visit studies [27,39,65]. Species were assumed to have similar detection probabilities because we kept the sampling radius relatively small (50 m); sampling points were located in similar habitats and vegetation cover was relatively low. This assumption is common to studies of urban bird communities [66–69]. Moreover, van Heezik and Seddon [64] reported that standardized, non-detectability-based point counts could provide useful information on the structure and relative abundance of bird communities in urban areas. Therefore, the use of the same methodology allowed for valid comparisons between the two study areas.

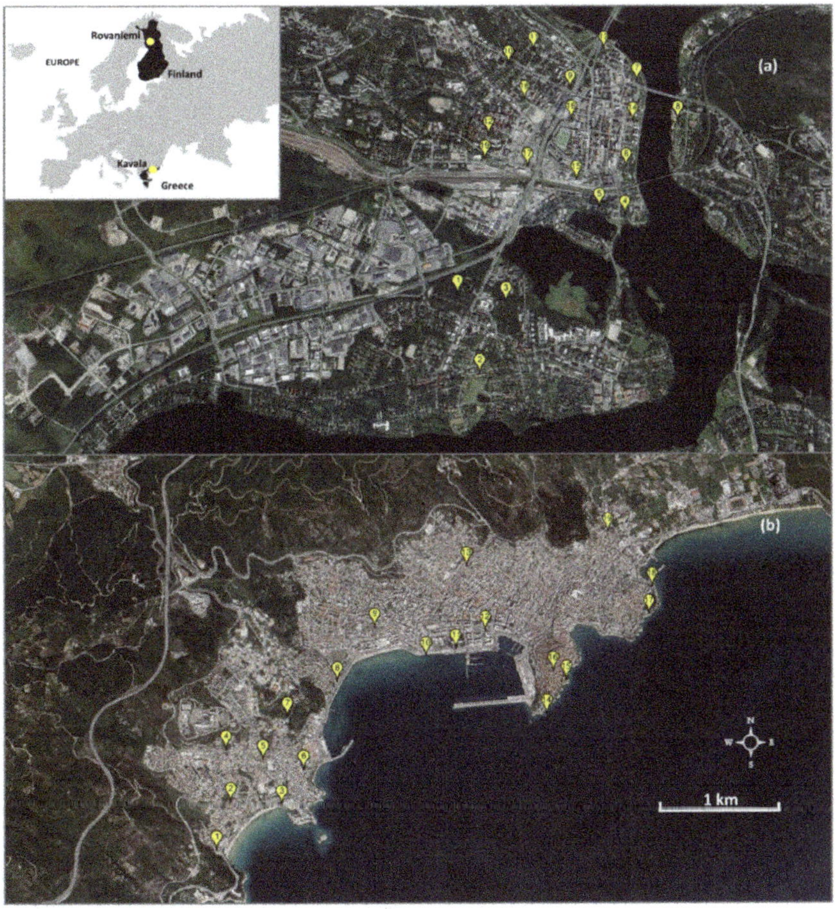

Figure 1. Maps of (**a**) Rovaniemi, Finland, and (**b**) Kavala, Greece, showing the location of the survey stations (n = 19). Scale applies to both maps. Main map: Google Earth; Image Landsat/Copernicus, Data SIO, NOAA, U.S. Navy, NGA, GEBCO; ©2020 Google, Image©2021 TerraMetrics, Image©2021 Maxar Technologies. Inset map: GinkgoMaps.

The presence of every individual bird as well as any indication of breeding activity within the survey station was noted (singing males, territorial behavior, nest construction or provisioning, occupied nests, etc.). This method is not appropriate for the assessment of the abundance of raptors, aerial feeders and crepuscular species [63], and therefore such species were not used in our analyses, with the exception of the Northern House Martin (*Delichon urbica*) and the Barn Swallow (*Hirundo rustica*) in Kavala and the Common Swift (*Apus apus*) and the Sand Martin (*Riparia riparia*) in Rovaniemi. These species are important features of the urban landscape and were counted only when involved in direct breeding activities (e.g., entering into the nest hole) within the 50 m survey point radius.

At the European level, species conservation status was taken from [70]. At the national level, conservation status was taken from the 2019 Red List of Finnish Species [71] and the Red Data Book of Threatened Animals of Greece [72]. (See species classification in Table S1.)

2.3. Environmental Variables

Sixteen environmental variables were measured in both cities to test their effect on bird abundance and diversity (Table S2). Distance of the center of each green space from the city center and city edge (km), green space size (ha), perimeter (edge) length (km) and relative edge length, calculated as perimeter divided by area (km/km^2), were measured from aerial maps. Land cover was measured in the field as the percentage of tree, shrub, bare ground and impervious (gray cover) surface within a radius of 50 m of every survey station. Tree cover was estimated separately for the coniferous and deciduous trees. Mean noise levels (dBA) were measured twice at each survey station, in the morning (just before the initial bird survey started), before daily traffic starts, and at midday, when daily traffic is at its maximum. Noise measurements were recorded for 5 min during each session (76 sessions in total), by using a portable noise meter in Kavala (Model Nova 42, Pulsar Instruments^{TM}) and an environmental multimeter in Rovaniemi (MASTECH® MS6300 Environment Multimeter), directed towards the nearest road at the breast height. Car traffic (number of cars per minute) and pedestrian traffic (number of people per minute) were also recorded within a radius of 50 m of every survey station during a 5 min period before the initial bird surveys started. We used the number of floors of the buildings surrounding each green space as an index of the matrix type. The number of floors was calculated from the four main compass directions in the field, and the average value was used later in analyses.

Although Kavala is more densely populated than Rovaniemi, green space size (Kavala, mean 1.5 $\pm$ 3.8 (SD) ha, range 0.1–17.0 ha; Rovaniemi, 1.7 $\pm$ 1.9 ha, 0.1–8.5 ha; Mann–Whitney U-test $z = 2.410$, $p = 0.063$ after Bonferroni correction) and matrix type indices such as distance of green spaces from the city center (Kavala, 1.5 $\pm$ 0.9 km, 0.1–3.1 km; Rovaniemi, 0.9 $\pm$ 0.7 km, 0.3–2.6 km; $z = -2.015$, $p = 0.075$) and number of floors (Kavala, 2.9 $\pm$ 1.6, 1–6; Rovaniemi, 2.6 $\pm$ 1.2, 1–4; $z = 0.150$, $p = 0.998$) did not significantly differ between the cities, suggesting that green spaces could be compared.

2.4. Data Analysis

The Outlying Mean Index (hereafter OMI) analysis [73], a two-table ordination technique, was used to explore the influence of environmental variables on the bird communities of Kavala and Rovaniemi. The first table included 26 species for the Kavala and 15 species for the Rovaniemi dataset (after excluding species occurring in <5% of all counts; see Table S1). The second table included the 16 environmental variables measured in both cities (see Table S2). In contrast to other multivariate methods, OMI analysis gives equal weight to species-rich and species-poor sites, makes no assumption about the shape of species response curves to the environmental gradients (linear or unimodal) and its interpretations are robust to multicollinearity among the environmental variables [73]. The OMI analysis provides an inertia estimate representing the total variance of the environmental table weighted by the species distribution profile. This variability is decomposed into three niche parameters, namely OMI, tolerance and residual tolerance (expressed as percentages of inertia). OMI measures the marginality of species, or the distance between the average environmental conditions used by a species and the mean environmental conditions of the sampling units of the study area. A high marginality indicates that a species is found under atypical environmental conditions within the study area, whereas a low marginality indicates that there is no difference between the overall environmental conditions and those where the species is found. Tolerance is a measure of niche breadth. High tolerance values indicate that the species is distributed along a variety of environmental conditions (generalist species), while low values imply that the species is distributed along a more limited range of environmental conditions (specialist species). Residual tolerance indicates the variance in species niche not considered. Niche parameters were estimated with the function *niche* of the *ade4* R package [74]. The statistical significance of the marginality of each species was tested by a Monte Carlo random permutation test with 10,000 permutations, applying the Holm correction for multiple testing. Pearson's correlation coefficients between the environmental variables and OMI axes were computed with the *rcorr* function of the *Hmisc* R package [75].

We classified species in conservation priority categories depending on niche specialization and conservation status. Species that were both specialized and threatened at the European or national levels were classified as first conservation priority. Species that were either specialized or threatened at the European or national levels were classified as second conservation priority.

All statistical analyses were performed in program R 4.0.2 [76].

3. Results

3.1. Niche Analysis for the Breeding Bird Community of Kavala

OMI analysis retained the first two axes, which accounted for 79.83% of the total inertia (axis 1: 56.34%; axis 2: 23.49%). OMI values varied greatly, ranging from 1.80% (Northern House Martin) to 87.20% of inertia (Eurasian Tree Sparrow (*Passer montanus*)) (Table 1, Figure 2a). Analysis showed that 12 of 26 species examined (46.2%) showed significant deviation from the mean habitat condition, indicating marginal niches. The remaining 14 species (53.8%) had low OMI index values, indicating non-marginal niches, typical of the local average urban niche. Moreover, the average marginality of all species, the criterion optimized by OMI analysis, was significantly different from the mean habitat condition ($p < 0.001$). Tolerance values revealed species occupying narrow niches (tolerance <20% of inertia) and species with broader niches (tolerance >30% of inertia). However, the generally high residual tolerance, ranging from 22.20% to 81.30% of inertia, indicated that additional important factors that affect species distribution and resource use should be considered.

The first OMI axis was significantly positively correlated with tree cover, coniferous tree cover, green space size and perimeter, and significantly negatively correlated with morning and midday noise levels, car and pedestrian traffic, gray cover and building height (Table 2, Figure 2c). The second OMI axis was significantly positively correlated with gray cover and relative edge length and significantly negatively correlated with car and pedestrian traffic, green space size and perimeter, tree cover and coniferous tree cover. A large group of species was positively correlated with the first axis and negatively correlated with the second axis, species characteristic of green spaces with higher tree cover and coniferous tree cover and larger size and perimeter (Figure 2a,c; Common Cuckoo (*Cuculus canorus*), European Bee-eater (*Merops apiaster*), Syrian Woodpecker (*Dendrocopos syriacus*), Eurasian Blackbird (*Turdus merula*), Sardinian Warbler (*Sylvia melanocephala*), Willow Warbler (*Phylloscopus trochilus*), Common Nightingale (*Luscinia megarhynchos*), Great Tit (*Parus major*), Eurasian Blue Tit (*Cyanistes caeruleus*), Eurasian Jay (*Garrulus glandarius*), Common Chaffinch (*Fringilla coelebs*), European Greenfinch (*Chloris chloris*), European Serin (*Serinus serinus*)). The Feral Pigeon (*Columba livia*), Collared Dove (*Streptopelia decaocto*), Northern House Martin, Eurasian Magpie (*Pica pica*), Western Jackdaw (*Corvus monedula*),

Hooded Crow (*Corvus cornix*), Common Starling (*Sturnus vulgaris*) and European Goldfinch (*Carduelis carduelis*) formed a group of species generally negatively correlated with both axes, being common in green spaces with high disturbance (noise and traffic load), surrounded by high buildings. The Eurasian Tree Sparrow was the species most positively correlated with the second axis, preferring edge habitats with generally high gray and bare ground cover.

Table 1. Niche parameters of the bird species of Kavala and Rovaniemi. Inertia: total variability; OMI: outlying mean index or marginality (%); T: tolerance index (%), RT: residual tolerance index (%).

Species Name	Species Code	Kavala				Rovaniemi			
		Inertia	OMI (%)	T (%)	RT (%)	Inertia	OMI (%)	T (%)	RT (%)
Black-headed Gull	Chrrid	-	-	-	-	12.1	79.0	3.9	17.1
Feral Pigeon	Colliv	23.6	37.0 *	36.4	26.6	-	-	-	-
Collared Dove	Strdec	15.1	5.3	24.8	69.9	-	-	-	-
Common Cuckoo	Cuccan	27.8	41.8 *	31.7	26.6	-	-	-	-
European Bee-eater	Merapi	40.1	47.0 *	21.3	31.7	-	-	-	-
Syrian Woodpecker	Densyr	32.8	39.5 *	29.1	31.3	-	-	-	-
Northern House Martin	Delurb	16.1	1.8	23.9	74.3	-	-	-	-
Barn Swallow	Hirrus	9.6	9.2	9.5	81.3	-	-	-	-
White Wagtail	Motalb	10.9	42.7	0.7	56.6	16.8	6.8	10.3	82.9
Eurasian Blackbird	Turmer	28.6	39.5 *	33.0	27.5	-	-	-	-
Fieldfare	Turpil	-	-	-	-	15.5	10.8	22.4	66.8
Sardinian Warbler	Sylmel	23.1	29.4	25.4	45.2	-	-	-	-
Willow Warbler	Phytro	32.6	45.0 *	30.4	24.7	19.7	15.3	15.8	68.9
Olivaceous Warbler	Idupal	14.9	25.1	0.0	74.9	-	-	-	-
Common Nightingale	Lusmeg	19.1	22.6	16.1	61.4	-	-	-	-
Great Tit	Parmaj	24.7	11.1	34.9	54.0	17.5	6.6	25.9	67.5
Eurasian Blue Tit	Cyacae	32.6	36.2 *	20.6	43.2	23.1	38.0 *	8.0	54.1
Eurasian Magpie	Picpic	13.1	58.1 *	5.0	36.9	25.9	64.7 *	4.9	30.5
Eurasian Jay	Gargla	34.2	41.7 *	33.5	24.8	-	-	-	-
Western Jackdaw	Cormon	17.4	33.1 *	14.4	52.4	19.1	43.5 *	21.9	34.6
Hooded Crow	Corcor	15.7	6.0	26.4	67.6	17.3	34.7 *	14.6	50.7
Common Starling	Stuvul	13.5	21.7	17.3	61.0	-	-	-	-
House Sparrow	Pasdom	14.8	11.2	26.1	62.8	14.6	11.9	16.6	71.5
Eurasian Tree Sparrow	Pasmon	13.8	87.2 *	0.3	12.5	11.9	20.9	11.4	67.7
Common Chaffinch	Fricoe	22.3	20.2 *	33.8	46.0	16.2	22.4 *	26.0	51.6
European Goldfinch	Carcar	17.3	4.3	15.4	80.3	-	-	-	-
European Greenfinch	Carchl	21.2	8.9	34.2	56.9	17.0	42.2 *	14.1	43.6
Redpoll	Acafla	-	-	-	-	19.5	38.4	17.9	43.6
European Serin	Serser	21.7	60.7	17.1	22.2	-	-	-	-
Eurasian Siskin	Spispi	-	-	-	-	12.4	12.8	4.3	82.9

* Significant OMI values (Monte Carlo random tests with 10,000 permutations; $p < 0.05$ after Holm correction).

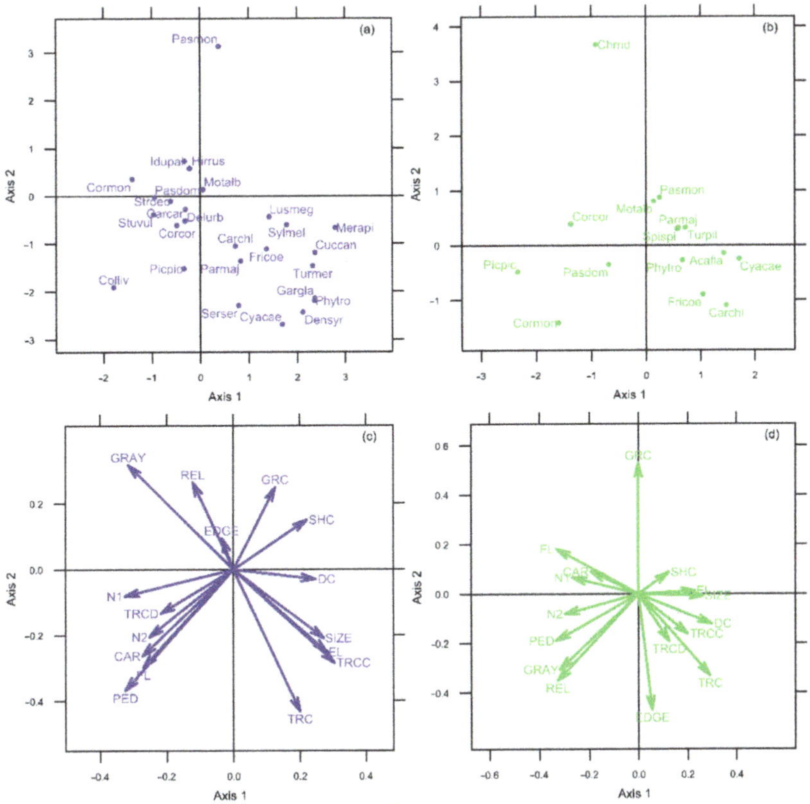

Figure 2. OMI analysis plots for Kavala (**a**,**c**) and Rovaniemi (**b**,**d**) datasets. The origin of the plots represents the mean environmental conditions. Species' position on the factorial plane reveals the degree of difference of their ecological niche to the mean environmental conditions and their correlation with each axis (Kavala, (**a**); Rovaniemi, (**b**)). The length of the arrows of environmental variables reflects their relative importance, and the direction of the arrow indicates correlations among variables and with each axis (Kavala, **c**; Rovaniemi, **d**). Refer to Table 2 for the species codes and to Table 1 for the environmental variable codes.

3.2. Niche Analysis for the Breeding Bird Community of Rovaniemi

OMI analysis also retained the first two axes, which accounted for 82.67% of the total inertia (axis 1: 70.57%; axis 2: 12.10%). OMI values highly varied from 6.60% (Great Tit) to 79.00% of inertia (Black-headed Gull (*Chroicocephalus ridibundus*)) (Table 1, Figure 2b). Analysis showed that 6 of 15 species examined (40.0%) showed significant deviation from the mean habitat condition, indicating marginal niches. The remaining nine species (60.0%) had low OMI index values, indicating non-marginal niches, typical of the local average urban niche. In addition, the average marginality of all species, the criterion optimized by OMI analysis, was significantly different from the mean habitat condition ($p < 0.001$). The relatively low tolerance values (3.90–26.00% of inertia) revealed that the 15 species of the Rovaniemi assemblage occupied relatively narrow niches. The considerable residual tolerance values (17.10–82.90% of inertia) suggested that there are other factors important for explaining species distribution and niche breadth.

Table 2. Pearson's correlation coefficients of the environmental variables with the first two axes of OMI analysis.

Environmental Variables	Code	Kavala		Rovaniemi	
		Axis1	Axis2	Axis1	Axis2
Tree cover (%)	TRC	0.538 *	−0.799 ***	0.554 *	−0.578 **
Shrub cover (%)	SHC	0.395	0.362	0.269	0.043
Bare ground cover (%)	GRC	0.302	0.393	0.120	0.933 ***
Gray cover (%)	GRAY	−0.783 ***	0.591 **	−0.754 ***	−0.509 *
Deciduous tree cover (%)	TRCD	−0.437	−0.275	0.252	−0.323
Coniferous tree cover (%)	TRCC	0.726 ***	−0.510 *	0.349	−0.356
Green space size (ha)	SIZE	0.696 ***	−0.551 *	0.712 **	0.069
Green space perimeter (km)	EL	0.666 **	−0.590 **	0.658 **	0.221
Relative edge length (km/km^2)	REL	−0.369	0.617 **	−0.745 ***	−0.538 *
Distance from city edge (km)	EDGE	0.045	0.062	0.035	−0.766 ***
Distance from city center (km)	DC	0.440	−0.035	0.788 ***	−0.011
Building height (floors/building)	FL	−0.576 **	−0.366	−0.734 ***	0.077
Morning noise level (dBA)	N1	−0.872 ***	−0.111	−0.688 **	0.029
Midday noise level (dBA)	N2	−0.699 ***	−0.417	−0.746 ***	−0.246
Car traffic (cars/min)	CAR	−0.679 ***	−0.492 *	−0.530 *	−0.172
Pedestrian traffic (people/min)	PED	−0.644 **	−0.593 **	−0.727 ***	−0.213

* $p < 0.05$; ** $p < 0.01$; *** $p < 0.001$.

The first OMI axis was significantly positively correlated with tree cover, green space size and perimeter and distance from the city center, and significantly negatively correlated with morning and midday noise levels, car and pedestrian traffic, gray cover, relative edge length and building height (Table 2, Figure 2d). The second OMI axis was significantly positively correlated with bare ground cover and significantly negatively correlated with distance from the city edge, tree cover, gray cover and relative edge length. The Fieldfare (*Turdus pilaris*), Willow Warbler, Great Tit, Eurasian Blue Tit, Common Chaffinch, European Greenfinch, Redpoll (*Acanthis flammea*) and Eurasian Siskin (*Spinus spinus*) were positively correlated with the first axis, species characteristic of green spaces with higher tree cover, larger size and perimeter, and further from the city center (Figure 2b,d). The Eurasian Magpie, Western Jackdaw, Hooded Crow and House Sparrow (*Passer domesticus*) were negatively correlated with the first axis, mostly occupying greenspaces with high noise levels, traffic load, gray cover, relative edge length and building height. The Eurasian Tree Sparrow, White Wagtail (*Motacilla alba*) and, more strongly, the Black-headed Gull were positively correlated with the second axis, preferring open green spaces.

3.3. Conservation Status and Priority

Six species had an unfavorable conservation status within Europe (Table S1). Three were recorded in both cities (Willow Warbler, Eurasian Tree Sparrow, House Sparrow), while the other three were recorded only in Kavala (Northern House Martin, Barn Swallow, Common Starling). At the national level, five species were threatened in Finland: the endangered House Sparrow and European Greenfinch, the vulnerable Black-headed Gull and the near-threatened White Wagtail and Eurasian Magpie. The conservation status of urban bird species recorded in Kavala has not been evaluated by the Red Data Book of Threatened Animals of Greece.

In Kavala, one exploiter (Eurasian Tree Sparrow) and one adapter (Willow Warbler) were the first conservation priority species. In Rovaniemi, one exploiter (Eurasian Magpie) and one adapter (European Greenfinch) were the first conservation priority species.

Most species in both communities were assigned to second conservation priority status. In Kavala, seven exploiters (Northern House Martin, Barn Swallow, Common Starling, House Sparrow, Feral Pigeon, Eurasian Magpie, Western Jackdaw) and seven adapters (Eurasian Jay, Common Chaffinch, Eurasian Blue Tit, Eurasian Blackbird, Common Cuckoo, European Bee-eater, Syrian Woodpecker) were the second conservation priority species. In Rovaniemi, six exploiters (House Sparrow, Eurasian Tree Sparrow, Black-headed Gull,

White Wagtail, Western Jackdaw, Hooded Crow) and three adapters (Common Chaffinch, Eurasian Blue Tit, Willow Warbler,) were the second conservation priority species.

4. Discussion

4.1. Niche Specialization and Habitat Conditions

The bird species breeding in Kavala's and Rovaniemi's urban core areas occupied different niches of varying marginality and breadth, indicating varying responses to urban environmental conditions. Birds in both Kavala and Rovaniemi could be broadly assigned to either exploiter or adapter status in relation to their response to local environmental variables [24,25]. The high niche marginality observed in urban bird communities might be partially explained by the differential use of resources by exploiters (high use of gray and low of green infrastructure) and adapters (low use of gray and high of green infrastructure) in the studied urban core areas.

Several species were recorded only in one of the two bird communities. The Collared Dove, European Bee-eater, Syrian Woodpecker, Sardinian Warbler, Olivaceous Warbler (*Iduna pallida*), Common Nightingale, European Goldfinch and European Serin are southernly distributed species that do not breed in Rovaniemi, while species such as the Redpoll do not breed in Greece, and others breed very rarely, mostly in mountainous areas (Fieldfare and Eurasian Siskin) [77]. The Northern House Martin, Barn Swallow and Feral Pigeon were abundant in Kavala, while they were not recorded in the study sites of Rovaniemi despite that they belong in the breeding bird community of the city. Feral Pigeon numbers were recently reduced in Rovaniemi, from several hundreds to several dozen due to viral infection (J. Jokimäki, unpublished data). Martins and swallows are not abundant in Rovaniemi, as 90% of buildings were destroyed in World War II [78], and modern architecture offers few nesting opportunities, especially where balconies are enclosed with windows in response to cold weather.

Eleven species were common in the urban core areas of Kavala and Rovaniemi, in both behaving either as exploiters, preferring green spaces with high gray cover, anthropogenic disturbance and further from the city limits (e.g., Western Jackdaw, House Sparrow, Eurasian Magpie) or as adapters, preferring larger green spaces with high vegetation cover, further from the city center (e.g., Common Chaffinch, European Greenfinch, Eurasian Blue Tit). Furthermore, four of the six specialists of Rovaniemi were also specialists in Kavala. Although Kavala and Rovaniemi are located in different biogeographical areas, both bird communities showed similar responses to urbanization and especially bird species common to both communities were making similar use of local environmental variables and also showed similar niche specialization. These findings agree with findings from previous studies. Urban adapters seek in cities conditions similar to their natural habitats, while urban exploiters are able to make use of novel conditions, which tend to be similar in cities around the world (e.g., nest sites in buildings, food waste), and thus become abundant [19–22,24,25,49].

4.2. Conservation Management Implications

A considerable number of species were threatened at the European or national levels or had specialized niches and were therefore assigned to conservation priority status. Other studies also found that European urban core areas [29] and Australian cities [28] are locally important for threatened bird species. Threatened species that were observed in Kavala and Rovaniemi were also recorded in other European cities, more frequently the Eurasian Tree and House Sparrows, Northern House Martin and Barn Swallow [29]. Our findings also suggested that birds, even exploiter species, might mainly depend on a limited set of critical environmental variables in urban core areas. Potential change in the availability of these environmental variables might render the urban environment inhospitable for certain species. This process could be illustrated with the example of exploiters abundant in Kavala, such as the narrow-niche Northern House Martin and Barn Swallow, which have severely declined in Rovaniemi due to changes in habitat conditions and were not recorded during our surveys. Such outcomes emphasize the need for conservation actions.

In both communities, adapters were more abundant in larger green spaces with high vegetation cover, especially tree cover, and low rates of urbanization and disturbance. Most conservation priority adapter species nest in trees (Common Chaffinch, European Greenfinch, Eurasian Jay), in holes in trees (Eurasian Blue Tit, Syrian Woodpecker), in shrubs (Eurasian Blackbird) or in shrubs and on the ground (Willow Warbler). The enlargement of green spaces and the increase in tree and shrub cover would benefit adapter species. Particular attention should be given to the retention of mature trees and dense shrubberies. Both of these elements are often scarce in the urban landscape or removed for aesthetic and safety reasons. However, they are vital for cavity nesters the former and for birds that nest in shrubs or on the ground under shrubs the latter. Decreased shrub cover increases nest visibility and consequently nest losses from both visually searching avian nest predators (e.g., Eurasian Magpie [79]) and mammalian predators (e.g., Red Fox (*Vulpes vulpes*), Pine Marten (*Martes foina*), free-ranging dogs and cats [16]). These shrub or ground-nesting bird species were also intolerant of impervious areas and disturbance. Therefore, recreational activities with their associated infrastructure (e.g., paths, kiosks, playgrounds) should be kept at a minimum and arranged at the periphery of green spaces.

Urban exploiters, such as the Eurasian Tree and House Sparrows, Feral Pigeon, Northern House Martin and Barn Swallow, are closely associated with built areas, taking advantage of the suitable sites for nesting they offer (i.e., windowsills, wall ledges, roof-tops). However, modern architecture makes buildings unsuitable for nesting for many species [80], and therefore architects must incorporate suitable nesting sites when designing buildings. This is especially urgent for Rovaniemi, were the modern architecture has decreased the nesting possibilities of the House Sparrow, Northern House Martin and Barn Swallow. The closing of balconies with windows has restricted access of martins and swallows to build their nests in wall corners of balconies (J. Jokimäki, unpublished data) and the use of new types of roof tiles with fewer holes or cracks might reduce nest site availability for sparrows [81]. In addition, secondary cavity-nesting species can be helped by erecting artificial nest-boxes [32,82,83].

Several other species that are tolerant of human disturbance and able to exploit anthropogenic food sources use natural substrates to nest, such as trees (Eurasian Magpie, Hooded Crow) or cavities therein (Common Starling, Western Jackdaw) and riverbanks (Black-headed Gull). Furthermore, grasslands and lawns are important foraging sites for species such as the Black-headed Gull and sparrows. These further emphasize the importance of preserving specific elements of the urban landscape, both natural and anthropogenic, for the conservation of bird species, even in urban core areas.

5. Conclusions

We analyzed the niche of bird species nesting in the urban core areas of Kavala and Rovaniemi, based on a set of 16 local environmental variables. This analysis involved describing niche characteristics and identifying niche specialization and differences and similarities between the cities. Specialization in niche position was high in both bird communities, with species generally preferring either larger, more vegetated green spaces with lower disturbance, or smaller, more built and disturbed green spaces. Species present in both communities occupied similar niches. This analysis allowed for determining species of conservation priority, also considering their conservation status at the European and national levels. Next, we proposed species-specific conservation actions that would allow for the protection of these species, but also other with similar ecological requirements. In doing so, we also secure the maintenance of diverse urban bird communities.

Several species that were observed in only one green space were not included in the analysis, with some of them being threatened at the European or national levels (Common Swift, Sand Martin, Redwing (*Turdus iliacus*), Spotted Flycatcher (*Muscicapa striata*) [70,71]). These species should be also considered a conservation priority and future research should explore their ecological requirements and the possibilities for their survival in urban core areas. There is a lack of information on the national conservation status of the bird species observed in Kavala [72]. It is important that threatened species would be classified as conservation priorities upon availability of such information. Future studies should also include other local environmental variables that might have important effects on urban birds, such as nest predation [15,79]), microclimate [31], artificial light at night [84] and the availability of holes in mature and dead standing trees [85]. In addition, although the generally small differences in the measured environmental variables between the cities indicated small differences in ecological conditions that allowed for comparisons, studies including variables of the wider landscape matrix would give further insights. Our study did not contain cities from Central Europe, and therefore the study does not represent an average situation in Europe. For example, some typical Central European parks birds (such as the Wood Pigeon (*Columba palumbus*), Eurasian Blackcap (*Sylvia atricapilla*), Eurasian Robin (*Erithacus rubecula*), Common Chiffchaff (*Phylloscopus collybita*), and Eurasian Wren (*Troglodytes troglodytes*) [86–88]) were missing from our samples. We encourage researchers from Central Europe as well as other continents to perform corresponding urban niche studies, also by using other groups than birds.

Supplementary Materials: The following are available online at https://www.mdpi.com/article/10.3390/su13116327/s1, Table S1: Common and scientific names, total abundance (sum of individuals in all survey stations) and incidence (number of stations a species was observed) of the bird species recorded in more than 5% of the survey stations in the green spaces of Kavala (26 species) and Rovaniemi (15 species); Table S2: Environmental variables associated with the urban green spaces of Kavala and Rovaniemi.

Author Contributions: Conceptualization, V.L., J.J. and V.J.K.; investigation, J.J., M.-L.K.-J. and E.V.; methodology, V.L., J.J., M.-L.K.-J. and V.J.K.; software, V.L.; validation, V.L. and J.J.; formal analysis, V.L.; resources, V.L., J.J., M.-L.K.-J., E.V. and V.J.K.; data curation, V.L.; writing—original draft preparation.; V.L.; writing—reviewing and editing, J.J., M.-L.K.-J., E.V. and V.J.K.; visualization, V.L.; supervision, V.L. and J.J.; project administration, V.L. All authors have read and agreed to the published version of the manuscript.

Funding: This research received no external funding.

Institutional Review Board Statement: Ethical review and approval were waived for this study because it did not involve approaching or handling animals.

Informed Consent Statement: Not applicable.

Data Availability Statement: The data presented in this study are available on reasonable request from the corresponding author.

Acknowledgments: The Kavala project was part of the "Planning and managing urban green spaces" postgraduate program of the Department of Forest and Natural Environment Sciences, International Hellenic University. We are grateful to three anonymous reviewers, whose comments and suggestions helped greatly improve the manuscript.

Conflicts of Interest: The authors declare no conflict of interest. V.L. served as a guest editor to the Special Issue "Wildlife Conservation: Managing Resources for a Sustainable World" in which this article has been included. However, he was not in any way involved in the review process, except as an author of the article.

References

1. Hutchinson, G.E. Concluding remarks. *Cold Spring Harb. Symp. Quant. Biol.* **1957**, *22*, 145–159. [CrossRef]
2. Whittaker, R.H.; Levin, S.A.; Root, R.B. Niche, habitat and ecotope. *Am. Nat.* **1973**, *107*, 321–338. [CrossRef]
3. Lehmann, S. Growing Biodiverse Urban Futures: Renaturalization and Rewilding as Strategies to Strengthen Urban Resilience. *Sustainability* **2021**, *13*, 2932. [CrossRef]
4. Lepczyk, C.A.; Warren, P.S. *Urban Bird Ecology and Conservation*; University of California Press: Berkeley, CA, USA, 2012.
5. Xu, X.; Xie, Y.; Qi, K.; Luo, Z.; Wang, X. Detecting the response of bird communities and biodiversity to habitat loss and fragmentation due to urbanization. *Sci. Total Environ.* **2018**, *624*, 1561–1576. [CrossRef]

6. Chace, J.F.; Walsh, J.J. Urban effects on native avifauna: A review. *Landsc. Urban Plan.* **2006**, *7*, 46–69. [CrossRef]
7. Pellissier, V.; Cohen, M.; Boulay, A.; Clergeau, P. Birds are also sensitive to landscape composition and configuration within the city centre. *Landsc. Urban Plan.* **2012**, *104*, 181–188. [CrossRef]
8. Jones, G.A.; Sieving, K.E. Sieving Intercropping sunflower in organic vegetables to augment bird predators of arthropods. *Agric. Ecosyst. Environ.* **2006**, *117*, 171–177. [CrossRef]
9. Klein, A.M.; Vaissière, B.E.; Cane, J.H.; Steffan-Dewenter, I.; Cunningham, S.A.; Kremen, C.; Tscharntke, T. Importance of pollinators in changing landscapes for world crops. *Proc. R. Soc. B Biol. Sci.* **2007**, *274*, 303–313. [CrossRef]
10. Whelan, C.J.; Sekercioglu, C.H.; Wenny, D.G. Why birds matter: From economic ornithology to ecosystem services. *J. Ornithol.* **2015**, *156*, 227–238. [CrossRef]
11. Fuller, R.A.; Irvine, K.N.; Devine-Wright, P.; Warren, P.H.; Gaston, K.J. Psychological benefits of greenspace increase with biodiversity. *Biol. Lett.* **2007**, *3*, 390–394. [CrossRef]
12. Cameron, R.W.F.; Brindley, P.; Mears, M.; McEwan, K.; Ferguson, F.; Sheffield, D.; Jorgensen, A.; Riley, J.; Goodrick, J.; Ballard, L.; et al. Where the wild things are! Do urban green spaces with greater avian biodiversity promote more positive emotions in humans? *Urban Ecosyst.* **2020**, *23*, 301–317. [CrossRef]
13. Randler, C.; Tryjanowski, P.; Jokimäki, J.; Kaisanlahti-Jokimäki, M.-L.; Staller, N. SARS-CoV2 (COVID-19) Pandemic lockdown influences nature-based recreational activity: The case of birders. *Int. J. Environ. Res. Public Health* **2020**, *17*, 7310. [CrossRef]
14. Grimm, N.B.; Faeth, S.H.; Golubiewski, N.E.; Redman, C.L.; Wu, J.G.; Bai, X.M.; Briggs, J.M. Global change and the ecology of cities. *Science* **2008**, *319*, 756–760. [CrossRef] [PubMed]
15. Jokimäki, J.; Suhonen, J.; Benedetti, Y.; Diaz, M.; Kaisanlahti-Jokimäki, M.-L.; Morelli, F.; Pérez-Contreras, T.; Rubio, E.; Sprau, P.; Tryjanowski, P.; et al. Land-sharing vs. land-sparing urban development modulate predator-prey interactions in Europe. *Ecol. Appl.* **2020**, *30*, 02049. [CrossRef] [PubMed]
16. Loss, S.R.; Will, T.; Marra, P.P. The impact of domestic cats on wildlife of the United States. *Nat. Commun.* **2013**, *4*, 1396. [CrossRef] [PubMed]
17. Pollack, L.; Ondrasek, N.R.; Calisi, R. Urban health and ecology: The promise of an avian biomonitoring tool. *Curr. Zool.* **2017**, *63*, 205–212. [CrossRef]
18. Jiguet, F.; Sunnen, L.; Prévot, A.-C.; Princé, K. Urban pigeons losing toes due to human activities. *Biol. Conserv.* **2019**, *240*, 108241. [CrossRef]
19. MacGregor-Fors, I.; Ortega-Álvarez, R. Fading from the forest: Bird community shifts related to urban park site-specific and landscape traits. *Urban For. Urban Green.* **2011**, *10*, 239–246. [CrossRef]
20. Schneider, S.C.; Fischer, J.D.; Miller, J.R. Two-sided edge responses of avian communities in an urban landscape. *Urban Ecosyst.* **2015**, *18*, 539–551. [CrossRef]
21. Leveau, L.M.; Leveau, C.M.; Villegas, M.; Cursach, J.A.; Suazo, C.G. Bird communities along urbanization gradients: A comparative analysis among three neo-tropical cities. *Ornitol. Neotrop.* **2017**, *28*, 77–87.
22. Tzortzakaki, O.; Kati, V.; Kassara, C.; Tietze, D.T.; Giokas, S. Seasonal patterns of urban bird diversity in a Mediterranean coastal city: The positive role of open green spaces. *Urban Ecosyst.* **2018**, *21*, 27–39. [CrossRef]
23. Kontsiotis, V.J.; Valsamidis, E.; Liordos, V. Organization and differentiation of breeding bird communities across a forested to urban landscape. *Urban For. Urban Green.* **2019**, *38*, 242–250. [CrossRef]
24. Blair, R.B. Land use and avian species diversity along an urban gradient. *Ecol. Appl.* **1996**, *6*, 506–519. [CrossRef]
25. McKinney, M.L. Urbanization as a major cause of biotic homogenization. *Biol. Conserv.* **2006**, *127*, 247–260. [CrossRef]
26. Fernández-Juricic, E.; Jokimäki, J. A habitat island approach to conserving birds in urban landscapes-case studies from southern and northern Europe. *Biodivers. Conserv.* **2001**, *10*, 2023–2043. [CrossRef]
27. Jokimäki, J.; Suhonen, J. Effects of urbanization on the breeding bird species richness in Finland: A biogeographical comparison. *Ornis Fenn.* **1993**, *70*, 71–77.
28. Ives, C.D.; Lentini, P.E.; Threlfall, C.G.; Ikin, K.; Shanahan, D.F.; Garrard, G.E.; Bekessy, S.A.; Fuller, R.A.; Mumaw, L.; Rayner, L.; et al. The importance of cities for threatened species. *Glob. Ecol. Biogeogr.* **2016**, *25*, 117–126. [CrossRef]
29. Jokimäki, J.; Suhonen, J.; Kaisanlahti-Jokimäki, M.-L. Urban core areas are important for species conservation: A European-level analysis of breeding bird species. *Landsc. Urban Plan.* **2018**, *178*, 73–81. [CrossRef]
30. Miller, J.R.; Hobbs, R.J. Conservation where people live and work. *Conserv. Biol.* **2002**, *16*, 330–337. [CrossRef]
31. Gilbert, O.L. *The Ecology of Urban Habitats*; Chapman and Hall: London, UK, 1989.
32. Jokimäki, J. Occurrence of breeding bird species in urban parks: Effects of park structure and broad-scale variables. *Urban Ecosyst.* **1999**, *3*, 21–34. [CrossRef]
33. Carbó-Ramírez, P.; Zuria, I. The value of small urban green spaces for birds in a Mexican city. *Landsc. Urban Plan.* **2011**, *100*, 213–222. [CrossRef]
34. Strohbach, M.W.; Lerman, S.B.; Warren, P.S. Are small greening areas enhancing bird diversity? Insights from community-driven greening projects in Boston. *Landsc. Urban Plan.* **2013**, *114*, 69–79. [CrossRef]
35. Peris, S.; Montelongo, T. Birds and small urban parks: A study in a high plateau city. *Turk. J. Zool.* **2014**, *38*, 316–325. [CrossRef]
36. Jasmani, Z.; Ravn, H.P.; van den Bosch, C.C.K. The influence of small urban parks characteristics on bird diversity: A case study of Petaling Jaya, Malaysia. *Urban Ecosyst.* **2017**, *20*, 227–243. [CrossRef]

37. Amaya-Espinel, J.D.; Hostetler, M.; Henríquez, C.; Bonacic, C. The influence of building density on Neotropical bird communities found in small urban parks. *Landsc. Urban Plan.* **2019**, *190*, 103578. [CrossRef]
38. Clergeau, P.; Jokimäki, J.; Savard, J.P.L. Are urban bird communities influenced by the bird diversity of adjacent landscapes? *J. Appl. Ecol.* **2001**, *38*, 1122–1134. [CrossRef]
39. Jokimäki, J.; Kaisanlahti-Jokimäki, M.-L. Spatial similarity of urban bird communities: A multiscale approach. *J. Biogeogr.* **2003**, *8*, 1183–1193. [CrossRef]
40. Shwartz, A.; Muratet, A.; Simon, L.; Julliard, R. Local and management variables outweigh landscape effects in enhancing the diversity of different taxa in a big metropolis. *Biol. Conserv.* **2013**, *157*, 285–292. [CrossRef]
41. Callaghan, C.T.; Major, R.E.; Lyons, M.B.; Martin, J.M.; Kingsford, R.T. The effects of local and landscape habitat attributes on bird diversity in urban greenspaces. *Ecosphere* **2018**, *9*, e02347. [CrossRef]
42. Muñoz-Pedreros, A.; González-Urrutia, M.; Encina-Montoya, F.; Norambuena, H.V. Effects of vegetation strata and human disturbance on bird diversity in green areas in a city in southern Chile. *Avian Res.* **2018**, *9*, 38. [CrossRef]
43. Titoko, R.; Lowry, J.H.; Osborne, T.; Naikatini, A.; Comely, J.; Riley, R. Relationship of bird richness, abundance and assemblage to the built environment in a small island tropical urban setting: A Suva, Fiji case study. *Urban Ecosyst.* **2019**, *22*, 709–719. [CrossRef]
44. Yang, X.; Tan, X.; Chen, C.; Wang, Y. The influence of urban park characteristics on bird diversity in Nanjing, China. *Avian Res.* **2020**, *11*, 45. [CrossRef]
45. Da Silva, B.F.; Pena, J.C.; Viana-Junior, A.B.; Vergne, M.; Pizo, M.A. Noise and tree species richness modulate the bird community inhabiting small public urban green spaces of a Neotropical city. *Urban Ecosyst.* **2021**, *24*, 71–81. [CrossRef]
46. MacGregor-Fors, I.; Schondube, J.E. Gray vs. green urbanization: Relative importance of urban features for urban bird communities. *Basic Appl. Ecol.* **2011**, *12*, 372–381. [CrossRef]
47. Ortega-Álvarez, R.; MacGregor-Fors, I. Living in the big city: Effects of urban land-use on bird community structure, diversity, and composition. *Landsc. Urban Plan.* **2009**, *90*, 189–195. [CrossRef]
48. Ferenc, M.; Sedláček, O.; Fuchs, R.; Dinetti, M.; Fraissinet, M.; Storch, D. Are cities different? Patterns of species richness and beta diversity of urban bird communities and regional species assemblages in Europe. *Glob. Ecol. Biogeogr.* **2014**, *23*, 479–489. [CrossRef]
49. Jokimäki, J.; Suhonen, J.; Kaisanlahti-Jokimäki, M.-L. Urbanization and species occupancy frequency distribution patterns in core zone areas of European towns. *Eur. J. Ecol.* **2016**, *2*, 23–43. [CrossRef]
50. Ibáñez-Álamo, J.D.; Morelli, F.; Benedetti, Y.; Rubio, E.; Jokimäki, J.; Pérez-Contreras, T.; Sprau, P.; Suhonen, J.; Tryjanowski, P.; Kaisanlahti-Jokimäki, M.-L.; et al. Biodiversity within the city: Effects of land sharing and land sparing urban development on avian diversity. *Sci. Total Environ.* **2020**, *707*, 135477. [CrossRef] [PubMed]
51. Kark, S.; Iwaniuk, A.; Schalimtzek, A.; Banker, E. Living in the city: Can anyone become an 'Urban Exploiter'? *J. Biogeogr.* **2007**, *34*, 638–651. [CrossRef]
52. Croci, S.; Butet, A.; Clergeau, P. Does urbanization filter birds on the basis of their biological traits. *Condor* **2008**, *110*, 223–240. [CrossRef]
53. Evans, K.L.; Chaberlain, D.E.; Hatchwell, B.J.; Gregory, R.D.; Gaston, K.J. What makes an urban bird? *Glob. Chang. Biol.* **2011**, *17*, 32–44. [CrossRef]
54. Callaghan, C.T.; Major, R.E.; Wilshire, J.H.; Martin, J.M.; Kingsford, R.T.; Cornwell, W.K. Generalists are the most urban-tolerant of birds: A phylogenetically controlled analysis of ecological and life history traits using a novel continuous measure of bird responses to urbanization. *Oikos* **2019**, *128*, 845–858. [CrossRef]
55. Callaghan, C.T.; Benedetti, Y.; Wilshire, J.H.; Morelli, F. Avian trait specialization is negatively associated with urban tolerance. *Oikos* **2020**, *129*, 1541–1551. [CrossRef]
56. Shugart, H.H.; Patten, B.C. Niche Quantification and the Concept of Niche Pattern. In *Systems Analysis and Simulation Ecology*; Patten, B.C., Ed.; Academic Press: New York, NY, USA, 1972; pp. 283–327.
57. Gregory, R.D.; Gaston, K.J. Explanations of commonness and rarity in British breeding birds, separating resource use and resource availability. *Oikos* **2000**, *88*, 5515–5526. [CrossRef]
58. Shultz, S.; Bradbury, R.; Evans, K.L.; Gregory, R.; Blackburn, T. Brain size and resource specialization predict long-term population trends in British birds. *Proc. R. Soc. B* **2005**, *272*, 2305–2311. [CrossRef]
59. Hurlbert, A.H.; White, E.P. Ecological correlates of geographical range occupancy in North American birds. *Glob. Ecol. Biogeogr.* **2007**, *16*, 764–773. [CrossRef]
60. Rannap, R.; Lõhmus, A.; Briggs, L. Niche position, but not niche breadth, differs in two coexisting amphibians having contrasting trends in Europe. *Divers. Distrib.* **2009**, *15*, 692–700. [CrossRef]
61. Greek Statistical Authority. Population Census 2011. Available online: http://www.statistics.gr/portal/page/portal/ESYE/PAGE-census2011 (accessed on 5 January 2021).
62. Statistics Finland. Rovaniemi Population, January 2021. Available online: https://pxnet2.stat.fi/PXWeb/pxweb/en/StatFin/StatFin__vrm__vamuu/statfin_vamuu_pxt_11lj.px/table/tableViewLayout1/ (accessed on 5 January 2021).
63. Bibby, C.; Burgess, N.; Hill, D.; Mustoe, S. *Bird Census Techniques*, 3rd ed.; Academic Press: London, UK, 2000.
64. Van Heezik, Y.; Seddon, P.J. Accounting for detectability when estimating avian abundance in an urban area. *N. Z. J. Ecol.* **2012**, *36*, 391–397.

65. Sólymos, P.; Lele, S.R.; Bayne, E. Conditional likelihood approach for analyzing single visit abundance survey data in the presence of zero inflation and detection error. *Environmetrics* **2012**, *23*, 197–205. [CrossRef]
66. Green, D.M.; Baker, M.G. Urbanization impacts on habitat and bird communities in a Sonoran desert ecosystem. *Landsc. Urban Plan.* **2003**, *36*, 225–239. [CrossRef]
67. Donnelly, R.; Marzluff, J.M. Importance of reserve size and landscape context to urban bird conservation. *Conserv. Biol.* **2004**, *18*, 733–745. [CrossRef]
68. Litteral, J.; Wu, J. Urban landscape matrix affects avian diversity in remnant vegetation fragments: Evidence from the Phoenix metropolitan region, USA. *Urban Ecosyst.* **2012**, *15*, 939–959. [CrossRef]
69. Rico-Silva, J.F.; Cruz-Trujillo, E.J.; Colorado, Z.G.J. Influence of environmental factors on bird diversity in greenspaces in an Amazonian city. *Urban Ecosyst.* **2021**, *24*, 365–374. [CrossRef]
70. BirdLife International. *European Birds of Conservation Concern: Populations, Trends and National Responsibilities*; BirdLife International: Cambridge, UK, 2017.
71. Hyvärinen, E.; Juslén, A.; Kemppainen, E.; Uddström, A.; Liukko, U.-M. *The 2019 Red List of Finnish Species*; Ministry of the Environment; Finnish Environment Institute: Helsinki, Finland, 2019.
72. Legakis, A.; Maragou, P. *The Red Data Book of Threatened Animals of Greece*; Hellenic Zoological Society: Athens, Greece, 2009.
73. Dolédec, S.; Chessel, D.; Gimaret-Carpentier, C. Niche separation in community analysis: A new method. *Ecology* **2000**, *81*, 2914–2927. [CrossRef]
74. Dray, S.; Dufour, A.B. The ade4 Package: Implementing the Duality Diagram for Ecologists. *J. Stat. Softw.* **2007**, *22*, 1–20. [CrossRef]
75. Harrell, F.E., Jr.; Dupont, C. Hmisc: Harrell Miscellaneous—R Package Version 4.4-1. 2020. Available online: https://CRAN.R-project.org/package=Hmisc (accessed on 5 January 2021).
76. R Core Team. *R: A Language and Environment for Statistical Computing*; R Foundation for Statistical Computing: Vienna, Austria, 2020. Available online: https://www.R-project.org/ (accessed on 5 January 2021).
77. Keller, V.; Herrando, S.; Voříšek, P.; Franch, M.; Kipson, M.; Milanesi, P.; Martí, D.; Anton, M.; Klvaňová, A.; Kalyakin, M.V.; et al. *European Breeding Bird Atlas 2: Distribution, Abundance and Change*; European Bird Census Council: Beek, The Netherlands; Lynx Edicions: Barcelona, Spain, 2020.
78. Ahvenainen, J. *Rovaniemen Historia II 1632–1960*; Kirjapaino Oy: Kuopio, Finland, 1970.
79. Jokimäki, J.; Huhta, E. Artificial nest predation and abundance of birds along an urban gradient. *Condor* **2000**, *102*, 838–847. [CrossRef]
80. Murgui, E. Breeding habitat selection in the House martin *Delichon urbica* in the city of Valencia (Spain). *Acta Ornithol.* **2002**, *37*, 75–83. [CrossRef]
81. Moudra, L.; Zasadil, P.; Moudry, V.; Salek, M. What makes new housing development unsuitable for house sparrows (*Passer domesticus*)? *Landsc. Urban Plan.* **2018**, *169*, 124–130. [CrossRef]
82. Gaston, K.J.; Warren, P.H.; Thompson, K.; Smith, R.M. Urban domestic gardens (IV): The extent of the resource and its associated features. *Biodivers. Conserv.* **2005**, *14*, 3327–3349. [CrossRef]
83. Balaji, S. Artificial nest box for house sparrow: An apt method to save the dwindling species in an urban environment. *Int. J. Biodivers. Conserv.* **2014**, *6*, 194–198. [CrossRef]
84. Ciach, M.; Fröhlich, A. Habitat type, food resources, noise and light pollution explain the species composition, abundance and stability of a winter bird assemblage in an urban environment. *Urban Ecosyst.* **2017**, *20*, 547–559. [CrossRef]
85. Murgui, E. Factors influencing the bird community of urban wooded streets along an annual cycle. *Ornis Fenn.* **2007**, *84*, 66–77.
86. Ferenc, M.; Sedláček, O.; Fuchs, R. How to improve urban greenspace for woodland birds: Site and local-scale determinants of bird species richness. *Urban Ecosyst.* **2014**, *17*, 625–640. [CrossRef]
87. Tryjanowski, P.; Morelli, F.; Mikula, P.; Krištín, A.; Indykiewicz, P.; Grzywaczewski, G.; Kronenberg, J.; Jerzak, L. Bird diversity in urban green space: A large-scale analysis of differences between parks and cemeteries in Central Europe. *Urban For. Urban Green.* **2017**, *27*, 264–271. [CrossRef]
88. Witt, K.; Mitschke, A.; Luniak, M. A comparison of common breeding bird populations in Hamburg, Berlin and Warsaw. *Acta Ornithol.* **2005**, *40*, 139–146. [CrossRef]

MDPI
St. Alban-Anlage 66
4052 Basel
Switzerland
www.mdpi.com

Sustainability Editorial Office
E-mail: sustainability@mdpi.com
www.mdpi.com/journal/sustainability

Disclaimer/Publisher's Note: The statements, opinions and data contained in all publications are solely those of the individual author(s) and contributor(s) and not of MDPI and/or the editor(s). MDPI and/or the editor(s) disclaim responsibility for any injury to people or property resulting from any ideas, methods, instructions or products referred to in the content.

www.ingramcontent.com/pod-product-compliance
Lightning Source LLC
Chambersburg PA
CBHW041950260926
51850CB00019B/2286
* 9 7 8 3 7 2 5 8 0 9 7 4 5 *